Discrete Mathematics

RICHARD JOHNSONBAUGH

DePaul University, Chicago

DISCRETE MATHEMATICS

REVISED EDITION

Macmillan Publishing Company

New York

Collier Macmillan Publishers

London

Macmillan Publishing Company
866 Third Avenue, New York, New York 10022

Collier Macmillan Canada, Inc.

Library of Congress Cataloging-in-Publication Data

Johnsonbaugh, Richard,
  Discrete mathematics.

  Bibliography: p.
  Includes index.
  1. Electronic data processing—Mathematics.
I. Title.
QA76.9.M35J63   1986   511.3   85-15459
ISBN 0-02-360720-3

Printing: 2 3 4 5 6 7 8   Year: 6 7 8 9 0 1 2 3 4

ISBN 0-02-360720-3

Preface

This book is intended for a one-semester or a one- or two-quarter introductory course in discrete mathematics. Although the formal mathematics prerequisites are minimal (calculus is *not* required), a certain level of sophistication is necessary. The recommended computer science prerequisite is one programming course using a higher-level language so that the examples drawn from computer science will be more meaningful.

Courses in discrete mathematics have been recommended for mathematics majors (see [Recommendations, 1981]), for secondary teachers of mathematics (see [Recommendations, 1982]), and for computer science majors (see [Curriculum, 1968, 1979]). The increased interest in discrete mathematics is principally attributable to the rise of computer science; however, discrete mathematics is also important in many other fields, such as operations research, engineering, and economics. Besides its applicability, discrete mathematics provides an ideal framework for developing problem-solving skills.

The topics treated in this book—logic, relations, functions, algorithms, mathematical induction, elementary combinatorics (counting methods and graph theory), elementary Boolean algebra, and introductory automata theory—reflect my view of what material should be treated in an introductory course in discrete mathematics. I believe that topics such as monoids, applied group theory, and Polya's theory of enumeration belong in a more advanced course. There is more than enough material in this book for a one-semester or a one-quarter course so that it is possible to tailor the book to the needs of a particular audience. (In two quarters, all of the material can be covered.)

The main changes in this revised edition are the addition of an appendix on logic, the integration of sections on relations and equivalence relations (which were in the appendix in the original edition) into Chapter 1, the rewriting of some sections to improve readability, and the increase in the number of problems and hints.

Introductory material is found in Chapter 1 and in the appendix. As much or as little of the introductory material can be used as needed, depending on the backgrounds of the students. I recommend covering, at the minimum, Sections 1.2 (Relations), 1.3 (Equivalence Relations), 1.5 (Algorithms), 1.6 (Complexity of Algorithms), and 1.7 (Mathematical Induction). Even if the students have seen some of these topics before, most can profit by a review. Furthermore, these sections make a nice starting point and help to set the tone of the course.

The accompanying figure shows the logical relationship among the chapters

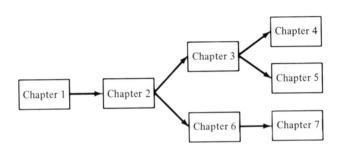

of the book. Within chapters, any of the following sections can be omitted without loss of continuity: 3.4, 3.6, 4.3–4.7, 5.4, and 5.5. A *Solutions Manual* is available from the publisher.

In writing this book, I have been guided by the following goals, which I believe are appropriate for an introductory course in discrete mathematics:

1. Stress problem-solving techniques rather than abstract "theorem/proof" mathematics.
2. Motivate the material through the use of examples.
3. Use an algorithmic approach.
4. Emphasize the interplay between mathematics and computer science.

Among the features included to help attain these goals are

1. An introduction to hash functions.
2. An elementary discussion of algorithms and the complexity of algorithms in Chapter 1.
3. Examples that illustrate how recurrence relations are used to analyze algorithms.
4. Comments throughout the book about the complexity of algorithms.
5. Analysis of algorithms.
6. A discussion using binary trees of the optimal time for sorting.

7. A section on Petri nets—a recent graph model of concurrent processing systems.
8. A treatment of Boolean algebras that emphasizes the relation of Boolean algebras to combinatorial circuits.
9. An approach to automata emphasizing modeling.
10. Computer exercises at the end of each chapter.

Consistent with the goal of emphasizing problem solving, this edition contains over 1800 exercises. Exercises felt to be more challenging than average are marked with a star. An exercise marked with an "H" has a hint or solution in the back of the book. Five exercises are clearly identified as requiring calculus. No calculus concepts are used in the main body of the text and, except for these five exercises, no calculus is needed to solve the exercises. Ends of proofs are marked with the symbol ■.

ACKNOWLEDGMENTS

In writing this book, I have been assisted by many persons including Gregory Bachelis, Wayne State University; Robert Busby, Drexel University; David G. Cantor, University of California–Los Angeles; Tim Carroll, Bloomsburg State College; Robert Crawford, Western Kentucky University; Henry D'Angelo, Boston University; Jerry Delazzer, DePaul University; Br. Michael Driscoll, C.F.C., DePaul University; Carl E. Eckberg, San Diego State University; Susanna Epp, DePaul University; Kevin Phelps, Georgia Institute of Technology; James H. Stoddard, Montclair State College; Michael Sullivan, Chicago State University; and Edward J. Williams, Ford Motor Company, Engineering Computer Center.

In preparing this revised edition, in addition to numerous anonymous reviewers, I found the comments of Martin Kalin, DePaul University, Jerrold Grossman, Oakland University, and Donald E. G. Malm, Oakland University, particularly useful.

I am indebted to the Department of Computer Science and Information Systems at DePaul University and its chairman, Helmut Epp, for providing time and encouragement for the development of both the original book and this revised edition.

I appreciate the assistance of the people at Macmillan, including Bill Winschief, Wayne Yuhasz, Susan Saltrick, and Elaine Wetterau. Gary W. Ostedt, Executive Editor, was of immense help in preparing this revised edition. Gary provided encouragement, insightful reviewers, an enthusiasm for discrete mathematics, and an incredible amount of knowledge about mathematics texts.

R. J.

Contents

CHAPTER 1

Introduction

Discrete mathematics deals mainly with the analysis of finite collections of objects, in contrast to continuous mathematics, which is concerned with infinite processes. Sorting is an example of a problem that uses discrete mathematics: How does one arrange a finite set of objects in order? Problems in classical physics belong to the world of continuous mathematics, calculus in particular. One example is: Find the force of water at every point against a dam. Computer science is finite in nature—computers have finite memories, instructions are executed at finite time intervals, programs are finite, and so on—thus computer science finds discrete mathematics extremely useful. However, the applications of discrete mathematics are by no means limited to computer science; operations research, business, engineering, economics, chemistry, political science, and biology, among others, find discrete mathematics an indispensable tool.

The present chapter consists of introductory material. Section 1.1 concerns sets. Sections 1.2 and 1.3 deal with relations. Section 1.4 discusses functions. Algorithms, a subject with which we will be concerned throughout the book, are treated in Sections 1.5 and 1.6. In the last section (Section 1.7) we discuss mathematical induction, a proof technique that we will use repeatedly. Additional introductory material (on logic and matrices) can be found in the appendixes at the back of the book.

Sets

The concept of set is basic to all of mathematics and mathematical applications. A **set** is simply any collection of objects. If a set is finite and not too large, we can describe it by listing the elements in it. For example, the equation

$$A = \{1, 2, 3, 4\} \qquad (1.1.1)$$

describes a set A made up of the four elements 1, 2, 3, and 4. A set is determined by its elements and not by any particular order in which the elements might be listed. Thus A might just as well be specified as

$$A = \{1, 3, 4, 2\}.$$

The elements making up a set are assumed to be distinct and although for some reason we may have duplicates in our list, only one occurrence of each element is in the set. For this reason we may also describe the set A defined in (1.1.1) as

$$A = \{1, 2, 2, 3, 4\}.$$

If a set is a large finite set or an infinite set, we can describe it by listing a property necessary for membership. For example, the equation

$$B = \{x \mid x \text{ is a positive, even integer}\} \qquad (1.1.2)$$

describes the set B made up of all positive, even integers; that is, B consists of the integers 2, 4, 6, and so on. The vertical bar "$\mid$" is read "such that." Equation (1.1.2) would be read, "B equals the set of all x such that x is a positive, even integer." Here the property necessary for membership is "is a positive, even integer." Note that the property appears after the vertical bar.

If X is a finite set, we let

$$|X| = \text{number of elements in } X.$$

Given a description of a set X such as (1.1.1) or (1.1.2) and an element x, we can determine whether or not x belongs to X. If the members of X are listed as in (1.1.1), we simply look to see whether or not x appears in the listing. In a description such as (1.1.2), we check to see whether the element x has the property listed. If x is in the set X, we write $x \in X$ and if x is not in X, we write $x \notin X$. For example, if $x = 1$, then $x \in A$, but $x \notin B$, where A and B are given by equations (1.1.1) and (1.1.2).

The set with no elements is called the **empty** (or **null** or **void**) **set** and is denoted $\varnothing$. Thus $\varnothing = \{ \ \}$.

Two sets X and Y are **equal** and we write $X = Y$ if X and Y have the same elements. To put it another way, $X = Y$ if whenever $x \in X$, then $x \in Y$ and whenever $x \in Y$, then $x \in X$.

EXAMPLE 1.1.1. If

$$A = \{x \mid x^2 + x - 6 = 0\}, \qquad B = \{2, -3\},$$

then $A = B$.

Suppose that X and Y are sets. If every element of X is an element of Y, we say that X is a **subset** of Y and write $X \subseteq Y$.

EXAMPLE 1.1.2. If

$$C = \{1, 3\} \qquad \text{and} \qquad A = \{1, 2, 3, 4\},$$

then C is a subset of A.

Any set X is a subset of itself, since any element in X is in X. If X is a subset of Y and X does not equal Y, we say that X is a **proper subset** of Y. The empty set is a subset of every set (see Exercise 63). The set of all subsets (proper or not) of a set X, denoted $\mathcal{P}(X)$, is called the **power set** of X. In Section 1.7 (Theorem 1.7.5) we will show that if $|X| = n$, then $|\mathcal{P}(X)| = 2^n$.

EXAMPLE 1.1.3. If $A = \{a, b, c\}$, the members of $\mathcal{P}(A)$ are

$$\varnothing, \quad \{a\}, \quad \{b\}, \quad \{c\}, \quad \{a, b\}, \quad \{a, c\}, \quad \{b, c\}, \quad \{a, b, c\}.$$

All but $\{a, b, c\}$ are proper subsets of A. For this example,

$$|A| = 3, \qquad |\mathcal{P}(A)| = 2^3 = 8.$$

Given two sets X and Y, there are various ways to combine X and Y to form a new set. The set

$$X \cup Y = \{x \mid x \in X \text{ or } x \in Y\}$$

is called the **union** of X and Y. The union consists of all elements belonging to either X or Y (or both).

The set

$$X \cap Y = \{x \mid x \in X \text{ and } x \in Y\}$$

is called the **intersection** of X and Y. The intersection consists of all elements belonging to both X and Y.

Sets X and Y are **disjoint** if $X \cap Y = \varnothing$. A collection of sets $\mathcal{S}$ is said to be **pairwise disjoint** if whenever X and Y are distinct sets in $\mathcal{S}$, X and Y are disjoint.

The set

$$X - Y = \{x \mid x \in X \text{ and } x \notin Y\}$$

is called the **difference** (or **relative complement**). The difference $X - Y$ consists of all elements in X that are not in Y.

EXAMPLE 1.1.4. If $A = \{1, 3, 5\}$ and $B = \{4, 5, 6\}$, then

$$A \cup B = \{1, 3, 4, 5, 6\}$$
$$A \cap B = \{5\}$$
$$A - B = \{1, 3\}$$
$$B - A = \{4, 6\}.$$

EXAMPLE 1.1.5. The sets

$$\{1, 4, 5\} \quad \text{and} \quad \{2, 6\}$$

are disjoint. The collection of sets

$$\mathcal{S} = \{\{1, 4, 5\}, \{2, 6\}, \{3\}, \{7, 8\}\} \qquad (1.1.3)$$

is pairwise disjoint.

Sometimes we are dealing with sets all of which are subsets of a set U. This set U is called a **universal set** or a **universe**. The set U must be explicitly given or inferred from the context. Given a universal set U and a subset X of U, the set $U - X$ is called the **complement** of X and is written $\overline{X}$.

EXAMPLE 1.1.6. Let $A = \{1, 3, 5\}$. If U, a universal set, is specified as $U = \{1, 2, 3, 4, 5\}$, then $\overline{A} = \{2, 4\}$. If, on the other hand, a universal set is specified as $U = \{1, 3, 5, 7, 9\}$, then $\overline{A} = \{7, 9\}$. The complement obviously depends on the universe in which we are working.

We define the union of an arbitrary family $\mathcal{S}$ of sets to be those elements x belonging to at least one set X in $\mathcal{S}$. Formally,

$$\cup \mathcal{S} = \{x \mid x \in X \text{ for some } X \in \mathcal{S}\}.$$

Similarly, we define the intersection of an arbitrary family $\mathcal{S}$ of sets to be those elements x belonging to every set X in $\mathcal{S}$. Formally,

$$\cap \mathcal{S} = \{x \mid x \in X \text{ for all } X \in \mathcal{S}\}.$$

If

$$\mathcal{S} = \{A_1, A_2, \ldots, A_n\}$$

we write

$$\bigcup \mathcal{S} = \bigcup_{i=1}^{n} A_i, \qquad \bigcap \mathcal{S} = \bigcap_{i=1}^{n} A_i$$

and if

$$\mathcal{S} = \{A_1, A_2, \ldots\}$$

we write

$$\bigcup \mathscr{S} = \bigcup_{i=1}^{\infty} A_i, \qquad \bigcap \mathscr{S} = \bigcap_{i=1}^{\infty} A_i.$$

EXAMPLE 1.1.7 If

$$A_n = \{n, n+1, \ldots\}$$

and

$$\mathscr{S} = \{A_1, A_2, \ldots\},$$

then

$$\bigcup_{i=1}^{\infty} A_i = \bigcup \mathscr{S} = \{1, 2, \ldots\}$$

$$\bigcap_{i=1}^{\infty} A_i = \bigcap \mathscr{S} = \varnothing.$$

A partition of a set X divides X into nonoverlapping subsets. More formally, a collection of nonempty sets $\mathscr{S}$ is said to be a **partition** of the set X if $\mathscr{S}$ is pairwise disjoint and

$$\bigcup \mathscr{S} = X.$$

EXAMPLE 1.1.8 The collection of sets (1.1.3) is a partition of $\{1, 2, 3, 4, 5, 6, 7, 8\}$.

At the beginning of this section we pointed out that a set is an unordered collection of elements; that is, a set is determined by its elements and not by any particular order in which the elements are listed. Sometimes, however, we do want to take order into account. An **ordered pair** of elements, written (a, b), is considered distinct from the ordered pair (b, a), unless, of course, $a = b$. To put it another way, $(a, b) = (c, d)$ if and only if $a = c$ and $b = d$. If X and Y are sets, we let $X \times Y$ denote the set of all ordered pairs (x, y) where $x \in X$ and $y \in Y$. We call $X \times Y$ the **Cartesian product** of X and Y.

EXAMPLE 1.1.9. If $X = \{1, 2, 3\}$ and $Y = \{a, b\}$, then

$$X \times Y = \{(1, a), (1, b), (2, a), (2, b), (3, a), (3, b)\}$$
$$Y \times X = \{(a, 1), (b, 1), (a, 2), (b, 2), (a, 3), (b, 3)\}$$
$$X \times X = \{(1, 1), (1, 2), (1, 3), (2, 1), (2, 2), (2, 3),$$
$$(3, 1), (3, 2), (3, 3)\}$$
$$Y \times Y = \{(a, a), (a, b), (b, a), (b, b)\}.$$

Example 1.1.9 shows that, in general, $X \times Y \neq Y \times X$. Notice that $|X \times Y| = |X| \cdot |Y|$.

EXAMPLE 1.1.10.   A restaurant serves four appetizers

$$r = \text{ribs}, \quad n = \text{nachos}, \quad s = \text{shrimp}, \quad f = \text{fried cheese}$$

and three main courses

$$c = \text{chicken}, \quad b = \text{beef}, \quad t = \text{trout}.$$

If we let $A = \{r, n, s, f\}$ and $M = \{c, b, t\}$, the Cartesian product $A \times M$ lists the 12 possible dinners consisting of one appetizer and one main course.

Ordered lists need not be restricted to two elements. An **n-tuple,** written $(a_1, a_2, \ldots , a_n)$, takes order into account:

$$(a_1, a_2, \ldots , a_n) = (b_1, b_2, \ldots , b_n)$$

if and only if

$$a_1 = b_1, a_2 = b_2, \ldots , a_n = b_n.$$

The Cartesian product of sets $X_1, X_2, \ldots , X_n$ is defined to be the set of all n-tuples $(x_1, x_2, \ldots , x_n)$, where $x_i \in X_i$ for $i = 1, \ldots , n$.

EXAMPLE 1.1.11.   If

$$X = \{1, 2\}, Y = \{a, b\}, Z = \{\alpha, \beta\},$$

then

$$X \times Y \times Z = \{(1, a, \alpha), (1, a, \beta), (1, b, \alpha), (1, b, \beta), (2, a, \alpha),$$
$$(2, a, \beta), (2, b, \alpha), (2, b, \beta)\}.$$

Notice that in Example 1.1.11, $|X \times Y \times Z| = |X| \cdot |Y| \cdot |Z|$. In general, we have

$$|X_1 \times X_2 \times \cdots \times X_n| = |X_1| \cdot |X_2| \cdot \ldots \cdot |X_n|.$$

EXAMPLE 1.1.12.   If A is a set of appetizers, M is a set of main courses, and D is a set of desserts, the Cartesian product $A \times M \times D$ lists all possible dinners consisting of one appetizer, one main course, and one dessert.

EXERCISES

In Exercises 1–16, let the universe be the set $U = \{1, 2, 3, \ldots , 10\}$. Let $A = \{1, 4, 7, 10\}$, $B = \{1, 2, 3, 4, 5\}$, and $C = \{2, 4, 6, 8\}$. List the elements of each set.

1H.†	$A \cup B$	**2.**	$B \cap C$
3H.	$A - B$	**4.**	$B - A$
5H.	$\overline{A}$	**6.**	$U - C$
7H.	$\overline{U}$	**8.**	$A \cup \varnothing$
9H.	$B \cap \varnothing$	**10.**	$A \cup U$
11H.	$B \cap U$	**12.**	$A \cap (B \cup C)$
13H.	$\overline{B} \cap (C - A)$	**14.**	$(A \cap B) - C$
15H.	$\overline{A \cap B} \cup C$	**16.**	$(A \cup B) - (C - B)$

In Exercises 17–20, let $X = \{1, 2\}$ and $Y = \{a, b, c\}$. List the elements in each set.

17H.	$X \times Y$	**18.**	$Y \times X$
19H.	$X \times X$	**20.**	$Y \times Y$

In Exercises 21–24, let $X = \{1, 2\}$, $Y = \{a\}$, and $Z = \{\alpha, \beta\}$. List the elements of each set.

21H.	$X \times Y \times Z$	**22.**	$X \times Y \times Y$
23H.	$X \times X \times X$	**24.**	$Y \times X \times Y \times Z$

In Exercises 25–28, list all partitions of the set.

25H.	$\{1\}$	**26.**	$\{1, 2\}$
27H.	$\{a, b, c\}$	**28.**	$\{a, b, c, d\}$

In Exercises 29–32, answer true or false.

29H.	$\{x\} \subseteq \{x\}$	**30.**	$\{x\} \in \{x\}$
31H.	$\{x\} \in \{x, \{x\}\}$	**32.**	$\{x\} \subseteq \{x, \{x\}\}$

In Exercises 33–37, determine whether each pair of sets is equal.

33H. $\{1, 2, 3\}, \{1, 3, 2\}$

34. $\{1, 2, 2, 3\}, \{1, 2, 3\}$

35H. $\{1, 1, 3\}, \{3, 3, 1\}$

36. $\{x \mid x^2 + x = 2\}, \{1, -2\}$

37H. $\{x \mid x$ is a real number and $0 < x \le 2\}, \{1, 2\}$

38. List the members of $\mathcal{P}(\{a, b\})$. Which are proper subsets of $\{a, b\}$?

39H. List the members of $\mathcal{P}(\{a, b, c, d\})$. Which are proper subsets of $\{a, b, c, d\}$?

†An exercise marked with "H" has a hint or solution in the back of the book.

40. If X has 10 members, how many members does $\mathscr{P}(X)$ have? How many proper subsets does X have?

41H. If X has n members, how many proper subsets does X have?

42. If X and Y are nonempty sets and $X \times Y = Y \times X$, what can we conclude about X and Y?

In each of Exercises 43–62, write true if the statement is true; otherwise, give a counterexample. The sets X, Y, and Z are subsets of a universal set U. Assume that the universe for Cartesian products is $U \times U$.

43H. For any sets X and Y, either X is a subset of Y or Y is a subset of X.

44. $\overline{X \cap Y} = \overline{X} \cap \overline{Y}$ for all sets X and Y

45H. $X \cap (Y - Z) = (X \cap Y) - (X \cap Z)$ for all sets X, Y, and Z

46. $X \cap Y = Y \cap X$ for all sets X and Y

47H. $(X - Y) \cap (Y - X) = \varnothing$ for all sets X and Y

48. $\overline{\overline{X}} = X$ for any set X

49H. $X - (Y \cup Z) = (X - Y) \cup Z$ for all sets X, Y, and Z

50. $\overline{X - Y} = \overline{Y - X}$ for all sets X and Y

51H. $X \cup \varnothing = X$ for any set X

52. $\overline{U} = \varnothing$

53H. $\overline{\overline{X \cap Y}} \subseteq X$ for all sets X and Y

54. $X \cap \overline{X} = \varnothing$ for any set X

55H. $\overline{X \cup Y} = \overline{X} \cap \overline{Y}$ for all sets X and Y

56. $(X \cap Y) \cup (Y - X) = X$ for all sets X and Y

57H. $X \times (Y \cup Z) = (X \times Y) \cup (X \times Z)$ for all sets X, Y, and Z

58. $\overline{X \times Y} = \overline{X} \times \overline{Y}$ for all sets X and Y

59H. $X \times (Y - Z) = (X \times Y) - (X \times Z)$ for all sets X, Y, and Z

60. $X - (Y \times Z) = (X - Y) \times (X - Z)$ for all sets X, Y, and Z

61H. $X \cap (Y \times Z) = (X \cap Y) \times (X \cap Z)$ for all sets X, Y, and Z

62. $X \times \varnothing = \varnothing$ for any set X

63H. Show that for any set X, $\varnothing \subseteq X$.

For each condition in Exercises 64–67, what relation must hold between sets A and B?

64. $A \cap B = A$ **65H.** $A \cup B = A$

66. $\overline{A} \cap U = \varnothing$ **67H.** $\overline{A \cap B} = \overline{B}$

The **symmetric difference** of two sets A and B is the set

$$A \Delta B = (A \cup B) - (A \cap B).$$

68. If $A = \{1, 2, 3\}$ and $B = \{2, 3, 4, 5\}$, find $A \Delta B$.

69H. Describe the symmetric difference of sets A and B in words.

70. Given a universe U, describe $A \Delta A$, $A \Delta \overline{A}$, $U \Delta A$, and $\varnothing \Delta A$.

71H. Show that

$$|A \cup B| = |A| + |B| - |A \cap B|.$$

72. Find a formula for $|A \cup B \cup C|$ similar to the formula of Exercise 71. Show that your formula holds for all sets A, B, and C.

73H. Let C be a circle and let $\mathscr{D}$ be the set of all diameters of C. What is $\cap \mathscr{D}$?

74. Let P denote the set of positive integers. For $i \geq 2$, define

$$X_i = \{ik \mid k \geq 2, k \in P\}.$$

Describe $P - \bigcup_{i=2}^{\infty} X_i$.

1.2

Relations

A **relation** can be thought of as a table that lists the relationship of elements to other elements (see Table 1.2.1). Table 1.2.1 shows which students are taking which courses. For example, Bill is taking Computer Science and Art and Mary is taking Mathematics. In the terminology of relations, we would say that Bill is related to Computer Science and Art and that Mary is related to Mathematics.

Another way to specify a relation is to list the rows of the table as ordered pairs. Abstractly, we *define* a relation to be a set of ordered pairs. In this setting, we consider the first element of the ordered pair to be related to the second element of the ordered pair.

DEFINITION 1.2.1. A (*binary*) *relation R from a set X to a set Y* is a subset of the Cartesian product $X \times Y$. If $(x, y) \in R$, we write xRy and say that *x is related to y*. In case $X = Y$, we call R a (*binary*) *relation on X*.

TABLE 1.2.1 Relation of Students to Courses

Student	Course
Bill	CompSci
Mary	Math
Bill	Art
Beth	History
Beth	CompSci
Dave	Math

The set

$$\{x \in X \mid (x, y) \in R \text{ for some } y \in Y\}$$

is called the *domain* of R. The set

$$\{y \in Y \mid (x, y) \in R \text{ for some } x \in X\}$$

is called the *range* of R.

If a relation is given as a table, the domain consists of the members of the first column and the range consists of the numbers of the second column.

EXAMPLE 1.2.2.   If we let

$$X = \{\text{Bill, Mary, Beth, Dave}\}$$

and

$$Y = \{\text{CompSci, Math, Art, History}\},$$

our relation R of Table 1.2.1 can be written

$$R = \{(\text{Bill, CompSci}), (\text{Mary, Math}), (\text{Bill, Art}),$$
$$(\text{Beth, History}), (\text{Beth, CompSci}), (\text{Dave, Math})\}.$$

Since (Beth, History) $\in R$, we may write Beth R History. The domain (first column) of R is the set X and the range (second column) of R is the set Y.

Example 1.2.2 shows that a relation can be given by simply specifying which ordered pairs belong to the relation. Our next example shows that sometimes it is possible to define a relation by giving a rule for membership in the relation.

EXAMPLE 1.2.3.   Let

$$X = \{2, 3, 4\} \qquad \text{and} \qquad Y = \{3, 4, 5, 6, 7\}.$$

If we define a relation R from X to Y by

$$(x, y) \in R \text{ if } x \text{ divides } y \text{ (with zero remainder)},$$

we obtain

$$R = \{(2, 4), (2, 6), (3, 3), (3, 6), (4, 4)\}.$$

If we rewrite R as a table, we obtain

X	Y
2	4
2	6
3	3
3	6
4	4

The domain of R is the set $\{2, 3, 4\}$ and the range of R is the set $\{3, 4, 6\}$.

EXAMPLE 1.2.4. Let R be the relation on $X = \{1, 2, 3, 4\}$ defined by $(x, y) \in R$ if $x \leq y$, $x, y \in X$. Then

$$R = \{(1, 1), (1, 2), (1, 3), (1, 4), (2, 2), (2, 3), (2, 4),$$
$$(3, 3), (3, 4), (4, 4)\}.$$

The domain and range of R are both equal to X.

An informative way to picture a relation on a set is to draw its **digraph**. (Digraphs are discussed in more detail in Chapter 3. For now, we mention digraphs only in connection with relations.) To draw the digraph of a relation on a set X, we first draw dots or **vertices** to represent the elements of X. In Figure 1.2.1, we have drawn four vertices to represent the elements of the set X of Example 1.2.4. Next, if the element (x, y) is in the relation, we draw an arrow (called a **directed edge**) from x to y. In Figure 1.2.1, we have drawn directed edges to represent the members of the relation R of Example 1.2.4. Notice that an element of the form (x, x) in a relation corresponds to a directed edge from x to x. Such an edge is called a **loop**. There is a loop at every vertex in Figure 1.2.1.

EXAMPLE 1.2.5. The relation R on $X = \{a, b, c, d\}$ given by the digraph of Figure 1.2.2 is

$$R = \{(a, a), (b, c), (c, b), (d, d)\}.$$

We next define several useful properties of relations.

DEFINITION 1.2.6. A relation R on a set X is called *reflexive* if $(x, x) \in R$ for every $x \in X$.

EXAMPLE 1.2.7. The relation R on $X = \{1, 2, 3, 4\}$ of Example 1.2.4 is reflexive because for each element $x \in X$, $(x, x) \in R$; specifically, $(1, 1)$, $(2, 2)$, $(3, 3)$, and $(4, 4)$ are each in R. The digraph of a reflexive relation has a loop at every vertex. Notice that the digraph of the reflexive relation of Example 1.2.4 (see Figure 1.2.1) has a loop at every vertex.

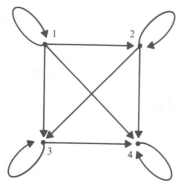

FIGURE 1.2.1

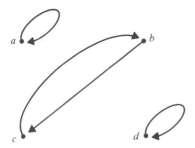

FIGURE 1.2.2

EXAMPLE 1.2.8. The relation R on $X = \{a, b, c, d\}$ of Example 1.2.5 is not reflexive. For example, $b \in X$, but $(b, b) \notin R$. That this relation is not reflexive can also be seen by looking at its digraph (see Figure 1.2.2); vertex b does not have a loop.

DEFINITION 1.2.9. A relation R on a set A is called *symmetric* if for all $(x, y) \in R$, we have $(y, x) \in R$.

EXAMPLE 1.2.10. The relation of Example 1.2.5 is symmetric because for all $(x, y) \in R$, we have $(y, x) \in R$. For example, (b, c) is in R and (c, b) is also in R. The digraph of a symmetric relation has the property that whenever there is a directed edge from v to w, there is also a directed edge from w to v. Notice that the digraph of the relation of Example 1.2.5 (see Figure 1.2.2) has the property that for every directed edge from v to w, there is also a directed edge from w to v.

EXAMPLE 1.2.11. The relation of Example 1.2.4 is not symmetric. For example, $(2, 3) \in R$, but $(3, 2) \notin R$. The digraph of this relation (see Figure 1.2.1) has a directed edge from 2 to 3, but there is no directed edge from 3 to 2.

DEFINITION 1.2.12. A relation R on a set A is called *antisymmetric* if for all $(x, y) \in R$ with $x \neq y$, we have $(y, x) \notin R$.

EXAMPLE 1.2.13. The relation of Example 1.2.4 is antisymmetric because for all $(x, y) \in R$ with $x \neq y$, we have $(y, x) \notin R$. For example, $(1, 2) \in R$, but $(2, 1) \notin R$. The digraph of an antisymmetric relation has the property that between any two vertices there is at most one directed edge. Notice that the digraph of the relation of Example 1.2.4 (see Figure 1.2.1) has at most one directed edge between each pair of vertices.

EXAMPLE 1.2.14. The relation of Example 1.2.5 is not antisymmetric because both (b, c) and (c, b) are in R. Notice that in the digraph of the relation of Example 1.2.5 (see Figure 1.2.2) there are two directed edges between b and c.

EXAMPLE 1.2.15. If a relation R has no members of the form (x, y) with $x \neq y$, then R is antisymmetric; the condition of Definition 1.2.12 is vacuously satisfied. For example, the relation

$$R = \{(a, a), (b, b), (c, c)\}$$

on $X = \{a, b, c\}$ is antisymmetric. The digraph of R shown in Figure 1.2.3 has at most one directed edge between each pair of vertices. Notice that R is also reflexive and symmetric. This example shows that "antisymmetric" is not the same as "not symmetric."

DEFINITION 1.2.16. A relation R on a set X is called *transitive* if for all (x, y), $(y, z) \in R$, we have $(x, z) \in R$.

EXAMPLE 1.2.17. The relation of Example 1.2.4 is transitive because for all (x, y), $(y, z) \in R$, we have $(x, z) \in R$. For example, $(1, 2)$, $(2, 3)$ are in R and $(1, 3)$ is also in R. The digraph of a transitive relation has the property that whenever there are directed edges from x to y and y to z, there is also a directed edge from x to z. Notice that the digraph of the relation (see Figure 1.2.1) of Example 1.2.4 has this property.

EXAMPLE 1.2.18. The relation of Example 1.2.5 is not transitive. For example, (b, c) and (c, b) are in R, but (b, b) is not in R. Notice that in the digraph (see Figure 1.2.2) of the relation of Example 1.2.5 there are directed edges from b to c and from c to b, but there is no directed edge from b to b.

DEFINITION 1.2.19. A relation R on a set X is called a *partial order* if R is reflexive, antisymmetric, and transitive.

EXAMPLE 1.2.20. The relations of Examples 1.2.4 and 1.2.15 are partial orders. The relation of Example 1.2.5 is not a partial order.

Given a relation R from X to Y we may define a relation from Y to X by reversing the order of each ordered pair in R. The formal definition follows.

DEFINITION 1.2.21. Let R be a relation from X to Y. The *converse* of R, denoted R^{-1}, is the relation from Y to X defined by

$$R^{-1} = \{(y, x) \mid (x, y) \in R\}.$$

EXAMPLE 1.2.22. The converse of the relation R of Example 1.2.3 is

$$R^{-1} = \{(4, 2), (6, 2), (3, 3), (6, 3), (4, 4)\}.$$

FIGURE 1.2.3

In words, we might describe this relation as "is divisible by."

If we have a relation R_1 from X to Y and a relation R_2 from Y to Z, we can form a relation from X to Z by applying first relation R_1 and then relation R_2. The resulting relation is denoted $R_2 \circ R_1$. Notice the order in which the relations are written. The formal definition follows.

DEFINITION 1.2.23. Let R_1 be a relation from X to Y and R_2 be a relation from Y to Z. The *composition* of R_1 and R_2, denoted $R_2 \circ R_1$, is the relation from X to Z defined by

$$R_2 \circ R_1 = \{(x, z) \mid (x, y) \in R_1 \text{ and } (y, z) \in R_2 \text{ for some } y \in Y\}.$$

EXAMPLE 1.2.24. The composition of the relations

$$R_1 = \{(1, 2), (1, 6), (2, 4), (3, 4), (3, 6), (3, 8)\}$$

and

$$R_2 = \{(2, u), (4, s), (4, t), (6, t), (8, u)\}$$

is

$$R_2 \circ R_1 = \{(1, u), (1, t), (2, s), (2, t), (3, s), (3, t), (3, u)\}.$$

EXERCISES

In Exercises 1–4, write the relation as a set of ordered pairs.

1H.

8840	Hammer
9921	Pliers
452	Paint
2207	Carpet

2.

a	3
b	1
b	4
c	1

3H.

Sally	Math
Ruth	Physics
Sam	Econ

4.

a	a
b	b

In Exercises 5–8, write the relation as a table.

5H. $R = \{(a, 6), (b, 2), (a, 1), (c, 1)\}$

6. $R = \{(\text{Roger, Music}), (\text{Pat, History}), (\text{Ben, Math}), (\text{Pat, PolySci})\}$.

7H. The relation R on $\{1, 2, 3, 4\}$ defined by $(x, y) \in R$ if $x^2 \geq y$.

8. The relation R from the set X of states beginning with the letter "M" to the set Y of cities defined by $(S, C) \in X \times Y$ if C is the capital of S.

In Exercises 9–12, draw the digraph of the relation.

9H. The relation of Exercise 4 on $\{a, b, c\}$.

10. The relation $R = \{(1, 2), (2, 1), (3, 3), (1, 1), (2, 2)\}$ on $X = \{1, 2, 3\}$.

11H. The relation $R = \{(1, 2), (2, 3), (3, 4), (4, 1)\}$ on $\{1, 2, 3, 4\}$.

12. The relation of Exercise 7.

In Exercises 13–16, write the relation as a set of ordered pairs.

13H.

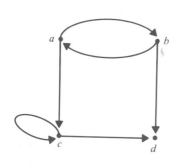

14.

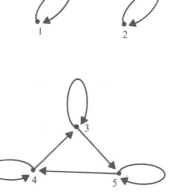

15H.

16.

17H. Find the domain and range of each relation in Exercises 1–16.

18. Find the converse (as a set of ordered pairs) of each relation in Exercises 1–16.

Exercises 19–24 refer to the relation R on the set $\{1, 2, 3, 4, 5\}$ defined by the rule $(x, y) \in R$ if 3 divides $x - y$.

19H. List the elements of R.

20. List the elements of R^{-1}.

21H. Find the domain of R.

22. Find the range of R.

23H. Find the domain of R^{-1}.

24. Find the range of R^{-1}.

25H. Repeat Exercises 19–24 for the relation R on the set $\{1, 2, 3, 4, 5\}$ defined by the rule $(x, y) \in R$ if $x + y \leq 6$.

26. Repeat Exercises 19–24 for the relation R on the set $\{1, 2, 3, 4, 5\}$ defined by the rule $(x, y) \in R$ if $x = y - 1$.

27H.   Is the relation of Exercise 25 reflexive; symmetric; antisymmetric; transitive; a partial order?

28.   Is the relation of Exercise 26 reflexive; symmetric; antisymmetric; transitive; a partial order?

In Exercises 29–34, determine whether each relation defined on the set of positive integers is reflexive, symmetric, antisymmetric, transitive, or a partial order.

29H. $(x, y) \in R$ if $x = y^2$.

30. $(x, y) \in R$ if $x > y$.

31H. $(x, y) \in R$ if $x \geq y$.

32. $(x, y) \in R$ if $x = y$.

33H. $(x, y) \in R$ if the greatest common divisor of x and y is 1.

34. $(x, y) \in R$ if 3 divides $x - y$.

35H.   Let X be a nonempty set. Define a relation on $\mathcal{P}(X)$, the power set of X, as $(A, B) \in R$ if $A \subseteq B$. Is this relation reflexive; symmetric; antisymmetric; transitive; a partial order?

36.   Suppose that R_i is a partial order on X_i, $i = 1, 2$. Show that R is a partial order on $X_1 \times X_2$ if we define

$$(x_1, x_2)R(x_1', x_2') \qquad \text{if } x_1 R_1 x_1' \text{ and } x_2 R_2 x_2'.$$

37H.   Let R_1 and R_2 be the relations on $\{1, 2, 3, 4\}$ given by

$$R_1 = \{(1, 1), (1, 2), (3, 4), (4, 2)\}$$
$$R_2 = \{(1, 1), (2, 1), (3, 1), (4, 4), (2, 2)\}.$$

List the elements of $R_1 \circ R_2$ and $R_2 \circ R_1$.

Give examples of relations on $\{1, 2, 3, 4\}$ having the properties specified in Exercises 38–42.

38.   Reflexive, symmetric, not transitive

39H.   Reflexive, not symmetric, not transitive

40.   Reflexive, antisymmetric, not transitive

41H.   Not reflexive, symmetric, not antisymmetric, transitive

42.   Not reflexive, not symmetric, transitive

Let R and S be relations on X. Determine whether each statement in Exercises 43–58 is true or false. If the statement is false, give a counterexample.

43H.   If R and S are transitive, then $R \cup S$ is transitive.

44.   If R and S are transitive, then $R \cap S$ is transitive.

45H. If R and S are transitive, then $R \circ S$ is transitive.

46. If R is transitive, then R^{-1} is transitive.

47H. If R and S are reflexive, then $R \cup S$ is reflexive.

48. If R and S are reflexive, then $R \cap S$ is reflexive.

49H. If R and S are reflexive, then $R \circ S$ is reflexive.

50. If R is reflexive, then R^{-1} is reflexive.

51H. If R and S are symmetric, then $R \cup S$ is symmetric.

52. If R and S are symmetric, then $R \cap S$ is symmetric.

53H. If R and S are symmetric, then $R \circ S$ is symmetric.

54. If R is symmetric, then R^{-1} is symmetric.

55H. If R and S are antisymmetric, then $R \cup S$ is antisymmetric.

56. If R and S are antisymmetric, then $R \cap S$ is antisymmetric.

57H. If R and S are antisymmetric, then $R \circ S$ is antisymmetric.

58. If R is antisymmetric, then R^{-1} is antisymmetric.

59H. What is wrong with the following argument, which supposedly shows that any relation R on X which is symmetric and transitive is reflexive?

> Let $x \in X$. Using symmetry, we have (x, y) and (y, x) both in R. Since $(x, y), (y, x) \in R$, by transitivity we have $(x, x) \in R$. Therefore, R is reflexive.

1.3

Equivalence Relations

Suppose that we have a set X of 10 balls, each of which is either red, blue, or green (see Figure 1.3.1). If we divide the balls into sets R, B, and G according to color, the family $\{R, B, G\}$ is a partition of X. (Recall that in Section 1.1, we defined a partition of a set X to be a pairwise disjoint collection $\mathcal{S}$ of subsets of X such that $X = \cup \mathcal{S}$.)

A partition can be used to define a relation. If $\mathcal{S}$ is a partition of X we may define xRy to mean that for some set $S \in \mathcal{S}$, both x and y belong to S. For the example of Figure 1.3.1, the relation obtained could be described as "is the

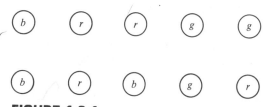

FIGURE 1.3.1

same color as." The next theorem shows that such a relation is always reflexive, symmetric, and transitive.

THEOREM 1.3.1. *Let $\mathcal{S}$ be a partition of a set X. Define xRy to mean that for some set S in $\mathcal{S}$, both x and y belong to S. Then R is reflexive, symmetric, and transitive.*

Proof. Let $x \in X$. Since $X = \cup \mathcal{S}$, $x \in S$ for some $S \in \mathcal{S}$. Thus xRx and R is reflexive.

Suppose that xRy. Then both x and y belong to some set $S \in \mathcal{S}$. Since both y and x belong to S, yRx and R is symmetric.

Finally, suppose that xRy and yRz. Then both x and y belong to some set $S \in \mathcal{S}$ and both y and z belong to some set $T \in \mathcal{S}$. If $S \neq T$, then y would belong to both S and T; but since $\mathcal{S}$ is a pairwise disjoint family, this is impossible. Thus $S = T$ and both x and z belong to S. Therefore, xRz and R is transitive. ∎

EXAMPLE 1.3.2. Consider the partition

$$\mathcal{S} = \{\{1, 3, 5\}, \{2, 6\}, \{4\}\}$$

of $X = \{1, 2, 3, 4, 5, 6\}$. The relation R on X given by Theorem 1.3.1 contains the ordered pairs $(1, 1)$, $(1, 3)$, and $(1, 5)$ because $\{1, 3, 5\}$ is in $\mathcal{S}$. The complete relation is

$$R = \{(1, 1), (1, 3), (1, 5), (3, 1), (3, 3), (3, 5), (5, 1), (5, 3), (5, 5),$$
$$(2, 2), (2, 6), (6, 2), (6, 6), (4, 4)\}.$$

Let $\mathcal{S}$ and R be as in Theorem 1.3.1. If $S \in \mathcal{S}$, we can regard the members of S as equivalent in the sense of the relation R. For this reason, relations that are reflexive, symmetric, and transitive are called **equivalence relations**. In the example of Figure 1.3.1, the relation is "is the same color as"; hence *equivalent* means "is the same color as." Each set in the partition consists of all the balls of a particular color.

DEFINITION 1.3.3. A relation that is reflexive, symmetric, and transitive on a set X is called an *equivalence relation on X*.

EXAMPLE 1.3.4. The relation R of Example 1.3.2 is an equivalence relation on $\{1, 2, 3, 4, 5, 6\}$ because of Theorem 1.3.1. We can also verify directly that R is reflexive, symmetric, and transitive.

The digraph of the relation R of Example 1.3.2 is shown in Figure 1.3.2. Again, we see that R is reflexive (there is a loop at every vertex), symmetric (for every directed edge from v to w, there is also a directed edge from w to v), and transitive (if there is a directed edge from x to y and a directed edge from y to z, there is a directed edge from x to z).

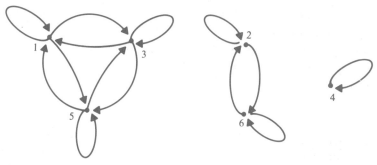

FIGURE 1.3.2

EXAMPLE 1.3.5. Consider the relation

$$R = \{(1, 1), (1, 3), (1, 5), (2, 2), (2, 4), (3, 1), (3, 3), (3, 5), (4, 2),$$
$$(4, 4), (5, 1), (5, 3), (5, 5)\}$$

on $\{1, 2, 3, 4, 5\}$. R is reflexive because $(1, 1), (2, 2), (3, 3), (4, 4), (5, 5)$ $\in R$. R is symmetric because whenever (x, y) is in R, (y, x) is also in R. Finally, R is transitive because whenever (x, y) and (y, z) are in R, (x, z) is also in R. Since R is reflexive, symmetric, and transitive, R is an equivalence relation on $\{1, 2, 3, 4, 5\}$.

EXAMPLE 1.3.6. The relation R of Example 1.2.4 is not an equivalence relation because R is not symmetric.

EXAMPLE 1.3.7. The relation R of Example 1.2.5 is not an equivalence relation because R is neither reflexive nor transitive.

EXAMPLE 1.3.8. The relation R of Example 1.2.15 is an equivalence relation because R is reflexive, symmetric, and transitive.

Given an equivalence relation on a set X, we can partition X by grouping equivalent members of X together. The next theorem gives the details.

THEOREM 1.3.9. *Let R be an equivalence relation on a set X. For each $a \in X$, let*

$$[a] = \{x \in X \mid xRa\}.$$

Then

$$\mathcal{S} = \{[a] \mid a \in X\}$$

is a partition of X.

Proof. Two facts need to be verified in order to deduce that $\mathcal{S}$ is a partition of X:

1. $X = \cup \, \mathcal{S}$.
2. $\mathcal{S}$ is a pairwise disjoint family.

Let $a \in X$. Since aRa, $a \in [a]$. Thus $X = \cup \mathcal{S}$, and fact 1 is established.

It remains to verify fact 2. We will first show that if aRb, then $[a] = [b]$. Suppose that aRb. Let $x \in [a]$. Then xRa. Since aRb and R is transitive, xRb. Therefore, $x \in [b]$ and $[a] \subseteq [b]$. The argument that $[b] \subseteq [a]$ is the same as that just given, but with the roles of a and b interchanged. Thus $[a] = [b]$.

We must show that $\mathcal{S}$ is a pairwise disjoint family. Suppose that $[a]$, $[b] \in \mathcal{S}$ with $[a] \neq [b]$. We must show that $[a] \cap [b] = \varnothing$. Suppose, by way of contradiction, that for some x, $x \in [a] \cap [b]$. Then xRa and xRb. Our result above shows that $[x] = [a]$ and $[x] = [b]$. Thus $[a] = [b]$, which is a contradiction. Therefore, $[a] \cap [b] = \varnothing$ and $\mathcal{S}$ is a pairwise disjoint family. ∎

DEFINITION 1.3.10.   Let R be an equivalence relation on a set X. The sets $[a]$ defined in Theorem 1.3.9 are called the *equivalence classes of X given by the relation R*.

EXAMPLE 1.3.11.   Consider the equivalence relation R of Example 1.3.2. The equivalence class $[1]$ containing 1 consists of all x such that $(x, 1) \in R$. Therefore,

$$[1] = \{1, 3, 5\}.$$

The remaining equivalence classes are found similarly:

$$[3] = [5] = \{1, 3, 5\}, \qquad [2] = [6] = \{2, 6\}, \qquad [4] = \{4\}.$$

EXAMPLE 1.3.12.   The equivalence classes appear quite clearly in the digraph of an equivalence relation. The three equivalence classes of the relation R of Example 1.3.2 appear in the digraph of R (shown in Figure 1.3.2) as the three subgraphs whose vertices are $\{1, 3, 5\}$, $\{2, 6\}$, and $\{4\}$. A subgraph G that represents an equivalence class is a largest subgraph of the original digraph having the property that for any vertices v and w in G, there is a directed edge from v to w. For example, if v, $w \in \{1, 3, 5\}$, there is a directed edge from v to w. Moreover, no additional vertices can be added to 1, 3, 5 so that the resulting vertex set has a directed edge between each pair of vertices.

EXAMPLE 1.3.13.   The equivalence classes for the equivalence relation of Example 1.3.5 are

$$[1] = [3] = [5] = \{1, 3, 5\}, \qquad [2] = [4] = \{2, 4\}.$$

EXAMPLE 1.3.14.   The equivalence classes for the equivalence relation of Example 1.2.15 are

$$[a] = \{a\}, \qquad [b] = \{b\}, \qquad [c] = \{c\}.$$

EXAMPLE 1.3.15. Let $X = \{1, 2, \ldots, 10\}$. Define xRy to mean that 3 divides $x - y$. We can readily verify that the relation R is reflexive, symmetric, and transitive. Thus R is an equivalence relation on X.

Let us determine the members of the equivalence classes. The equivalence class [1] consists of all x with $xR1$. Thus

$$[1] = \{x \in X \mid 3 \text{ divides } x - 1\}$$
$$= \{1, 4, 7, 10\}.$$

Similarly,

$$[2] = \{2, 5, 8\}$$
$$[3] = \{3, 6, 9\}.$$

These three sets partition X. Note that

$$[1] = [4] = [7] = [10]$$
$$[2] = [5] = [8]$$
$$[3] = [6] = [9].$$

For this relation, *equivalence* is "has the same remainder when divided by 3."

EXERCISES

In Exercises 1–8, determine whether the given relation is an equivalence relation on $\{1, 2, 3, 4, 5\}$. If the relation is an equivalence relation, list the equivalence classes. (In Exercises 5–8, x, $y \in \{1, 2, 3, 4, 5\}$).

1H. $\{(1, 1), (2, 2), (3, 3), (4, 4), (5, 5), (1, 3), (3, 1)\}$

2. $\{(1, 1), (2, 2), (3, 3), (4, 4), (5, 5), (1, 3), (3, 1), (3, 4), (4, 3)\}$

3H. $\{(1, 1), (2, 2), (3, 3), (4, 4)\}$

4. $\{(1, 1), (2, 2), (3, 3), (4, 4), (5, 5), (1, 5), (5, 1), (3, 5), (5, 3), (1, 3), (3, 1)\}$

5H. $\{(x, y) \mid 1 \leq x \leq 5, 1 \leq y \leq 5\}$

6. $\{(x, y) \mid 4 \text{ divides } x - y\}$

7H. $\{(x, y) \mid 3 \text{ divides } x + y\}$

8. $\{(x, y) \mid x \text{ divides } 2 - y\}$

In Exercises 9–14, list the members of the equivalence relation on $\{1, 2, 3, 4\}$ defined (as in Theorem 1.3.1) by the given partition. Also, find the equivalence classes [1], [2], [3], and [4].

9H. $\{\{1, 2\}, \{3, 4\}\}$ **10.** $\{\{1\}, \{2\}, \{3, 4\}\}$

11H. $\{\{1\}, \{2\}, \{3\}, \{4\}\}$ **12.** $\{\{1, 2, 3\}, \{4\}\}$

13H. $\{\{1, 2, 3, 4\}\}$ **14.** $\{\{1\}, \{2, 4\}, \{3\}\}$

Exercises 15–17 refer to the relation R defined on the set of eight-bit strings by b_1Rb_2 provided that the first four bits of b_1 and b_2 coincide.

15H. Show that R is an equivalence relation.

16. How many equivalence classes are there?

17H. List one member of each equivalence class.

18. Let

$$X = \{\text{San Francisco, Pittsburgh, Chicago, San Diego,}$$
$$\text{Philadelphia, Los Angeles}\}.$$

Define a relation R on X as xRy if x and y are in the same state.
(a) Show that R is an equivalence relation.
(b) List the equivalence classes of X.

19H. Show that if R is an equivalence relation on X, then

$$\text{domain } R = \text{range } R = X.$$

20. If an equivalence relation has only one equivalence class, what must the relation look like?

21H. If R is an equivalence relation on a set X and $|X| = |R|$, what must the relation look like?

22. By listing ordered pairs, give an example of an equivalence relation on $\{1, 2, 3, 4, 5, 6\}$ having exactly four equivalence classes.

23H. How many equivalence relations are there on the set $\{1, 2, 3\}$?

24. Let $X = \{1, 2, \ldots, 10\}$. Define a relation R on $X \times X$ by $(a, b)R(c, d)$ if $a + d = b + c$.
(a) Show that R is an equivalence relation on $X \times X$.
(b) List one member of each equivalence class of $X \times X$.

25H. Let $X = \{1, 2, \ldots, 10\}$. Define a relation R on $X \times X$ by $(a, b)R(c, d)$ if $ad = bc$.
(a) Show that R is an equivalence relation on $X \times X$.
(b) List one member of each equivalence class of $X \times X$.
(c) Describe the relation R in familiar terms.

26. Let R be a reflexive and transitive relation on X. Show that $R \cap R^{-1}$ is an equivalence relation on X.

27H. Let R_1 and R_2 be equivalence relations on X.
(a) Show that $R_1 \cap R_2$ is an equivalence relation on X.
(b) Describe the equivalence classes of $R_1 \cap R_2$ in terms of the equivalence classes of R_1 and the equivalence classes of R_2.

28. Suppose that $\mathscr{S}$ is a collection of subsets of a set X and $X = \cup \mathscr{S}$. (It is not assumed that the family $\mathscr{S}$ is pairwise disjoint.) Define xRy to mean that for some set $S \in \mathscr{S}$, both x and y are in S. Is R necessarily reflexive, symmetric, or transitive?

29H. Let S be a unit square including the interior as shown in the accompanying figure. Define a relation R on S by $(x, y)R(x', y')$ if $(x = x'$ and $y = y')$ or $(y = y'$ and $x = 0$ and $x' = 1)$ or $(y = y'$ and $x = 1$ and $x' = 0)$.

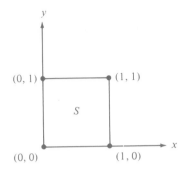

(a) Show that R is an equivalence relation on S.

(b) If points in the same equivalence class are glued together, how would you describe the figure formed?

30. Let S be a unit square including the interior (as in Exercise 29). Define a relation R' on S by $(x, y)R'(x', y')$ if $(x = x'$ and $y = y')$ or $(y = y'$ and $x = 0$ and $x' = 1)$ or $(y = y'$ and $x = 1$ and $x' =0)$ or $(x = x'$ and $y = 0$ and $y' = 1)$ or $(x = x'$ and $y = 1$ and $y' = 0)$. Let $R = R' \cup \{((0, 0), (1, 1)), ((0, 1), (1,0)), ((1, 0), (0, 1)), ((1, 1), (0, 0))\}$.

(a) Show that R is an equivalence relation on S.

(b) If points in the same equivalence class are glued together, how would you describe the figure formed?

Let R be a relation on a set X. Define

$$\rho(R) = R \cup \{(x, x) \mid x \in X\}$$
$$\sigma(R) = R \cup R^{-1}$$
$$R^n = R \circ R \circ R \circ \cdots \circ R \quad (n\ R's)$$
$$\tau(R) = \cup \{R^n \mid n = 1, 2, \ldots\}.$$

The relation $\tau(R)$ is called the **transitive closure** of R.

31H. For the relations R_1 and R_2 of Exercise 37, Section 1.2, find $\rho(R_i)$, $\sigma(R_i)$, $\tau(R_i)$, and $\tau(\sigma(\rho(R_i)))$ for $i = 1, 2$.

32. Show that $\rho(R)$ is reflexive.

33H. Show that $\sigma(R)$ is symmetric.

34. Show that $\tau(R)$ is transitive.

★35H.† Show that $\tau(\sigma(\rho(R)))$ is an equivalence relation containing R.

★36. Show that $\tau(\sigma(\rho(R)))$ is the smallest equivalence relation on X containing R; that is, show that if R' is an equivalence relation on X and $R' \supseteq R$, then $R' \supseteq \tau(\sigma(\rho(R)))$.

★37H. Show that R is transitive if and only if $\tau(R) = R$.

†A starred exercise indicates a problem of above-average difficulty.

In Exercises 38–44, write true if the statement is true for all relations R_1 and R_2 on an arbitrary set X; otherwise, give a counterexample.

38. $\rho(R_1 \cup R_2) = \rho(R_1) \cup \rho(R_2)$

39H. $\sigma(R_1 \cap R_2) = \sigma(R_1) \cap \sigma(R_2)$

40. $\tau(R_1 \cup R_2) = \tau(R_1) \cup \tau(R_2)$

41H. $\tau(R_1 \cap R_2) = \tau(R_1) \cap \tau(R_2)$

42. $\sigma(\tau(R_1)) = \tau(\sigma(R_1))$

43H. $\sigma(\rho(R_1)) = \rho(\sigma(R_1))$

44. $\rho(\tau(R_1)) = \tau(\rho(R_1))$

1.4

Functions

A **function** is a special kind of relation. Recall (see Definition 1.2.1) that a relation R from X to Y is a subset of the Cartesian product $X \times Y$ and that

$$\text{domain } R = \{x \in X \mid (x, y) \in R \text{ for some } y \in Y\}.$$

If f is a relation from X to Y, in order for f also to be a function, the domain of f must equal X and if (x, y) and (x, y') are in f, we must have $y = y'$.

DEFINITION 1.4.1.   A *function* f from X to Y is a relation from X to Y having the properties:

1. The domain of f is X.
2. If $(x, y), (x, y') \in f$, then $y = y'$.

A function from X to Y is sometimes denoted $f: X \to Y$.

EXAMPLE 1.4.2.   The relation

$$f = \{(1, a), (2, b), (3, a)\}$$

from $X = \{1, 2, 3\}$ to $Y = \{a, b, c\}$ is a function from X to Y. The domain of f is X and the range of f is $\{a, b\}$. [Recall (see Definition 1.2.1) that the range of a relation R is the set

$$\{y \in Y \mid (x, y) \in R \text{ for some } x \in X\}.]$$

EXAMPLE 1.4.3.   The relation

$$R = \{(1, a), (2, a), (3, b)\} \qquad (1.4.1)$$

from $X = \{1, 2, 3, 4\}$ to $Y = \{a, b, c\}$ is not a function from X to Y. Property 1 of Definition 1.4.1 is violated. The domain of R, $\{1, 2, 3\}$, is not equal to X. If (1.4.1) were regarded as a relation from $X' = \{1, 2, 3\}$ to $Y = \{a, b, c\}$, it would be a function from X' to Y.

EXAMPLE 1.4.4. The relation

$$R = \{(1, a), (2, b), (3, c), (1, b)\}$$

from $X = \{1, 2, 3\}$ to $Y = \{a, b, c\}$ is not a function from X to Y. Property 2 of Definition 1.4.1 is violated. We have $(1, a)$ and $(1, b)$ in R but $a \neq b$.

Given a function f from X to Y, according to Definition 1.4.1, for each element x of the domain X, there is exactly one $y \in Y$ with $(x, y) \in f$. This unique value y is denoted $f(x)$. In other words, $y = f(x)$ is another way to write $(x, y) \in f$.

EXAMPLE 1.4.5. For the function f of Example 1.4.2, we may write

$$f(1) = a, \qquad f(2) = b, \qquad f(3) = a.$$

Functions involving the **modulus operator** play an important role in mathematics and computer science.

DEFINITION 1.4.6. If x is a nonnegative integer and y is a positive integer, we define $x \bmod y$ to be the remainder when x is divided by y.

EXAMPLE 1.4.7

$$6 \bmod 2 = 0, \quad 5 \bmod 1 = 0, \quad 8 \bmod 12 = 8, \quad 199673 \bmod 2 = 1$$

EXAMPLE 1.4.8 Hash Functions. Suppose that we have cells in a computer memory indexed from 0 to 10 (see Figure 1.4.1). We wish to store and retrieve arbitrary nonnegative integers in these cells. One approach is to use a **hash function**. A hash function takes a data item to be stored or retrieved and computes the first choice for a location for the item. For example, for our problem, to store or retrieve the number n, we might take as the first choice for a location, $n \bmod 11$. Our hash function becomes

$$h(n) = n \bmod 11.$$

Figure 1.4.1 shows the result of storing 15, 558, 32, 132, 102, and 5, in this order, in initially empty cells.

Now suppose that we want to store 257. Since $h(257) = 4$, 257 should be stored at location 4; however, this position is already occupied. In this case we say that a **collision** has occurred. More precisely, a collision occurs for a hash function H if $H(x) = H(y)$, but $x \neq y$. To handle collisions, a **collision resolution policy** is required. One simple collision resolution policy is to find the next highest (with 0 assumed to follow 10) unoccupied cell. If we use this collision resolution policy, we would store 257 at location 6 (see Figure 1.4.1).

132			102	15	5	257		558		32
0	1	2	3	4	5	6	7	8	9	10

FIGURE 1.4.1

If we want to locate a stored value n, we compute $m = h(n)$ and begin looking at location m. If n is not at this position, we look in the next highest position (again, 0 is assumed to follow 10), if n is not in this position, we proceed to the next highest position, and so on. If we reach an empty cell or return to our original position, we conclude that n is not present; otherwise, we obtain the position of n.

DEFINITION 1.4.9. A function f from X to Y is said to be *one-to-one* (or *injective*) if for each $y \in Y$, there is at most one $x \in X$ with $f(x) = y$.

The condition given in Definition 1.4.9 for a function to be one-to-one is equivalent to: If $x, x' \in X$ and $f(x) = f(x')$, then $x = x'$.

Because the amount of potential data is usually so much larger than the available memory, hash functions are usually not one-to-one. In other words, most hash functions produce collisions.

EXAMPLE 1.4.10. The function

$$f = \{(1, b), (3, a), (2, c)\}$$

from $X = \{1, 2, 3\}$ to $Y = \{a, b, c, d\}$ is one-to-one.

EXAMPLE 1.4.11. The function of Example 1.4.2 is not one-to-one since $f(1) = a = f(3)$.

If the range of a function f is Y, the function is said to be **onto** Y.

DEFINITION 1.4.12. If f is a function from X to Y and the range of f is Y, f is said to be *onto* Y (or an *onto function* or a *surjective function*).

EXAMPLE 1.4.13. The function

$$f = \{(1, a), (2, c), (3, b)\}$$

from $X = \{1, 2, 3\}$ to $Y = \{a, b, c\}$ is one-to-one and onto Y.

EXAMPLE 1.4.14. The function f of Example 1.4.10 is *not* onto $Y = \{a, b, c, d\}$. It is onto $\{a, b, c\}$.

DEFINITION 1.4.15. A function that is both one-to-one and onto is called a *bijection*.

EXAMPLE 1.4.16. The function f of Example 1.4.13 is a bijection.

Suppose that f is a one-to-one, onto function from X to Y. It can be shown (see Exercise 61) that the converse relation

$$\{(y, x) \mid (x, y) \in f\}$$

is a function from Y to X. This new function, denoted f^{-1}, is called f **inverse**.

EXAMPLE 1.4.17. For the function f of Example 1.4.13, we have

$$f^{-1} = \{(a, 1), (c, 2), (b, 3)\}.$$

Since functions are special kinds of relations, we can form the **composition** of two functions. Specifically, suppose that g is a function from X to Y and f is a function from Y to Z. Given $x \in X$, we may apply g to determine a unique element $y = g(x) \in Y$. We may then apply f to determine a unique element $z = f(y) = f(g(x)) \in Z$. The resulting function from X to Z is called the composition of f with g and is denoted $f \circ g$.

EXAMPLE 1.4.18. Given

$$g = \{(1, a), (2, a), (3, c)\},$$

a function from $X = \{1, 2, 3\}$ to $Y = \{a, b, c\}$, and

$$f = \{(a, y), (b, x), (c, z)\},$$

a function from Y to $Z = \{x, y, z\}$, the composition function from X to Z is the function

$$f \circ g = \{(1, y), (2, y), (3, z)\}.$$

We next define some special kinds of functions.

DEFINITION 1.4.19. A *sequence of elements of X* is a function from $\{1, 2, \ldots\}$ to X.

EXAMPLE 1.4.20. If we define $f(n) = -2n$, we obtain the sequence

$$f(1) = -2, f(2) = -4, \ldots$$

of negative integers. We sometimes write the sequence

$$-2, -4, \ldots$$

or

$$f_1, f_2, \ldots$$

or

$$\{f_n\}$$

if the definition of $f(n) = f_n$ is given or can be easily inferred.

DEFINITION 1.4.21. A *string over X* (or a *finite sequence of elements of X*) is a function from $\{1, 2, \ldots, n\}$, for some positive integer n, to X.

EXAMPLE 1.4.22. Let $X = \{a, b, c\}$. If we let

$$f(1) = b, \quad f(2) = a, \quad f(3) = a, \quad f(4) = c,$$

we obtain a string over X. This string is written *baac*.

The order of the elements in a string is taken into account. For example, the string *baac* is distinct from the string *aacb*.

Repetitions in a string can be specified by superscripts. For example, the string *bbaaac* may be written b^2a^3c.

We regard the null function $\emptyset$ as a string with no elements. This string is called the **null string** and is denoted λ. We let X^* denote the set of all strings over X (including the null string).

If $s = f_1f_2 \cdots f_n$ is a string, we call n the **length** of s. The length of s is denoted $|s|$. For example,

$$|baac| = 4.$$

We define $|\lambda| = 0$.

If $s = f_1 \cdots f_m$ and $t = g_1 \cdots g_n$ are two strings, we define the **concatenation** of s and t to be the string

$$st = f_1 \cdots f_m g_1 \cdots g_n.$$

EXAMPLE 1.4.23. If $s = aab$ and $t = cabd$, then

$$st = aabcabd$$

$$ts = cabdaab$$

$$s\lambda = s = aab$$

$$\lambda s = s = aab.$$

A **binary operator** on a set X associates with each ordered pair of elements in X one element in X.

DEFINITION 1.4.24. A function from $X \times X$ into X is called a *binary operator* on X.

EXAMPLE 1.4.25. Let $X = \{1, 2, \ldots\}$. If we define

$$f(x, y) = x + y,$$

then f is a binary operator on X.

EXAMPLE 1.4.26. Let $X = \{a, b, c\}$. If we define

$$f(s, t) = st,$$

where s and t are strings over X and st is the concatenation of s and t, then f is a binary operator on X^*.

DEFINITION 1.4.27. A function from X into X is called a *unary operator* on X.

EXAMPLE 1.4.28. Let U be a universal set. If we define

$$f(X) = \overline{X}, \qquad X \subseteq U,$$

then f is a unary operator on $\mathcal{P}(U)$.

EXERCISES

Determine whether each relation in Exercises 1–5 is a function from $X = \{1, 2, 3, 4\}$ to $Y = \{a, b, c, d\}$. If it is a function, find its domain and range and determine if it is one-to-one or onto. If it is both one-to-one and onto, give the description of the inverse function as a set of ordered pairs and give the domain and range of the inverse function.

1H. $\{(1, a), (2, a), (3, c), (4, b)\}$

2. $\{(1, c), (2, a), (3, b), (4, c), (2, d)\}$

3H. $\{(1, c), (2, d), (3, a), (4, b)\}$

4. $\{(1, d), (2, d), (4, a)\}$

5H. $\{(1, b), (2, b), (3, b), (4, b)\}$

6.   Give an example of a function that is one-to-one, but not onto.

7H.   Give an example of a function that is onto, but not one-to-one.

8.   Give an example of a function that is neither one-to-one nor onto.

9H.   Given

$$g = \{(1, b), (2, c), (3, a)\},$$

a function from $X = \{1, 2, 3\}$ to $Y = \{a, b, c, d\}$, and

$$f = \{(a, x), (b, x), (c, z), (d, w)\},$$

a function from Y to $Z = \{w, x, y, z\}$, write $f \circ g$ as a set of ordered pairs.

10.   Given

$$f = \{(x, x^2) \mid x \in X\},$$

a function from $X = \{-5, -4, \ldots, 4, 5\}$ to the set of integers, write f as a set of ordered pairs. Is f one-to-one or onto?

11H.   How many functions are there from $\{1, 2\}$ into $\{a, b\}$? Which are one-to-one? Which are onto?

12.   Given

$$f = \{(a, b), (b, a), (c, b)\},$$

a function from $X = \{a, b, c\}$ to X:

(a) Write $f \circ f$ and $f \circ f \circ f$ as sets of ordered pairs.

(b) Define

$$f^n = f \circ f \circ \cdots \circ f$$

to be the *n*-fold composition of f with itself. Find f^9 and f^{623}.

13H.   Let f be the function from $X = \{0, 1, 2, 3, 4\}$ to X defined by

$$f(x) = 4x \bmod 5.$$

Write f as a set of ordered pairs. Is f one-to-one or onto?

14.   Let f be the function from $X = \{0, 1, 2, 3, 4, 5\}$ to X defined by

$$f(x) = 4x \bmod 6.$$

Write f as a set of ordered pairs. Is f one-to-one or onto?

★15H.   Let m and n be positive integers. Let f be the function from

$$X = \{0, 1, \ldots, m - 1\}$$

to X defined by

$$f(x) = nx \bmod m.$$

Find conditions on m and n that assure that f is one-to-one and onto.

For each hash function in Exercises 16–19, show how the data would be inserted in the order given in initially empty cells. Use the collision resolution policy of Example 1.4.8.

16. $h(x) = x \bmod 11$; cells indexed 0 to 10; data: 53, 13, 281, 743, 377, 20, 10, 796

17H. $h(x) = x \bmod 17$; cells indexed 0 to 16; data: 714, 631, 26, 373, 775, 906, 509, 2032, 42, 4, 136, 1028

18. $h(x) = x^2 \bmod 11$; cells and data as in Exercise 16

19H. $h(x) = (x^2 + x) \bmod 17$; cells and data as in Exercise 17

20.   Suppose that we store and retrieve data as described in Example 1.4.8. Will any problem arise if we delete data? Explain.

21H.   Suppose that we store data as described in Example 1.4.8 and that we never store more than 10 items. Will any problem arise when retrieving data if we stop searching when we encounter an empty cell? Explain.

22.   Suppose that we store data as described in Example 1.4.8 and retrieve data as described in Exercise 21. Will any problem arise if we delete data? Explain.

Let g be a function from X to Y and let f be a function from Y to Z. For each statement in Exercises 23–29, write true if the statement is true. If the statement is false, give a counterexample.

23H. If f is one-to-one, then $f \circ g$ is one-to-one.

24. If f and g are onto, then $f \circ g$ is onto.

25H. If f and g are one-to-one and onto, then $f \circ g$ is one-to-one and onto.

26. If $f \circ g$ is one-to-one, then f is one-to-one.

27H. If $f \circ g$ is one-to-one, then g is one-to-one.

28. If $f \circ g$ is onto, then f is onto.

29H. If $f \circ g$ is onto, then g is onto.

If f is a function from X to Y and $A \subseteq X$ and $B \subseteq Y$, we define

$$f(A) = \{f(x) \mid x \in A\}$$
$$f^{-1}(B) = \{x \in X \mid f(x) \in B\}.$$

We call $f^{-1}(B)$ the **inverse image** of B under f.

30. Let

$$g = \{(1, a), (2, c), (3, c)\}$$

be a function from $X = \{1, 2, 3\}$ to $Y = \{a, b, c, d\}$. Let $S = \{1\}$, $T = \{1, 3\}$, $U = \{a\}$, and $V = \{a, c\}$. Find $g(S)$, $g(T)$, $g^{-1}(U)$, and $g^{-1}(V)$.

★31H. Let f be a function from X to Y. Prove that f is one-to-one if and only if

$$f(A \cap B) = f(A) \cap f(B)$$

for all subsets A and B of X.

32. Let f be a function from X to Y. Define a relation R on X by

$$xRy \qquad \text{if } f(x) = f(y).$$

Show that R is an equivalence relation on X.

33H. Let f be a function from X onto Y. Let

$$\mathcal{S} = \{f^{-1}(\{y\}) \mid y \in Y\}.$$

[The definition of $f^{-1}(B)$, where B is a set, precedes Exercise 30.] Show that $\mathcal{S}$ is a partition of X. Describe an equivalence relation that gives rise to this partition.

34. Let R be an equivalence relation on a set A. Define a function f from A to the set of equivalence classes of A by the rule

$$f(x) = [x].$$

When do we have $f(x) = f(y)$?

35H. Let R be an equivalence relation on a set A. Suppose that g is a function from A into a set X having the property that if xRy, then $g(x) = g(y)$. Show that

$$h([x]) = g(x)$$

defines a function from the set of equivalence classes of A into X. [What needs to be shown is that h *uniquely* assigns a value to $[x]$; that is, if $[x] = [y]$, then $g(x) = g(y)$.]

★**36.**   Let f be a function from X to Y. Show that f is one-to-one if and only if whenever g is a one-to-one function from any set A to X, $f \circ g$ is one-to-one.

★**37H.**   Let f be a function from X to Y. Show that f is onto Y if and only if whenever g is a function from Y onto any set Z, $g \circ f$ is onto Z.

Let U be a universal set and let $X \subseteq U$. Define

$$C_X(x) = \begin{cases} 1 & \text{if } x \in X \\ 0 & \text{if } x \notin X. \end{cases}$$

We call C_X the **characteristic function** of X (in U).

38.   Show that $C_{X \cap Y}(x) = C_X(x)C_Y(x)$ for all $x \in U$.

39H.   Show that $C_{X \cup Y}(x) = C_X(x) + C_Y(x) - C_X(x)C_Y(x)$ for all $x \in U$.

40.   Show that $C_{\overline{X}}(x) = 1 - C_X(x)$ for all $x \in U$.

41H.   Show that $C_{X - Y}(x) = C_X(x)[1 - C_Y(x)]$ for all $x \in U$.

42.   Show that if $X \subseteq Y$, then $C_X(x) \le C_Y(x)$ for all $x \in U$.

43H.   Show that $C_{X \cup Y}(x) = C_X(x) + C_Y(x)$ for all $x \in U$, if and only if $X \cap Y = \varnothing$.

44.   Find a formula for $C_{X \triangle Y}$. ($X \triangle Y$ is the symmetric difference of X and Y. The definition is given before Exercise 68, Section 1.1.)

45H.   Show that the function f from $\mathcal{P}(U)$ to the set of characteristic functions in U defined by

$$f(X) = C_X$$

is one-to-one and onto.

46.   Let f be a characteristic function in X. Define a relation R on X by xRy if $f(x) = f(y)$. According to Exercise 32, R is an equivalence relation. What are the equivalence classes?

If X and Y are sets, we define X to be **equivalent** to Y if there is a one-to-one, onto function from X to Y.

47H.   Show that set equivalence is an equivalence relation.

48.   If X and Y are finite sets and X is equivalent to Y, what does this say about X and Y?

49.   Show that the sets $\{1, 2, \ldots\}$ and $\{2, 4, \ldots\}$ are equivalent.

★**50H.**   Show that for any set X, X is not equivalent to $\mathcal{P}(X)$, the power set of X.

51H.   Let X and Y be sets. Show that there is a one-to-one function from X to Y if and only if there is a function from Y onto X.

52. Let $X = \{0, 1\}$. List all strings of length 2 over X. List all strings of length 2 or less over X.

53H. A string s is a **substring** of t if there are strings u and v with $t = usv$. Find all substrings of the string *babc*.

A binary operator f on a set X is **commutative** if $f(x, y) = f(y, x)$ for all $x, y \in X$. In Exercises 54–58, state whether the given function f is a binary operator on the set X. If f is not a binary operator, state why. State whether each binary operator is commutative or not.

54. $f(x, y) = x + y, X = \{1, 2, \ldots\}$
55H. $f(x, y) = x - y, X = \{1, 2, \ldots\}$
56. $f(s, t) = st, X$ is the set of strings over $\{a, b\}$
57H. $f(x, y) = x/y, X = \{0, 1, 2, \ldots\}$
58. $f(x, y) = x^2 + y^2 - xy, X = \{1, 2, \ldots\}$

In Exercises 59 and 60, give an example of a unary operator (different from $f(x) = x$, for all x) on the given set.

59H. $\{\ldots, -2, -1, 0, 1, 2, \ldots\}$
60. The set of strings over $\{a, b\}$

61. Show that if f is a one-to-one, onto function from X to Y, then

$$\{(y, x) \mid (x, y) \in f\}$$

is a one-to-one, onto function from Y to X.

1.5

Algorithms

The solution of a problem by a computer requires a set of precise instructions. Informally, such a set of precise instructions is called an **algorithm**. Although examples of algorithms can be found throughout history going back at least as far as ancient Babylonia, it is the rise of computer science that has triggered the recent surge of interest in algorithms. It is proper to think of an algorithm as a set of instructions precisely enough stated so that a machine could carry out the instructions to solve the problem. In this section we describe in some detail what an algorithm is and illustrate the concepts by an example.

We begin with an example of an algorithm. Algorithm 1.5.1 determines the largest value in a finite numerical sequence.

ALGORITHM 1.5.1 Finding the Largest Value in a Finite Sequence. The algorithm searches a given sequence

$$S(1), S(2), \ldots, S(N)$$

for the largest value and returns it as LARGE.

1. [Initialization.] Set $I := 1$ and LARGE $:= S(1)$. (I gives the current element of the sequence under examination.)
2. [Find a larger value?] If $S(I) >$ LARGE, then LARGE $:= S(I)$. (If a larger value is found, update LARGE.)
3. [Terminate?] If $I = N$, then stop. (The largest value is LARGE.)
4. [Continue search.] $I := I + 1$. Go to step 2.

Algorithm 1.5.1 consists of a finite (namely four) sequence of steps for solving the problem of finding the largest value in a sequence of N numbers. Each of the four steps tells us precisely what action to take. Given a sequence of N numbers, if we follow the steps set forth in Algorithm 1.5.1, we will find the largest value.

The algorithm specification illustrated by Algorithm 1.5.1, which we will use throughout the book, is similar to that used in [Knuth, 1973 Vols. 1 and 3; 1981]. Each algorithm has a title followed by a brief description of the algorithm. Next are the numbered steps. The algorithm begins with step 1 and proceeds sequentially through the numbered steps unless there is an explicit instruction to do otherwise—such as "Go to step 2." Each step contains a brief description of its purpose in brackets. We will insert comments, in parentheses, any place we feel that additional details will aid the reader in understanding a part of a step.

The assignment operator is denoted $:=$ and thus, $A := B$ means that the value of the variable B is to be copied to the variable A, or equivalently, that the current value of the variable A is to be replaced by the value of the variable B. The equal sign "$=$" will only be used in a conditional statement such as that in step 3 in Algorithm 1.5.1. Here, if the condition $I = N$ is true, we will execute the instructions following "then"; otherwise, no further instructions will be executed in this step.

We will illustrate how Algorithm 1.5.1 would be applied to a particular sequence.

EXAMPLE 1.5.2. Show how Algorithm 1.5.1 finds the largest value in the sequence

$$S(1) = 2.3, \quad S(2) = 4.1.$$

We begin with the first step, which tells us to set

$$I := 1, \qquad \text{LARGE} := 2.3.$$

At step 2, we test the condition

$$S(I) > \text{LARGE},$$

which in our case is

$$2.3 > 2.3.$$

Since this condition is false, we ignore the rest of step 2 and proceed to step 3. We test the condition

$$I = N,$$

which in our case is

$$1 = 2.$$

Since this condition is false, we ignore the rest of step 3 and proceed to step 4. At this point, we set

$$I := 2$$

and proceed to step 2. The condition this time is

$$4.1 > 2.3.$$

Since this condition is true, we set

$$LARGE := S(2),$$

which in our case is

$$LARGE := 4.1.$$

At step 3, we test the condition

$$2 = 2,$$

which is true, so we stop. At this point,

$$LARGE := 4.1,$$

which is the largest value in the given sequence.

An **algorithm** is a finite set of instructions having the following characteristics:

1. *Precision.* The steps are precisely stated.
2. *Uniqueness.* The intermediate results of each step of execution are uniquely defined and depend only on the inputs and the results of the preceding steps.
3. *Finiteness.* The algorithm stops after finitely many instructions have been executed.
4. *Input.* The algorithm receives input.
5. *Output.* The algorithm produces output.
6. *Generality.* The algorithm applies to a set of inputs.

We consider next each of these properties, in turn, in more detail.

The steps of an algorithm must be precisely stated. The steps of Algorithm 1.5.1 are stated sufficiently precisely so that the algorithm can be converted to

```
10   INPUT N
20   DIM S(N)
30   FOR I = 1 TO N
40   INPUT S(I)
50   NEXT I
60   I = 1
70   LARGE = S(1)
80   IF S(I) > LARGE THEN LARGE = S(I)
90   IF I = N THEN PRINT LARGE: END
100  I = I + 1
110  GOTO 80
```

FIGURE 1.5.1

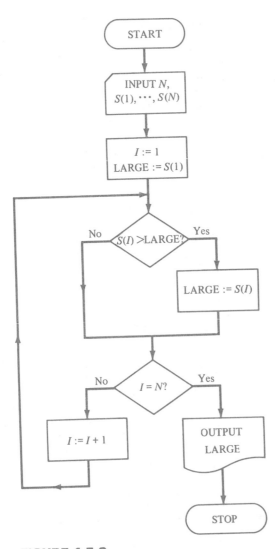

FIGURE 1.5.2

```
procedure FIND_LARGE (S, N);
  integer N, I;
  real S(N), LARGE;
  I := 1;
  LARGE := S(1);
  while I < = N do
    if S(I) > LARGE then LARGE := S(I);
    I := I + 1;
  endwhile;
  return (LARGE);
end FIND_LARGE.
```

FIGURE 1.5.3

a computer program. A representation of Algorithm 1.5.1 in the BASIC programming language is given in Figure 1.5.1. Lines 10 to 50 provide a method
of entering the sequence into the computer. Lines 60 and 70 correspond to
step 1. Line 80 corresponds to step 2. Line 90 corresponds to step 3. The
largest value LARGE is also output at line 90. Lines 100 and 110 correspond
to step 4.

Given values of the inputs, each intermediate step of an algorithm produces a unique result. The action to be taken at each of the four steps of Algorithm 1.5.1 is completely determined once the value of N and the values of
$S(1), \ldots, S(N)$ are given. If two persons independently applied Algorithm
1.5.1 to the same sequence, each would obtain the same intermediate results at
each step.

An algorithm stops after finitely many steps answering the given question.
Algorithm 1.5.1 stops after N iterations giving the result in the variable LARGE.

An algorithm receives input and produces output. Algorithm 1.5.1 receives,
as input, the sequence of N numbers

$$S(1), S(2), \ldots, S(N)$$

and outputs one value,

$$LARGE.$$

An algorithm must be general. Algorithm 1.5.1 finds the largest value in a
sequence of arbitrary length and there is no restriction on the numbers making
up the sequence. An algorithm that would find the largest value in a sequence
of length N, where $37 < N \le 54$ and each $S(I)$ satisfies $104.6 \le S(I) < 723.5$,
would be of little use.

There are many ways to specify algorithms. In Figure 1.5.2, Algorithm 1.5.1
is given as a flowchart and in Figure 1.5.3, Algorithm 1.5.1 is written in a
hypothetical computer language. A specification written in ordinary text is usually not sufficiently precise to be called an algorithm.

Our description of what an algorithm is will suffice for our needs in this book.
However, it should be noted that it is possible to give a precise, mathematical
definition of algorithm (see Section 7.6).

EXERCISES

1H.   Write an algorithm, in the style of Algorithm 1.5.1, which finds the smallest element in a sequence of numbers

$$S(1), \ldots, S(N).$$

2.   Write an algorithm, in the style of Algorithm 1.5.1, which finds the index J of the first occurrence of the largest element in a sequence of numbers

$$S(1), \ldots, S(N).$$

(For example, if the sequence were

$$6.2 \quad 8.9 \quad 4.2 \quad 8.9,$$

the algorithm would return the value 2.)

3H.   Write an algorithm, in the style of Algorithm 1.5.1, which returns the index of the first occurrence of the value KEY in a sequence of strings

$$S(1), \ldots, S(N).$$

If KEY is not in the sequence, the algorithm returns the value 0. (For example, if the sequence were

$$\text{'MARY'} \quad \text{'JOE'} \quad \text{'MARK'} \quad \text{'RUDY'},$$

and KEY were 'MARK', the algorithm would return the value 3.)

4.   Write an algorithm, in the style of Algorithm 1.5.1, which finds the index of the first string that is out of alphabetical order in a sequence of strings

$$S(1), \ldots, S(N).$$

If all of the strings are in alphabetical order, the algorithm returns the value 0. (For example, if the sequence were

$$\text{'AMY'} \quad \text{'BRUNO'} \quad \text{'ELIE'} \quad \text{'DAN'} \quad \text{'ZEKE'},$$

the algorithm would return the value 4.)

5H.   Write an algorithm, in the style of Algorithm 1.5.1, which reverses the sequence

$$S(1), \ldots, S(N).$$

(For example, if the sequence were

$$\text{'AMY'} \quad \text{'BRUNO'} \quad \text{'ELIE'},$$

the algorithm would return the sequence

$$\text{'ELIE'} \quad \text{'BRUNO'} \quad \text{'AMY'}.)$$

6.   Write an algorithm, in the style of Algorithm 1.5.1, which tests whether the positive integer $N > 1$ is prime.

7H.   Write an algorithm, in the style of Algorithm 1.5.1, which finds the smallest prime number greater than the positive integer N.

8. Write the following algorithm in the style of Algorithm 1.5.1.

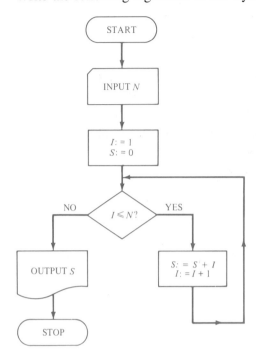

9H. Show how the algorithm of Exercise 8 executes if $N = 3$.

10. What does the algorithm of Exercise 8 compute?

11H. Write the following algorithm in the style of Algorithm 1.5.1. (The input values of M and N are positive integers.)

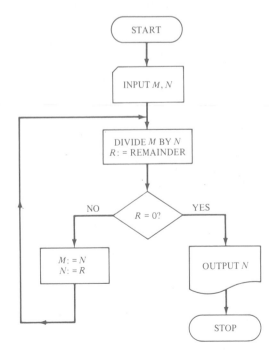

12. Show how the algorithm of Exercise 11 executes if $M = 6$ and $N = 10$.

13H. What does the algorithm of Exercise 11 compute?

14H. Give arguments to show that the algorithms of Exercises 8 and 11 terminate after a finite number of steps.

15. Consult a telephone book for the instructions for making a long-distance call. Which of the properties of an algorithm—precision, uniqueness, finiteness, input, output, generality—are present? Which properties are lacking?

16. Which of the properties of an algorithm—precision, uniqueness, finiteness, input, output, generality—do not hold for the following flowchart? (The input values of M and N are positive integers.)

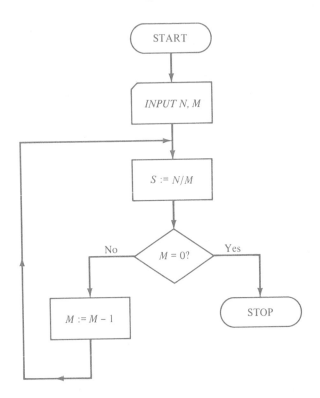

17H. Which of the properties of an algorithm—precision, uniqueness, finiteness, input, output, generality—do not hold for the following flowchart?

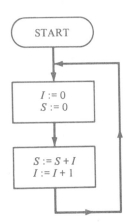

18H. Write the standard method of adding two positive integers, taught in elementary schools, as an algorithm in the style of Algorithm 1.5.1.

19. Write the standard method of multiplying two positive integers, taught in elementary schools, as an algorithm in the style of Algorithm 1.5.1.

20. Convert the quadratic formula for solving

$$ax^2 + bx + c = 0$$

to an algorithm in the style of Algorithm 1.5.1.

21. Represent the algorithms of Exercises 1–7 and 18–20 as flowcharts.

22. Write the algorithms of Exercises 1–8, 11, and 18–20 in a hypothetical computer language like that in Figure 1.5.3.

1.6

Complexity of Algorithms

Several questions can be asked about a particular algorithm:

1. Is it correct? That is, does it do what it claims to do when given valid input data?
2. How long does it take to run?
3. How much space is needed to execute the algorithm? That is, how much storage is needed to hold the data, arrays, program variables, and so on?

In this section we deal with the **complexity of algorithms**—questions 2 and 3.

A computer program, even though derived from a correct algorithm, might be useless for certain kinds of input because either the memory required exceeds that available or because the time needed to run the program is too great. Thus it is important to be able to estimate the time and space requirements of a given algorithm.

Suppose that we are given a set X of n elements, some labeled "red" and some labeled "black," and we want to find the number of subsets of X that contain at least one red item. Suppose we construct an algorithm that examines all subsets of X and counts those that contain at least one red item and then implement this algorithm as a computer program. We will see in Section 1.7 that a set which has n elements has 2^n subsets; thus the program would require at least 2^n units of time to execute. It does not matter what the units of time are—2^n grows so fast as n increases (see Table 1.6.1) that, except for small values of n, it would be infeasible to run the program.

Determining the performance parameters of a computer program is a difficult task and depends on a number of factors such as the computer used, the way the data are represented, and the compiler used. Although precise estimates of the efficiency of a program must take such factors into account, useful information can be obtained by analyzing the complexity of the underlying algorithm.

The time and space requirements of an algorithm are a function of the size

TABLE 1.6.1 Time to Execute an Algorithm if One Step Takes 1 Microsecond to Execute

Number of Steps to Termination for Input of Size n	Time to Execute if n =								
	3	6	9	12	50	100	1000	10^5	10^6
1	10^{-6} sec	10^{-6} sec	10^{-6} sec	10^{-6} sec	10^{-6} sec	10^{-6} sec	10^{-6} sec	10^{-6} sec	10^{-6} sec
$\lg \lg n$	10^{-6} sec	10^{-6} sec	2×10^{-6} sec	2×10^{-6} sec	2×10^{-6} sec	3×10^{-6} sec	3×10^{-6} sec	4×10^{-6} sec	4×10^{-6} sec
$\lg n$	2×10^{-6} sec	3×10^{-6} sec	3×10^{-6} sec	4×10^{-6} sec	6×10^{-6} sec	7×10^{-6} sec	10^{-5} sec	2×10^{-5} sec	2×10^{-5} sec
n	3×10^{-6} sec	6×10^{-6} sec	9×10^{-6} sec	10^{-5} sec	5×10^{-5} sec	10^{-4} sec	10^{-3} sec	0.1 sec	1 sec
$n \lg n$	5×10^{-6} sec	2×10^{-5} sec	3×10^{-5} sec	4×10^{-5} sec	3×10^{-4} sec	7×10^{-4} sec	10^{-2} sec	2 sec	20 sec
n^2	9×10^{-6} sec	4×10^{-5} sec	8×10^{-5} sec	10^{-4} sec	3×10^{-3} sec	0.01 sec	1 sec	3 hrs	12 days
n^3	3×10^{-5} sec	2×10^{-4} sec	7×10^{-4} sec	2×10^{-3} sec	0.13 sec	1 sec	16.7 min	32 yr	$31,710$ yr
2^n	8×10^{-6} sec	6×10^{-5} sec	5×10^{-4} sec	4×10^{-3} sec	36 yr	4×10^{16} yr	3×10^{287} yr	3×10^{30089} yr	3×10^{301016} yr

of the data that are input to the algorithm. We will let n denote the size of the input. For example, for Algorithm 1.5.1, n would be the number of elements in the sequence S. Thus the complexity problem becomes: Given an algorithm and an input of size n, determine the time and space required to execute the algorithm.

We must decide how to measure the time and space required by an algorithm. We could measure the space requirements by determining the amount of memory required to hold the variables, array elements, and so on. We could measure the time requirements by counting the number of instructions executed. Alternatively, we could use a cruder time estimate, such as the number of times each loop is executed or, if the principal activity of an algorithm is making comparisons, as might happen in a sorting routine, we might count the number of comparisons. Usually, we are interested in general estimates since, as we have already observed, the actual performance of a program implementation of an algorithm is dependent on many factors.

We are usually interested in the complexity of an algorithm for input selected from a particular class. For example, we might select input of size n that gives the worst time or space performance. Similarly, we might select input of size n that gives the best time or space performance. Such cases are called, respectively, **worst case** and **best case**. Another important case is **average case**—the average time or space performance over all inputs of size n.

EXAMPLE 1.6.1. The worst-case, best-case, and average-case times for Algorithm 1.5.1, assuming that n is the length of the input sequence and that the execution time is the number of iterations of the loop, are all n, since the loop is always executed n times.

For the purposes of estimating the time or space requirements of an algorithm and of comparing algorithms that perform the same function, it is often useful to have upper estimates of the performance parameters rather than precise values of these parameters. Such estimates are often written using the ''big oh'' notation.

DEFINITION 1.6.2. Let f and g be functions on $\{1, 2, 3, \ldots\}$. We write

$$f(n) = O(g(n))$$

and say that $f(n)$ is of *order at most* $g(n)$ if there exists a positive constant C such that

$$|f(n)| \leq C\,|g(n)|$$

for all but finitely many positive integers n.

EXAMPLE 1.6.3. Since

$$n^2 + n + 3 \leq 3n^2 + 3n^2 + 3n^2 = 9n^2,$$

we may take $C = 9$ in Definition 1.6.2, to obtain

$$n^2 + n + 3 = O(n^2).$$

EXAMPLE 1.6.4.   If we replace each integer $1, 2, \ldots, n$ by n in the sum $1 + 2 + \cdots + n$, the sum does not decrease and we have

$$1 + 2 + \cdots + n \le n + n + \cdots + n = n \cdot n = n^2. \quad (1.6.1)$$

It follows that

$$1 + 2 + \cdots + n = O(n^2).$$

EXAMPLE 1.6.5.   If k is a positive integer, we have

$$1^k + 2^k + \cdots + n^k \le n^k + n^k + \cdots + n^k = n \cdot n^k = n^{k+1};$$

hence

$$1^k + 2^k + \cdots + n^k = O(n^{k+1}).$$

EXAMPLE 1.6.6.   Since

$$|3n^3 + 6n^2 - 4n + 2| \le 3n^3 + 6n^2 + 4n + 2$$
$$\le 6n^3 + 6n^3 + 6n^3 + 6n^3$$
$$= 24n^3$$

it follows that

$$3n^3 + 6n^2 - 4n + 2 = O(n^3).$$

The method of Example 1.6.6 can be used to show that a polynomial in n of degree k is $O(n^k)$.

THEOREM 1.6.7.   *Let*

$$a_k n^k + a_{k-1} n^{k-1} + \cdots + a_1 n + a_0$$

be a polynomial in n of degree k. Then

$$a_k n^k + a_{k-1} n^{k-1} + \cdots + a_1 n + a_0 = O(n^k).$$

Proof.   Let

$$C = \max \{|a_k|, |a_{k-1}|, \ldots, |a_1|, |a_0|\}.$$

Then

$$|a_k n^k + a_{k-1} n^{k-1} + \cdots + a_1 n + a_0|$$
$$\le |a_k| n^k + |a_{k-1}| n^{k-1} + \cdots + |a_1| n + |a_0|$$
$$\le Cn^k + Cn^{k-1} + \cdots + Cn + C$$
$$\le Cn^k + Cn^k + \cdots + Cn^k + Cn^k$$
$$= (k + 1)Cn^k.$$

Therefore,

$$a_k n^k + a_{k-1} n^{k-1} + \cdots + a_1 n + a_0 = O(n^k). \qquad \blacksquare$$

We will next define the order of an algorithm.

DEFINITION 1.6.8. If an algorithm requires $f(n)$ units of time to terminate for an input of size n and

$$f(n) = O(g(n)),$$

we say that the *time required by the algorithm* is of order at most $g(n)$ or that the *time required by the algorithm* is $O(g(n))$. Similarly, if an algorithm requires $f(n)$ units of memory during execution for an input of size n and

$$f(n) = O(g(n)),$$

we say that the *space required by the algorithm* is of order at most $g(n)$ or that the *space required by the algorithm* is $O(g(n))$.

EXAMPLE 1.6.9. Suppose that a particular algorithm is known to require exactly

$$n^2 + n + 3$$

units of memory for an input of size n. We showed in Example 1.6.3 that

$$n^2 + n + 3 = O(n^2).$$

Thus the space required by this algorithm is $O(n^2)$.

EXAMPLE 1.6.10. Determine, in "big oh" notation, the best-case, worst-case, and average-case times required to execute the following algorithm. Assume that the input size is N and that the run time of the algorithm is the number of comparisons made at step 3. Also, assume that the $N + 1$ possibilities of KEY being at any particular position in the sequence or not being in the sequence are equally likely.

ALGORITHM 1.6.11   Searching an Unordered Sequence.   Given the sequence

$$S(1), S(2), \ldots, S(N)$$

and a value KEY, this algorithm finds the first occurrence of KEY and returns its location in I. If KEY is not found, the algorithm returns the value 0.
   1. [Initialize.] $I := 1$.
   2. [Not found?] If $I > N$, then set $I := 0$ and terminate. (Search not successful.)
   3. [Test.] If KEY $= S(I)$, then terminate. (Search successful.)
   4. [Continue search.] $I := I + 1$. Go to step 2.
   The best case can be analyzed as follows. If $S(1) =$ KEY, step 3 is

executed once. Thus the best case of Algorithm 1.6.11 has a run time that is

$$O(1).$$

The worst-case analysis of Algorithm 1.6.11 is analyzed as follows. If KEY is not in the sequence, step 3 will be executed N times so the worst case of Algorithm 1.6.11 has a run time that is

$$O(N).$$

Finally, consider the average run time of Algorithm 1.6.11. If KEY is found at the Ith position, step 3 is executed I times and if KEY is not in the sequence, step 3 is executed N times. Thus the average number of times step 3 is executed is

$$\frac{(1 + 2 + \cdots + N) + N}{N + 1} \leq \frac{N^2 + N}{N + 1} \qquad \text{by (1.6.1)}$$

$$= \frac{N(N + 1)}{N + 1} = N.$$

Thus the average case of Algorithm 1.6.11 has a run time that is

$$O(N).$$

For this algorithm, the average-case and worst-case run times are both $O(N)$.

When using the ''big oh'' notation to specify the performance of an algorithm, it is important to keep in mind that it gives only an upper estimate for the actual value of the performance parameter. For example, if we are told that the average-case run times of algorithms A and B are, respectively, $O(n^2)$ and $O(n^3)$, we may feel that algorithm A is superior. However, since we are only given upper estimates, we cannot be sure. For all we know, algorithm B is $O(n^2)$ also! Normally, one chooses the ''best'' function $g(n)$ to describe the order $O(g(n))$ of an algorithm.

Suppose that algorithms A and B have space requirements which are, respectively, $O(n)$ and $O(n^2)$. For a specific input, the size of the constants may be important. For example, suppose that for an input of size n, algorithms A and B require exactly $300n$ and $5n^2$ units of memory, respectively. For an input size of $n = 5$, algorithms A and B then require 1500 and 125 units of memory, respectively; and in this case, algorithm B is more efficient. Of course, for sufficiently large inputs, algorithm A is more efficient. Despite the cautionary remarks above, the ''big oh'' notation is very useful.

Certain forms occur so often that they are given special names, as shown in Table 1.6.2. In this table and throughout the text, lg will denote the logarithm to the base 2. The forms in Table 1.6.2, with the obvious exception of $O(n^m)$, are arranged so that if $O(f(n))$ is above $O(g(n))$, then $f(n) \leq g(n)$ for all but finitely many positive integers n. Thus, if algorithms A and B have run times which are $O(f(n))$ and $O(g(n))$, respectively, and, further, algorithms A and B

TABLE 1.6.2

"Big Oh" Form[a]	Name
$O(1)$	Constant
$O(\lg \lg n)$	Log log
$O(\lg n)$	Logarithmic
$O(n)$	Linear
$O(n \lg n)$	$n \log n$
$O(n^2)$	Quadratic
$O(n^3)$	Cubic
$O(n^m)$	Polynomial
$O(m^n)$, $m \geq 2$	Exponential
$O(n!)$	Factorial

[a]$\lg$ = log to the base 2; m is a fixed nonnegative integer.

require exactly $C_1 f(n)$ and $C_2 g(n)$ time units, respectively, and $O(f(n))$ is above $O(g(n))$ in Table 1.6.2, then algorithm A is more time efficient than algorithm B for sufficiently large inputs.

It is important to develop some feeling for the relative sizes of the functions in Table 1.6.2. In Figure 1.6.1 we have graphed some of these functions. Another way to develop some appreciation for the relative sizes of the functions $f(n)$ in Table 1.6.2 is to determine how long it would take an algorithm to

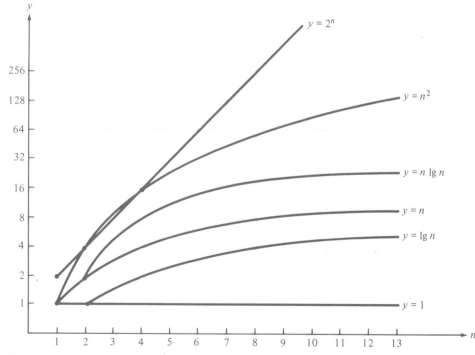

FIGURE 1.6.1

terminate whose run time is exactly $f(n)$. For this purpose, let us assume that we have a computer that can execute one step in 1 microsecond (10^{-6} sec). Table 1.6.1 shows the execution times, under this assumption, for various input sizes. Notice that it is feasible to implement an algorithm that requires 2^n steps for an input of size n only for very small input sizes. Algorithms requiring n^2 or n^3 steps also become infeasible, but for relatively larger input sizes. Also, notice the dramatic improvement that results when we move from n^2 steps to $n \lg n$ steps.

EXERCISES

Select the best "big oh" notation from Table 1.6.2 for each expression in Exercises 1–12.

1H. $6n + 1$ **2.** $2n^2 + 1$

3H. $6n^3 - 12n^2 - 1$ **4.** $3n^2 + 2n \lg n$

5H. $2 \lg n - 4n + 3n \lg n$ **6.** $6n^6 + n - 4$

7H. $2 + 4 + 6 + \cdots + 2n$ **8.** $(6n - 1)^2$

9H. $(6n + 4)(1 + \lg n)$ **10.** $\dfrac{(n + 1)(n - 3)}{n + 2}$

11H. $\dfrac{(n^2 + \lg n)(n - 1)}{n + n^2}$ **12.** $2 + 4 + 8 + 16 + \cdots + 2^n$

In Exercises 13–15, select the best "big oh" notation for $f(n) + g(n)$.

13H. $f(n) = O(1), g(n) = O(n^2)$

14. $f(n) = 6n^3 - 2n^2 + 4, g(n) = O(n \lg n)$

15H. $f(n) = O(n^{3/2}), g(n) = O(n^{5/2})$

16. Determine the exact number of steps executed by Algorithm 1.5.1 for the best case and the worst case for inputs of size n. Here one step is defined as one of the listed steps 1, 2, 3, or 4 in Algorithm 1.5.1.

In Exercises 17–22, select the best "big oh" notation as a function of N for the run time of each partial program. The run time is defined to be the number of times the instruction $X = X + 1$ is executed. ($\lfloor x \rfloor$ denotes the greatest integer less than or equal to x.)

17H.
```
FOR I = 1 TO N
    FOR J = 1 TO N
        X = X + 1
    NEXT J
NEXT I
```

18. FOR $I = 1$ TO N
  FOR $J = 1$ TO N
    FOR $K = 1$ TO N
      $X = X + 1$
    NEXT K
  NEXT J
NEXT I

19H. FOR $I = 1$ TO N
  FOR $J = 1$ TO I
    $X = X + 1$
  NEXT J
NEXT I

20. FOR $I = 1$ TO N
  FOR $J = 1$ TO $\lfloor (I + 1)/2 \rfloor$
    $X = X + 1$
  NEXT J
NEXT I

★21H. $J = N$
10 $X = X + 1$
  $J = \lfloor J/2 \rfloor$
  IF $J < 1$ THEN END
  GOTO 10

★22. $J = N$
10 FOR $I = 1$ TO J
    $X = X + 1$
  NEXT I
  IF $J < 1$ THEN END
  $J = \lfloor J/2 \rfloor$
  GOTO 10

23H. Suppose that $a > 1$ and that $f(n) = O(\log_a n)$. Show that $f(n) = O(\lg n)$.

24. Suppose that $g(n) > 0$ for $n = 1, 2, \ldots$. Show that $f(n) = O(g(n))$ if and only if there exists a positive constant c such that

$$|f(n)| \le c\, g(n).$$

for $n = 1, 2, \ldots$.

25H. Show that if

$$f(n) = O(h(n)) \qquad \text{and} \qquad g(n) = O(h(n)),$$

then

$$f(n) + g(n) = O(h(n)) \qquad \text{and} \qquad cf(n) = O(h(n)),$$

for any number c.

26. Show that $n! = O(n^n)$.

27H. Show that $2^n = O(n!)$.

★28. Show that $n \lg n = O(\lg (n!))$.

29H. Show that $\lg (n!) = O(n \lg n)$.

Define

$$f(n) = \Theta(g(n))$$

if there exist positive constants C_1 and C_2 such that

$$C_1|g(n)| \le |f(n)| \le C_2|g(n)|$$

for all but finitely many positive integers n.

30. Show that $2n - 1 = \Theta(n)$.

31H. Show that $3n^2 - 1 = \Theta(n^2)$.

32. Show that $(4n - 1)^2 = \Theta(n^2)$.

★33H. Show that $(2n - 1)(6n + 1)/(n - 1) = \Theta(n)$.

34. Determine a Θ notation for each expression in Exercises 1–12. (Θ is defined above.)

35H. Does

$$f(n) = O(g(n))$$

define an equivalence relation on the set of real-valued functions on $\{1, 2, \ldots\}$?

36. Does

$$f(n) = \Theta(g(n))$$

define an equivalence relation on the set of real-valued functions on $\{1, 2, \ldots\}$? (Θ is defined above.)

37. [Requires the integral]

(a) Show, by consulting the figure that

$$\frac{1}{2} + \frac{1}{3} + \cdots + \frac{1}{n} < \log_e n.$$

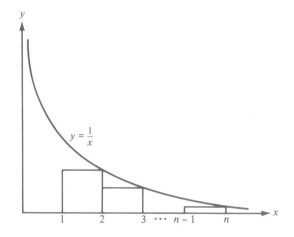

(b) Use part (a) to show that

$$1 + \frac{1}{2} + \frac{1}{3} + \cdots + \frac{1}{n} = O(\lg n).$$

38. [Requires the integral] Use an argument like that in Exercise 37 to show that

$$2^m + 3^m + \cdots + n^m < \frac{n^{m+1}}{m+1},$$

where m is a positive integer.

1.7

Mathematical Induction

Suppose that a sequence of blocks numbered 1, 2, . . . sits on an (infinitely) long table (see Figure 1.7.1) and that some blocks are marked with an "X." (All of the blocks visible in Figure 1.7.1 are marked.) Suppose that

The first block is marked. (1.7.1)

If all the blocks preceding the $(n + 1)$st block
are marked, then the $(n + 1)$st block is also marked. (1.7.2)

We will show that (1.7.1) and (1.7.2) imply that every block is marked by examining the blocks one-by-one.

Statement (1.7.1) explicitly states that block 1 is marked. Consider block 2. All of the blocks preceding block 2, namely block 1, are marked; thus, according to (1.7.2), block 2 is also marked. Consider block 3. All of the blocks preceding block 3, namely, blocks 1 and 2, are marked; thus, according to (1.7.2), block 3 is also marked. In this way we can show that every block is marked. For example, suppose we have verified that blocks 1–5 are marked, as shown in Figure 1.7.1. To show that block 6, which is not shown in Figure 1.7.1, is marked, we note that all the blocks that precede block 6 are marked, so by (1.7.2), block 6 is also marked.

The preceding example illustrates the **Principle of Mathematical Induction**.

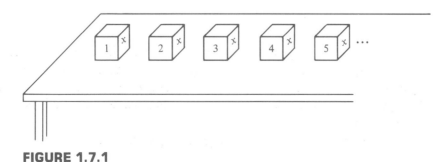

FIGURE 1.7.1

To show how mathematical induction can be used in a more profound way, let S_n denote the sum of the first n positive integers

$$S_n = 1 + 2 + 3 + \cdots + n. \tag{1.7.3}$$

Suppose that someone claims that

$$S_n = \frac{n(n + 1)}{2} \qquad \text{for } n = 1, 2, \ldots. \tag{1.7.4}$$

A sequence of statements is really being made, namely

$$S_1 = \frac{1(2)}{2} = 1$$

$$S_2 = \frac{2(3)}{2} = 3$$

$$S_3 = \frac{3(4)}{2} = 6$$

$$\vdots$$

Suppose that each true equation has an "$\times$" placed beside it [see (1.7.5)]. Since the first equation is true,

$$S_1 = \frac{1(2)}{2} \qquad \qquad \times$$

$$S_2 = \frac{2(3)}{2} \qquad \qquad \times$$

$$\vdots$$

$$S_{n-1} = \frac{(n - 1)n}{2} \qquad \qquad \times \tag{1.7.5}$$

$$S_n = \frac{n(n + 1)}{2} \qquad \qquad \times$$

$$S_{n+1} = \frac{(n + 1)(n + 2)}{2} \qquad \qquad ?$$

$$\vdots$$

it is marked. Now suppose we can show that if all the equations preceding a particular equation, say the $(n + 1)$st equation, are marked, then the $(n + 1)$st equation is also marked. Then, as in the example involving the blocks, all of the equations are marked; that is, all the equations are true and the formula (1.7.4) is verified.

We must show that if all of the equations preceding the $(n + 1)$st equation are true, then the $(n + 1)$st equation is also true. Assuming that all of the equations preceding the $(n + 1)$st equation are true, then, in particular, the nth equation is true and we have

$$S_{n+1} = 1 + 2 + \cdots + n + (n + 1)$$

$$= S_n + (n + 1)$$

$$= \frac{n(n + 1)}{2} + (n + 1)$$

$$= \frac{(n + 1)(n + 2)}{2}.$$

Therefore, the $(n + 1)$st equation is true. It follows that (1.7.4) is true for every positive integer n.

We next formally state the Principle of Mathematical Induction.

PRINCIPLE OF MATHEMATICAL INDUCTION. *Suppose that for each positive integer n we have a statement $S(n)$ that is either true or false. Suppose that*

$$S(1) \text{ is true;} \tag{1.7.6}$$

$$\text{if } S(i) \text{ is true, for all } i < n + 1, \text{ then } S(n + 1) \text{ is true.} \tag{1.7.7}$$

Then $S(n)$ is true for every positive integer n.

Condition (1.7.6) is sometimes called the **Basis Step** and condition (1.7.7) is sometimes called the **Inductive Step**. Hereafter, "induction" will mean "mathematical induction."

At this point, we illustrate the Principle of Mathematical Induction with another example.

EXAMPLE 1.7.1. Use induction to show that

$$n! \geq 2^{n-1} \qquad \text{for } n = 1, 2, \ldots . \tag{1.7.8}$$

Basis Step [Condition (1.7.6)]. We must show that (1.7.8) is true if $n = 1$. This is easily accomplished, since

$$1! = 1 \geq 1 = 2^{1-1}.$$

Inductive Step [Condition (1.7.7)]. We must show that if $i! \geq 2^{i-1}$ for $i = 1, \ldots, n$, then

$$(n + 1)! \geq 2^n. \tag{1.7.9}$$

Assume that $i! \geq 2^{i-1}$ for $i = 1, \ldots, n$. Then, in particular, for $i = n$, we have

$$n! \geq 2^{n-1}. \tag{1.7.10}$$

We can relate (1.7.9) and (1.7.10) by observing that

$$(n + 1)! = (n + 1)(n!).$$

Now

$$(n + 1)! = (n + 1)(n!)$$

$$\geq (n + 1)2^{n-1} \qquad \text{by (1.7.10)}$$

$$\geq 2 \cdot 2^{n-1} \qquad \text{since } n + 1 \geq 2$$

$$= 2^n.$$

Therefore, (1.7.9) is true. We have completed the Inductive Step.

Since the Basis Step and the Inductive Step have been verified, the Principle of Mathematical Induction tells us that (1.7.8) is true for every positive integer n.

A proof by induction is essentially an algorithm to compute the expression. For instance, in (1.7.3), the Basis Step tells us that

$$S_1 = 1. \tag{1.7.11}$$

The key observation in the Inductive Step is that

$$S_{n+1} = S_n + n + 1. \tag{1.7.12}$$

Equations (1.7.11) and (1.7.12) yield an algorithm for computing S_n for any n.

ALGORITHM 1.7.2 Computing the Sum 1 + 2 + 3 + $\cdots$ + N. Given a value of N, this algorithm computes the sum of the first N positive integers. The value of the sum is returned in S.
  1. [Initialize.] Set $S := 1$ and $M := 1$.
  2. [Done?] If $M = N$, terminate the algorithm with the sum in S.
  3. [Increment M.] $M := M + 1$.
  4. [Add the new term.] $S := S + M$. Go to step 2.

Notice that in step 1, $S := 1$ corresponds to (1.7.11), and that step 4 corresponds to (1.7.12).

To verify the Inductive Step (1.7.7), we assume that $S(i)$ is true for all $i < n + 1$ and then prove that $S(n + 1)$ is true. This formulation of mathematical induction is called the **strong form of mathematical induction**. Often, as was the case in the preceding examples, we can deduce $S(n + 1)$ assuming only $S(n)$. Indeed, the Inductive Step is often stated:

If $S(n)$ is true, then $S(n + 1)$ is true.

In these two formulations, the Basis Step is unchanged. It can be shown (see Exercise 40) that the two forms of mathematical induction are logically equivalent.

If we want to verify that the statements

$$S(n_0), S(n_0 + 1), \ldots ,$$

where $n_0 \neq 1$, are true, we must change the Basis Step to

$$S(n_0) \text{ is true.}$$

The Inductive Step is unchanged.

EXAMPLE 1.7.3. Use induction to show that if $a \neq 1$,

$$1 + a^1 + a^2 + \cdots + a^n = \frac{a^{n+1} - 1}{a - 1} \qquad (1.7.13)$$

for $n = 0, 1, \ldots$.

The sum on the left is called the **geometric sum**.

Basis Step. The Basis Step, which in this case is obtained by setting $n = 0$, is

$$1 = \frac{a^1 - 1}{a - 1},$$

which is true.

Inductive Step. Assume that the statement is true for n. Now

$$1 + a^1 + a^2 + \cdots + a^n + a^{n+1} = \frac{a^{n+1} - 1}{a - 1} + a^{n+1}$$

$$= \frac{a^{n+1} - 1}{a - 1} + \frac{a^{n+1}(a - 1)}{a - 1}$$

$$= \frac{a^{n+2} - 1}{a - 1}.$$

Since the modified Basis Step and the Inductive Step have been verified, the Principle of Mathematical Induction tells us that (1.7.13) is true for $n = 0$, $1, \ldots$.

Our final two examples show that induction is not limited to proving formulas for sums and verifying inequalities.

EXAMPLE 1.7.4. Use induction to show that $5^n - 1$ is divisible by 4 for $n = 1, 2, \ldots$.

Basis Step. If $n = 1$, $5^n - 1 = 5^1 - 1 = 4$, which is divisible by 4.

Inductive Step. Assume that $5^n - 1$ is divisible by 4. We must show that $5^{n+1} - 1$ is divisible by 4. To relate the $(n + 1)$st case to the nth case, we write

$$5^{n+1} - 1 = 5 \cdot 5^n - 1 = (5^n - 1) + 4 \cdot 5^n.$$

By assumption, $5^n - 1$ is divisible by 4 and, since $4 \cdot 5^n$ is divisible by 4, the sum

$$(5^n - 1) + 4 \cdot 5^n = 5^{n+1} - 1$$

is divisible by 4. Since the Basis Step and the Inductive Step have been verified, the Principle of Mathematical Induction tells us that $5^n - 1$ is divisible by 4 for $n = 1, 2, \ldots$.

We close by using induction to prove that the power set of a set with n elements has 2^n elements.

THEOREM 1.7.5. *If* $|X| = n$, *then*

$$|\mathscr{P}(X)| = 2^n. \tag{1.7.14}$$

Proof. The proof is by induction on n.

Basis Step. If $n = 0$, X is the empty set. The only subset of the empty set is the empty set itself; thus

$$|\mathscr{P}(X)| = 1 = 2^0 = 2^n.$$

Thus (1.7.14) is true for $n = 0$.

Inductive Step. Assume that (1.7.14) holds for n. Let X be a set with $|X| = n + 1$. Choose $x \in X$. We partition $\mathscr{P}(X)$ into two classes. The first class consists of those subsets of X that include x and the second class consists of those subsets of X that do not include x. If we list the subsets of X that do not include x

$$X_1, X_2, \ldots, X_k,$$

then

$$X_1 \cup \{x\}, X_2 \cup \{x\}, \ldots, X_k \cup \{x\}$$

is a list of the subsets of X that do include x. Thus the number of subsets of X that include x is equal to the number of subsets of X that do not include x. If $Y = X - \{x\}$, $\mathscr{P}(Y)$ consists precisely of those subsets of X that do not include x. Thus $|\mathscr{P}(X)| = 2|\mathscr{P}(Y)|$. Since $|Y| = n$, by the inductive assumption, $|\mathscr{P}(Y)| = 2^n$. Therefore,

$$|\mathscr{P}(X)| = 2|\mathscr{P}(Y)| = 2 \cdot 2^n = 2^{n+1}.$$

Thus (1.7.14) holds for $n + 1$ and the inductive step is complete. By the Principle of Mathematical Induction, (1.7.14) holds for all $n \geq 0$. ■

In Section 2.1 (see Example 2.1.2) we will give another proof of Theorem 1.7.5.

EXERCISES

In Exercises 1–11, using induction, verify that each equation is true for every positive integer n.

1H. $1 + 3 + 5 + \cdots + 2n - 1 = n^2$

2. $1 \cdot 2 + 2 \cdot 3 + 3 \cdot 4 + \cdots + n(n + 1) = \dfrac{n(n + 1)(n + 2)}{3}$

3. $1(1!) + 2(2!) + \cdots + n(n!) = (n + 1)! - 1$

4H. $1^2 + 2^2 + 3^2 + \cdots + n^2 = \dfrac{n(n + 1)(2n + 1)}{6}$

5. $1^2 - 2^2 + 3^2 - \cdots + (-1)^{n+1} n^2 = \dfrac{(-1)^{n+1} n(n + 1)}{2}$

6. $1^3 + 2^3 + 3^3 + \cdots + n^3 = \left[\dfrac{n(n + 1)}{2} \right]^2$

7H. $\dfrac{1}{1 \cdot 3} + \dfrac{1}{3 \cdot 5} + \dfrac{1}{5 \cdot 7} + \cdots + \dfrac{1}{(2n - 1)(2n + 1)} = \dfrac{n}{2n + 1}$

8. $\dfrac{1}{2 \cdot 4} + \dfrac{1 \cdot 3}{2 \cdot 4 \cdot 6} + \dfrac{1 \cdot 3 \cdot 5}{2 \cdot 4 \cdot 6 \cdot 8} + \cdots + \dfrac{1 \cdot 3 \cdot 5 \cdots (2n - 1)}{2 \cdot 4 \cdot 6 \cdots (2n + 2)}$

$$= \dfrac{1}{2} - \dfrac{1 \cdot 3 \cdot 5 \cdots (2n + 1)}{2 \cdot 4 \cdot 6 \cdots (2n + 2)}$$

9. $\dfrac{1}{2^2 - 1} + \dfrac{1}{3^2 - 1} + \cdots + \dfrac{1}{(n + 1)^2 - 1} = \dfrac{3}{4} - \dfrac{1}{2(n + 1)} - \dfrac{1}{2(n + 2)}$

★10H. $\cos x + \cos 2x + \cdots + \cos nx = \dfrac{\cos [(x/2)(n + 1)] \sin (nx/2)}{\sin (x/2)}$ provided that $\sin (x/2) \neq 0$.

★11. $1 \sin x + 2 \sin 2x + \cdots + n \sin nx =$

$$\dfrac{\sin [(n + 1)x]}{4 \sin^2 (x/2)} - \dfrac{(n + 1) \cos \left(\dfrac{2n + 1}{2} x \right)}{2 \sin (x/2)}$$

provided that $\sin (x/2) \neq 0$.

12. Write algorithms to compute each of the sums in Exercises 1–11.

In Exercises 13–18, using induction, verify the inequality.

13H. $\dfrac{1}{2n} \leq \dfrac{1 \cdot 3 \cdot 5 \cdots (2n - 1)}{2 \cdot 4 \cdot 6 \cdots (2n)}, \quad n = 1, 2, \ldots$

★14. $\dfrac{1 \cdot 3 \cdot 5 \cdots (2n - 1)}{2 \cdot 4 \cdot 6 \cdots (2n)} \leq \dfrac{1}{\sqrt{n + 1}}, \quad n = 1, 2, \ldots$

15. $2n + 1 \leq 2^n, \quad n = 3, 4, \ldots$

★16H. $2^n \geq n^2, \quad n = 4, 5, \ldots$

★17H. $(a_1 a_2 \cdots a_{2^n})^{1/2^n} \leq \dfrac{a_1 + a_2 + \cdots + a_{2^n}}{2^n}, \quad n = 1, 2, \ldots$ and the a_i are positive numbers.

18. $(1 + x)^n \geq 1 + nx,$ for $x \geq -1$ and $n = 1, 2, \ldots$.

In Exercises 19–22, use induction to prove the statement.

19H. $7^n - 1$ is divisible by 6, for $n = 1, 2, \ldots$

20. $11^n - 6$ is divisible by 5, for $n = 1, 2, \ldots$

21H. $6 \cdot 7^n - 2 \cdot 3^n$ is divisible by 4, for $n = 1, 2, \ldots$

★22. $3^n + 7^n - 2$ is divisible by 8, for $n = 1, 2, \ldots$

In Exercises 23 and 24, by experimenting with small values of n, guess a formula for the given sum; then use induction to verify your formula.

23H. $\dfrac{1}{1 \cdot 2} + \dfrac{1}{2 \cdot 3} + \cdots + \dfrac{1}{n(n + 1)}$

24. $4\left(\dfrac{1}{2 \cdot 3}\right) + 8\left(\dfrac{2}{3 \cdot 4}\right) + \cdots + 2^{n+1}\left[\dfrac{n}{(n + 1)(n + 2)}\right]$

25. Use induction to show that if $X_1, \ldots, X_n$ and X are sets, then
(a) $X \cap (X_1 \cup X_2 \cup \cdots \cup X_n) = (X \cap X_1) \cup (X \cap X_2) \cup \cdots \cup (X \cap X_n)$
(b) $\overline{X_1 \cap X_2 \cap \cdots \cap X_n} = \overline{X_1} \cup \overline{X_2} \cup \cdots \cup \overline{X_n}$

26. Determine the number of times the statement $X = X + 1$ is executed in the programs in Exercises 17–22, Section 1.6. Use induction to prove your answer.

27H. Use induction to show that n straight lines in the plane divide the plane into $(n^2 + n + 2)/2$ regions. Assume that no two lines are parallel and that no three lines have a common point.

28. Show that postage of 5 cents or more can be achieved by using only 2-cent and 5-cent stamps.

★29H. Show that postage of 24 cents or more can be achieved by using only 5-cent and 7-cent stamps.

The Egyptians of antiquity expressed a fraction as a sum of fractions whose numerators were 1. For example, $\frac{5}{6}$ might be expressed as

$$\frac{5}{6} = \frac{1}{2} + \frac{1}{3}.$$

We say that a fraction p/q, where p and q are positive integers, is in **Egyptian form** if

$$\frac{p}{q} = \frac{1}{n_1} + \frac{1}{n_2} + \cdots + \frac{1}{n_k}, \tag{1.7.15}$$

where $n_1, n_2, \ldots, n_k$ are distinct positive integers.

30. Show that the representation (1.7.15) need not be unique by representing $\frac{5}{6}$ in two different ways.

★31H. Show that the representation (1.7.15) is never unique.

★32. By completing the following steps, give a proof by induction on p to show that every fraction p/q with $0 < p/q < 1$ may be expressed in Egyptian form.
(a) Verify the Basis Step ($p = 1$).
(b) Suppose that $0 < p/q < 1$ and that all fractions i/q', with $1 \le i < p$ and q' arbitrary, can be expressed in Egyptian form. Choose the smallest positive integer n with $1/n \le p/q$. Show that

$$n > 1 \quad \text{and} \quad \frac{p}{q'} < \frac{1}{n - 1}.$$

(c) Show that if $p/q = 1/n$, the proof is complete.
(d) Assume that $1/n < p/q$. Let

$$p_1 = np - q \quad \text{and} \quad q_1 = nq.$$

Show that

$$\frac{p_1}{q_1} = \frac{p}{q} - \frac{1}{n}, \quad 0 < \frac{p_1}{q_1} < 1, \quad \text{and} \quad p_1 < p.$$

Conclude that

$$\frac{p_1}{q_1} = \frac{1}{n_1} + \frac{1}{n_2} + \cdots + \frac{1}{n_k}$$

with $n_1, n_2, \ldots, n_k$ distinct.
(e) Show that $p_1/q_1 < 1/n$.
(f) Show that

$$\frac{p}{q} = \frac{1}{n} + \frac{1}{n_1} + \cdots + \frac{1}{n_k}$$

and $n, n_1, \ldots, n_k$ are distinct.

33H. Use the method of Exercise 32 to find Egyptian forms of $\frac{3}{8}$, $\frac{5}{7}$, and $\frac{13}{19}$.

★34. Show that any fraction p/q, where p and q are positive integers, can be written in Egyptian form. (We are not assuming that $p/q < 1$.)

35H. Suppose that $S_n = (n + 2)(n - 1)$ is (incorrectly) proposed as a formula for

$$2 + 4 + \cdots + 2n.$$

(a) Show that the Inductive Step is satisfied but that the Basis Step fails.
★(b) If S'_n is an arbitrary expression that satisfies the Inductive Step, what form must S'_n assume?

36. What is wrong with the following argument, which allegedly shows that any two positive integers are equal?

BASIS STEP. If the maximum of two integers is 1, they are equal. (In this case, both are 1.)

INDUCTIVE STEP. Assume that if the maximum of two integers is n, they are equal. Suppose that the maximum of two integers, a and b, is $n + 1$.

Subtract 1 from each of a and b. The maximum of the altered integers, $a - 1$ and $b - 1$, is n. By induction, $a - 1 = b - 1$. But this implies that $a = b$.

Since we have verified the Basis Step and the Inductive Step by the Principle of Mathematical Induction, any two positive integers are equal!

37H. What is wrong with the following "proof" that any algorithm has a run time that is $O(n)$?

We must show that the time required for an input of size n is at most a constant times n.

BASIS STEP. Suppose that $n = 1$. If the algorithm takes C units of time for an input of size 1, the algorithm takes at most $C \cdot 1$ units of time. Thus the assertion is true for $n = 1$.

INDUCTIVE STEP. Assume that the time required for an input of size n is at most $C'n$ and that the time for processing an additional item is C''. Let C be the maximum of C' and C''. Then the total time required for an input of size $n + 1$ is at most

$$C'n + C'' \le Cn + C = C(n + 1).$$

The Inductive Step has been verified.

By induction, for input of size n, the time required is at most Cn. Therefore, the run time is $O(n)$.

★38H. Assume the form of mathematical induction where the Inductive Step is: If $S(n)$ is true, then $S(n + 1)$ is true. Prove the Well-Ordering Theorem for Positive Integers.

WELL-ORDERING THEOREM FOR POSITIVE INTEGERS. *If X is a nonvoid set of positive integers, X contains a least element.*

Assume that there is no positive integer less than 1 and that if n is a positive integer, there is no positive integer between n and $n + 1$.

★39H. Assume the Well-Ordering Theorem for Positive Integers (see Exercise 38). Prove the strong form of the Principle of Mathematical Induction.

★40H. Show that the strong form of the Principle of Mathematical Induction and the form of mathematical induction where the Inductive Step is "if $S(n)$ is true, then $S(n + 1)$ is true" are equivalent; that is, assume the strong form and prove the alternative form, then assume the alternative form and prove the strong form.

1.8

Notes

General references on discrete mathematics are [Arbib, 1981; Berztiss; Birkhoff; Bobrow; Dornhoff; Fisher; Gersting; Gill; Knuth, 1973 Vols. 1 and 3, 1981;

Levy; Lipschutz, 1976, 1982; Liu, 1985; Prather; Sahni; Stanat; Stone; Tucker; Tremblay; and Wand]. [Birkhoff; Dornhoff; Fisher; Gersting; and Gill] emphasize algebraic aspects of discrete mathematics while [Arbib, 1981; Levy; Lipschutz, 1976, 1982; Liu, 1985; Bobrow; Prather; Sahni; Stanat; Stone; and Tremblay] treat much the same topics as this text. [Berztiss] is about half discrete mathematics and half data structures. [Tucker] emphasizes counting, networks, and graph theory. [Wand] covers graph theory (briefly), induction, recursion, computability theory, logic, and program correctness.

The books by Knuth [1973 Vols. 1 and 3, 1981] are the first three books in a projected seven-volume set. The first half of Volume 1 introduces the concept of an algorithm and various mathematical topics, including mathematical induction. The second half of Volume 1 is devoted to data structures. Everyone should become familiar with these volumes as early as possible. They are classics in the area of algorithms. They are also among the finest examples of technical writing.

[Carberry] is a good introduction to computer science and can serve as a reference to many of the basic definitions. [Berztiss; Horowitz, 1976; and Standish] are references on data structures. [Grillo] is a microcomputer-based data structures text.

The reader wanting to study set theory, including relations and functions, in more detail could consult any of [Halmos, 1974; Lipschutz, 1964; or Stoll].

[Baase] is highly recommended as an introductory text on the analysis of algorithms. More advanced books on algorithms are [Aho; Horowitz, 1978; Knuth, 1973 Vols. 1 and 3, 1981; Nievergelt; and Reingold]. [McNaughton] contains a very thorough discussion on an introductory level of what an algorithm is. Knuth's expository article about algorithms ([Knuth, 1977]) is also recommended.

Finally, concerning the interplay between mathematics and computer science, [Knuth, 1974] is strongly recommended.

COMPUTER EXERCISES

In Exercises 1–4, assume that a set X of n elements is represented as an array A of size at least $n + 1$. The elements of X are listed consecutively in A starting in the first position and terminating with 0. Assume further that no set contains 0.

1. Write a program to represent the sets $X \cup Y$, $X \cap Y$, $X - Y$, and $X \Delta Y$, given the arrays A and B representing X and Y. (The definition of Δ is given before Exercise 68, Section 1.1.)

2. Assuming a universe represented as an array, write a program to represent the set $\overline{X}$, given the array A representing X.

3. Given an element E and the array A, which represents X, write a program that determines whether $E \in X$ or not.

4.   Given the array A representing X, write a program that lists all subsets of X.

5.   [*Project*] Find a way different from that given above to represent a set in a computer, as for example, by using linked lists (see [Carberry] or [Standish]). Redo Exercises 1–4 using your method of representing sets.

In Exercises 6 and 7, assume that A is an array, indexed from 1 to N, of real numbers. Consider A to be a function from $\{1, \ldots, N\}$ into the real numbers.

6.   Write a program that tests whether A is one-to-one.

7.   Write a program that tests whether A is onto a given set.

8.   Implement the algorithms of Exercises 1–8, 11, and 18–20 of Section 1.5 as programs.

9.   Implement Algorithm 1.6.11 as a program. Generate some random data and find the average number of iterations needed to locate a key.

10.   Verify the entries in Table 1.6.1.

11.   Compare the values of

$$1 + \frac{1}{2} + \cdots + \frac{1}{n} \quad \text{and} \quad \lg n$$

for $n = 10, 100, 1000$. (See Exercise 37, Section 1.6.)

12.   Write programs to compute the sums in Exercises 1–11, Section 1.7.

CHAPTER 2

Counting Methods and Recurrence Relations

In many discrete problems, we are confronted with the problem of counting. For example, in Section 1.6 we saw that in order to estimate the run time of an algorithm, we needed to count the number of times certain steps or loops were executed. Because of the importance of counting, a variety of useful aids, some quite sophisticated, have been developed. Sections 2.1–2.4 present an introduction to counting.

Recurrence relations are useful in certain counting problems. A recurrence relation relates the nth element in a sequence to its predecessors. We will see that recurrence relations arise naturally in the analysis of algorithms.

Readers wanting to explore these ideas further should consult the references listed in Section 2.8.

2.1

Basic Principles

Before stating the **First Counting Principle**, we will illustrate it by an example.

EXAMPLE 2.1.1. How many strings of length 2 can be formed using the letters ABC if repetitions are allowed?

We have three choices for the first letter and three choices for the second

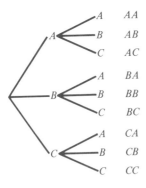

FIGURE 2.1.1

letter. It follows that there are $3 \cdot 3 = 9$ possible strings. The tree diagram of Figure 2.1.1 shows why we multiply 3 times 3—we have three groups of three objects. This example illustrates our first general principle.

FIRST COUNTING PRINCIPLE. *If an activity can be constructed in t successive steps and step 1 can be done in n_1 ways; step 2 can then be done in n_2 ways; . . . ; and step t can then be done in n_t ways, then the number of possible activities is $n_1 \cdot n_2 \cdot \cdots \cdot n_t$.*

In Example 2.1.1 the activity is constructing strings of length 2. The first step is "select the first letter" and the second step is "select the second letter." Thus $n_1 = 3$ and $n_2 = 3$ and by the First Counting Principle, the total number of activities is $3 \cdot 3 = 9$.

We may summarize the First Counting Principle by saying that we multiply together the numbers of ways of doing each step when an activity is constructed in successive steps.

EXAMPLE 2.1.2.   Use the First Counting Principle to show that a set $\{x_1, \ldots, x_n\}$ containing n elements has 2^n subsets.

A subset can be constructed in n successive steps: pick or do not pick x_1, pick or do not pick x_2, . . . , pick or do not pick x_n, Each step can be done in two ways. Thus the number of possible subsets is

$$2 \cdot 2 \cdot \cdots \cdot 2 = 2^n.$$

EXAMPLE 2.1.3.   How many different eight-bit strings are there? (A **bit** is either 0 or 1.)

An eight-bit string can be constructed in eight successive steps: select the first bit, select the second bit, . . . , select the eighth bit. Since there are two ways to select each bit by the First Counting Principle, the total number of eight-bit strings is

$$2 \cdot 2 \cdot 2 \cdot 2 \cdot 2 \cdot 2 \cdot 2 \cdot 2 = 2^8 = 256.$$

If a computer interprets an eight-bit string as a character, Example 2.1.3 shows that a total of 256 distinct characters can be represented.

Next, we will illustrate the Second Counting Principle by an example and then present the principle.

EXAMPLE 2.1.4. How many eight-bit strings begin either 101 or 111? An eight-bit string that begins 101 can be constructed in five successive steps: select the fourth bit, select the fifth bit, . . . , select the eighth bit. Since each of the five bits can be selected in two ways, by the First Counting Principle, there are

$$2 \cdot 2 \cdot 2 \cdot 2 \cdot 2 = 2^5 = 32$$

eight-bit strings that begin 101. The same argument can be used to show that there are 32 eight-bit strings that begin 111. Since there are 32 eight-bit strings that begin 101 and 32 eight-bit strings that begin 111, there are $32 + 32 = 64$ eight-bit strings that begin either 101 or 111.

In Example 2.1.4 we added the numbers of eight-bit strings (32 and 32) of each type to determine the final result. The **Second Counting Principle** tells us when to add to compute the total number of possibilities.

SECOND COUNTING PRINCIPLE. *Suppose that $X_1, \ldots, X_t$ are sets and that the ith set X_i has n_i elements. If $\{X_1, \ldots, X_t\}$ is a pairwise disjoint family, the number of possible elements that can be selected from X_1 or X_2 or $\cdots$ or X_t is*

$$n_1 + n_2 + \cdots + n_t.$$

(Equivalently, the union $\bigcup\limits_{i=1}^{t} X_i$ contains $n_1 + n_2 + \cdots + n_t$ elements.)

In Example 2.1.4 we could let X_1 denote the set of eight-bit strings that begin 101 and X_2 denote the set of eight-bit strings that begin 111. Since X_1 is disjoint from X_2, according to the Second Counting Principle, the number of eight-bit strings of either type, which is the number of elements in $X_1 \cup X_2$, is $32 + 32 = 64$.

We may summarize the Second Counting Principle by saying that we add the numbers of elements in each subset when the elements being counted can be decomposed into disjoint subsets.

EXAMPLE 2.1.5. How many eight-bit strings either begin 101 or have the fourth bit 1?

Arguing as in Example 2.1.4, we find that there are 32 eight-bit strings that begin 101. Similarly, there are 128 eight-bit strings whose fourth bit is 1. The total number of possibilities is not the sum, since the set of eight-bit strings that begin 101 is not disjoint from the set of eight-bit strings whose fourth bit is 1. (An eight-bit string can begin 1011.)

To solve this problem, we decompose the possibilities into disjoint sets:

$X_1 = \{x \mid x$ is an eight-bit string whose fourth bit is 1 and begins 101$\}$.

$X_2 = \{x \mid x$ is an eight-bit string whose fourth bit is 1 and does not begin 101$\}$.

$X_3 = \{x \mid x$ is an eight-bit string whose fourth bit is 0 and begins 101$\}$.

The union $X_1 \cup X_2 \cup X_3$ consists of the eight-bit strings to be counted.

There are 16 eight-bit strings that begin 1011 and there are 16 eight-bit strings that begin 1010. Thus X_1 and X_3 each contain 16 elements. Since there are 128 eight-bit strings whose fourth bit is 1 and 16 of these begin 101, there are $128 - 16 = 112$ elements in X_2. By the Second Counting Principle, there are

$$16 + 112 + 16 = 144$$

eight-bit strings that either begin 101 or have the fourth bit 1.

We close this section with an example that illustrates both counting principles.

EXAMPLE 2.1.6. A six-person committee composed of Alice, Ben, Connie, Dolph, Egbert, and Francisco is to select a chairperson, secretary, and treasurer.

(a) In how many ways can this be done?

(b) In how many ways can this be done if either Alice or Ben must be chairperson?

(c) In how many ways can this be done if Egbert must hold one of the offices?

(d) In how many ways can this be done if both Dolph and Francisco must hold office?

(a) We use the First Counting Principle. The officers can be selected in three successive steps: Select the chairperson, select the secretary, select the treasurer. The chairperson can be selected in six ways. Having selected the chairperson, the secretary can be selected in five ways. Having selected the chairperson and secretary, the treasurer can be selected in four ways. Therefore, the total number of possibilities is

$$6 \cdot 5 \cdot 4 = 120.$$

(b) Arguing as in (a), if Alice is chairperson, there are $5 \cdot 4 = 20$ ways to select the remaining officers. Similarly, if Ben is chairperson, there are 20 ways to select the remaining officers. Since these cases are disjoint, by the Second Counting Principle, there are

$$20 + 20 = 40$$

possibilities.

(c) [First solution.]   Arguing as in (a), if Egbert is chairperson, there are 20 ways to select the remaining officers. Similarly, if Egbert is secretary, there are 20 possibilities, and if Egbert is treasurer, there are 20 possibilities. Since these three cases are pairwise disjoint, by the Second Counting Principle, there are

$$20 + 20 + 20 = 60$$

possibilities.

[Second solution.]   Let us consider the activity of assigning Egbert and two others to offices to be made up of three successive steps: Assign Egbert an office, fill the highest remaining office, fill the last office. There are three ways to assign Egbert an office. Having assigned Egbert, there are five ways to fill the highest remaining office. Having assigned Egbert and filled the highest remaining office, there are four ways to fill the last office. By the First Counting Principle, there are

$$3 \cdot 5 \cdot 4 = 60$$

possibilities.

(d) Let us consider the activity of assigning Dolph, Francisco, and one other person to offices to be made up of three successive steps: Assign Dolph, assign Francisco, fill the remaining office. There are three ways to assign Dolph. Having assigned Dolph, there are two ways to assign Francisco. Having assigned Dolph and Francisco, there are four ways to fill the remaining office. By the First Counting Principle, there are

$$3 \cdot 2 \cdot 4 = 24$$

possibilities.

EXERCISES

1H.   A man has eight shirts, four pairs of pants, and five pairs of shoes. How many different outfits are possible?

In Exercises 2–10, two dice are rolled, one blue and one red.

2.   How many outcomes are possible?

3H.   How many outcomes give the sum of 4? no distinction between blue and red

4.   How many outcomes are doubles? (A double occurs when both dice show the same number.)

5H.   How many outcomes give the sum of 7 or the sum of 11?

6.   How many outcomes have the blue die showing 2?

7H.   How many outcomes have exactly one die showing 2?

8.   How many outcomes have at least one die showing 2?

9H.   How many outcomes have neither die showing 2?

10.   How many outcomes give an even sum?

In Exercises 11–13, suppose that there are 10 roads from Oz to Mid Earth and five roads from Mid Earth to Fantasy Island.

11H.   How many routes are there from Oz to Fantasy Island passing through Mid Earth?

12.   How many round trips are there of the form Oz–Mid Earth–Fantasy Island–Mid Earth–Oz?

13H.   How many round trips are there of the form Oz–Mid Earth–Fantasy Island–Mid Earth–Oz in which on the return trip we do not reverse the original route from Oz to Fantasy Island?

14.   How many different car license plates can be constructed if the licenses contain three letters followed by two digits if repetitions are allowed? if repetitions are not allowed?

15H.   How many eight-bit strings begin 1100?

16.   How many eight-bit strings begin and end with 1?

17H.   How many eight-bit strings have either the second or the fourth bit 1 (or both)?

18.   How many eight-bit strings have exactly one 1?

19H.   How many eight-bit strings have exactly two 1's?

20.   How many eight-bit strings have at least one 1?

21H.   How many eight-bit strings read the same from either end? (An example of such an eight-bit string is 01111110.)

In Exercises 22–28, a six-person committee composed of Alice, Ben, Connie, Dolph, Egbert, and Francisco is to select a chairperson, secretary, and treasurer.

22.   How many selections exclude Connie?

23H.   How many selections exclude Ben and Francisco?

24.   How many selections include Ben and Francisco?

25H.   How many selections exclude Ben or Francisco?

26.   How many selections include Ben or Francisco?

27H.   How many selections either include both Ben and Francisco or exclude both Ben and Francisco?

28.   How many selections are there in which either Alice is chairperson or she is not an officer?

In Exercises 29–36, the letters *ABCDE* are to be used to form strings of length 3.

29H. How many strings can be formed if we allow repetitions?

30. How many strings can be formed if we do not allow repetitions?

31H. How many strings begin with *A*, allowing repetitions?

32. How many strings begin with *A* if repetitions are not allowed?

33H. How many strings do not contain the letter *A*, allowing repetitions?

34. How many strings do not contain the letter *A* if repetitions are not allowed?

35H. How many strings contain the letter *A*, allowing repetitions?

36. How many strings contain the letter *A* if repetitions are not allowed?

Exercises 37–47 refer to the integers from 5 to 200, inclusive.

37H. How many numbers are there?

38. How many are even?

39H. How many are odd?

40. How many are divisible by 5?

41H. How many are greater than 72?

42. How many consist of distinct digits?

43H. How many contain the digit 7?

44. How many do not contain the digit 0?

45H. How many are greater than 101 and do not contain the digit 6?

46. How many have the digits in strictly increasing order? (Examples are 13, 147, 8.)

47H. How many are of the form xyz, where $0 \neq x < y$ and $y > z$?

48. (a) In how many ways can the months of the birthdays of five people be distinct?
(b) How many possibilities are there for the months of the birthdays of five people?
(c) In how many ways can at least two people from among five have their birthdays in the same month?

Exercises 49–54 refer to a set of five different computer science books, three different mathematics books, and two different art books.

49H. In how many ways can these books be arranged on a shelf?

50. In how many ways can these books be arranged on a shelf if all five computer science books are on the left and both art books are on the right?

51H. In how many ways can these books be arranged on a shelf if all five computer science books are on the left?

52. In how many ways can these books be arranged on a shelf if all books of the same discipline are grouped together?

★**53H.** In how many ways can these books be arranged on a shelf if the two art books are not together?

54. In how many ways can we select two books from different subjects?

55H. In one version of BASIC, a valid variable name consists of a string of one or two alphanumeric characters beginning with a letter (excepting FN, IF, ON, OR, and TO), optionally followed by one of %, !, #, or $. (An **alphanumeric** character is one of A to Z or 0 to 9.) How many valid BASIC variable names are there?

56. A valid FORTRAN identifier consists of a string of one to six alphanumeric characters beginning with a letter. How many valid FORTRAN identifiers are there?

57H. A password-free file specification in the TRSDOS operating system consists of a name followed by an optional extension followed by an optional drive specification. The name is a string of at most eight alphanumeric characters beginning with a letter. The extension is a slash (/) followed by a string of at most three alphanumeric characters beginning with a letter. A drive specification is :0, :1, :2, or :3. Consider an omitted drive specification and :0 to be identical. (EXAMPLE: EXERCISE/CH1:1.) How many password-free file specifications are there in TRSDOS?

58. If X is an n-element set and Y is an m-element set, how many functions are there from X to Y?

★**59H.** There are 10 copies of one book and one copy each of 10 other books. In how many ways can we select 10 books?

60. How many terms are there in the expansion of

$$(x + y)(a + b + c)(e + f + g)(h + i)?$$

★**61H.** How many subsets of a $(2n + 1)$-element set have n elements or less?

2.2

Permutations and Combinations

A **permutation** of objects involves ordering, whereas a **combination** does not take ordering into account. In this section we define these concepts precisely, give formulas for counting permutations and combinations, and give several examples.

DEFINITION 2.2.1. Given a set containing n (distinct) elements $X = \{x_1, \ldots, x_n\}$,

(a) a *permutation* of X is an ordering of the n elements $x_1, \ldots, x_n$;

(b) an *r-permutation* of X, where $r \leq n$, is an ordering of a subset of r elements of X;

(c) the number of r-permutations of a set of n distinct elements is denoted $P(n, r)$;

(d) an *r-combination* of X is an unordered selection of r elements of X (i.e., an r-element subset of X);

(e) the number of r-combinations of a set of n distinct elements is denoted $C(n, r)$ or $\binom{n}{r}$.

EXAMPLE 2.2.2. Examples of permutations of $X = \{a, b, c\}$ are

$$abc, \quad acb, \quad bac.$$

Examples of 2-permutations of X are

$$ab, \quad ba, \quad ca.$$

Examples of 2-combinations of X are

$$\{a, b\}, \quad \{a, c\}, \quad \{b, c\}.$$

We can use the First Counting Principle to compute $P(n, r)$, the number of r-permutations of an n-element set.

THEOREM 2.2.3. *The number of r-permutations of a set of n distinct objects is*

$$P(n, r) = n(n - 1)(n - 2) \cdots (n - r + 1).$$

Proof. We are to count the number of ways to order r elements selected from a set having n elements. The first element can be selected in n ways. Having selected the first element, the second element can be selected in $n - 1$ ways. We continue selecting elements until, having selected the $(r - 1)$st element, we select the rth element. This last element can be chosen in $n - r + 1$ ways. By the First Counting Principle, these selections can be made in

$$n(n - 1)(n - 2) \cdots (n - r + 1)$$

ways. ∎

EXAMPLE 2.2.4. According to Theorem 2.2.3, the number of 2-permutations of $X = \{a, b, c\}$ is

$$P(3, 2) = 3 \cdot 2 = 6.$$

These six 2-permutations are

$$ab, \quad ac, \quad ba, \quad bc, \quad ca, \quad cb.$$

By Theorem 2.2.3, the number of permutations of an n-element set is

$$P(n, n) = n(n - 1) \cdots 2 \cdot 1 = n!.$$

We may also write $P(n, r)$ in terms of factorials

$$P(n, r) = n(n - 1) \cdots (n - r + 1)$$

$$= \frac{n(n - 1) \cdots (n - r + 1)(n - r) \cdots 2 \cdot 1}{(n - r) \cdots 2 \cdot 1}$$

$$= \frac{n!}{(n - r)!} \tag{2.2.1}$$

EXAMPLE 2.2.5. In how many ways can we select a chairperson, vice-chairperson, secretary, and treasurer from a group of 10 persons?

We need to count the number of orderings of four persons selected from a group of 10, since an ordering picks (uniquely) a chairperson (first pick), a vice-chairperson (second pick), a secretary (third pick), and a treasurer (fourth pick). By Theorem 2.2.3, the solution is

$$P(10, 4) = 10 \cdot 9 \cdot 8 \cdot 7 = 5040.$$

We could also have solved Example 2.2.5 by appealing directly to the First Counting Principle.

EXAMPLE 2.2.6. How many permutations of the letters *ABCDEF* contain the letters *DEF* together in any order?

First, we will count the number of permutations of the letters *ABCDEF* that contain the pattern *DEF*. To guarantee the presence of the pattern *DEF*, these three letters must be kept together in this order. The remaining letters *A*, *B*, and *C* can be placed arbitrarily. We can think of constructing permutations of the letters *ABCDEF* that contain the pattern *DEF* by permuting four tokens—one labeled *DEF* and the others labeled *A*, *B*, and *C* (see Figure 2.2.1). Thus the number of permutations of the letters *ABCDEF* that contain the pattern *DEF* is the number

$$4! = 24$$

of permutations of four objects.

We can solve the given problem by a two-step procedure: select an ordering of the letters *DEF*; construct a permutation of *ABCDEF* containing the given ordering of the letters *DEF*. The first step can be done in $3! = 6$ ways and, by our argument above, the second step can be done in 24 ways. Thus the number of permutations of the letters *ABCDEF* containing the letters *DEF* together in any order is

$$6 \cdot 24 = 144.$$

FIGURE 2.2.1

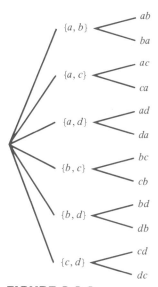

FIGURE 2.2.2

We can construct all *r*-permutations of an *n*-element set *X* in two successive steps: select an *r*-combination of *X* (an unordered subset of *r* items), and order it. For example, to construct a 2-permutation of {*a*, *b*, *c*, *d*}, we can first select a 2-combination and then order it. Figure 2.2.2 shows how all 2-permutations of {*a*, *b*, *c*, *d*} are obtained in this way. The First Counting Principle tells us that the number of *r*-permutations $P(n, r)$ is the product of the number of *r*-combinations $C(n, r)$ and the number of orderings of *r* elements *r*!. That is,

$$P(n, r) = C(n, r)r!.$$

Therefore,

$$C(n, r) = \frac{P(n, r)}{r!}.$$

Our next theorem summarizes this result and gives some alternative ways to write $C(n, r)$.

THEOREM 2.2.7. *The number of r-combinations of a set of n distinct objects is*

$$C(n, r) = \frac{P(n, r)}{r!} = \frac{\frac{n!}{(n-r)!}}{r!} = \frac{n!}{(n-r)!} \cdot \frac{1}{r!} = \frac{n!}{(n-r)!r!}$$

$$= \frac{n(n - 1) \cdots (n - r + 1)}{r!}$$

$$= \frac{n!}{(n - r)!r!}$$

Proof. The proof of the first equation is given just before the statement of the theorem. The other forms of the equation follow from Theorem 2.2.3 and equation (2.2.1).   ∎

EXAMPLE 2.2.8.   In how many ways can we select a committee of three from a group of 10 persons?

Since a committee is an unordered group of people, the answer is

$$C(10, 3) = \frac{10 \cdot 9 \cdot 8}{3!} = 120.$$

EXAMPLE 2.2.9.   In how many ways can we select a committee of two men and three women from a group of five men and six women?

As in Example 2.2.8, we find that the two men can be selected in

$$C(5, 2) = 10$$

ways and that the three women can be selected in

$$C(6, 3) = 20$$

ways. The committee can be constructed in two successive steps: select the men, select the women. By the First Counting Principle, the total number of committees is

$$10 \cdot 20 = 200.$$

EXAMPLE 2.2.10.   How many eight-bit strings contain exactly four 1's?

An eight-bit string containing four 1's is uniquely determined once we tell which bits are 1. But this can be done in

$$C(8, 4) = 70$$

ways.

EXAMPLE 2.2.11

(a) How many (unordered) five-card poker hands, selected from an ordinary 52-card deck, are there?

(b) How many poker hands contain four aces?

(c) How many poker hands contain four of a kind, that is, four cards of the same denomination?

(a) The answer is given by the combination formula

$$C(52, 5) = 2,598,960.$$

(b) The answer is 48 since, if a hand contains four aces, the fifth card may be selected in 48 ways.

(c) A hand containing four of a kind can be constructed in two successive steps: select the denomination, select the fifth card. The first step can be done in 13 ways and the second step can be done in 48 ways. By the First Counting Principle, the number of hands containing four of a kind is

$$13 \cdot 48 = 624.$$

EXERCISES

In Exercises 1–5, let $X = \{a, b, c, d\}$.

1H. Compute the number of 3-permutations of X.

2. List the 3-permutations of X.

3H. Compute the number of 3-combinations of X.

4. List the 3-combinations of X.

5H. Show the relationship between the 3-permutations and the 3-combinations of X by drawing a picture like that in Figure 2.2.2.

6. In how many ways can we select a chairperson, vice-chairperson, and recorder from a group of 11 persons?

7H. In how many ways can we select a committee of three from a group of 11 persons?

8. How many strings can be formed by ordering the letters *APPLIED*?

Exercises 9–13 refer to a club consisting of six men and seven women.

9H. In how many ways can we select a committee of three men and four women?

10. In how many ways can we select a committee of four persons which has at least one woman?

11H. In how many ways can we select a committee of four persons that has at most one man?

12. In how many ways can we select a committee of four persons that has persons of both sexes?

13H. In how many ways can we select a committee of four persons so that Mabel and Ralph do not serve together?

14. In how many ways can we select a committee of four Republicans, three Democrats, and two Independents from a group of 10 Republicans, 12 Democrats, and four Independents?

15H. How many eight-bit strings contain exactly three 0's?

16. How many eight-bit strings contain three 0's in a row and five 1's?

★17H. How many eight-bit strings contain at least two 0's in a row?

In Exercises 18–26, find the number of (unordered) five-card poker hands, selected from an ordinary 52-card deck, having the properties indicated.

18. Containing all spades

19H. All of the same suit

20. Containing cards of exactly two suits

21H.   Containing cards of all suits

22.   Of the form A2345 of the same suit

23H.   Consecutive and of the same suit. (Assume that the ace is the lowest denomination.)

24.   Consecutive. (Assume that the ace is the lowest denomination.)

25H.   Containing three of one denomination and two of another denomination

26.   Containing two of one denomination, two of another denomination, and one of a third denomination

27H.   Find the number of (unordered) 13-card bridge hands selected from an ordinary 52-card deck.

28.   How many bridge hands are all of the same suit?

29H.   How many bridge hands contain exactly two suits?

30.   How many bridge hands contain all four aces?

31H.   How many bridge hands contain five spades, four hearts, three clubs, and one diamond?

32.   How many bridge hands contain five of one suit, four of another suit, three of another suit, and one of another suit?

33H.   How many bridge hands contain four cards of three suits and one card of the fourth suit?

34.   How many bridge hands contain no face cards? (A face card is one of 10, J, Q, K, A.)

35H.   What is wrong with the following argument, which purports to show that $4C(39, 13)$ bridge hands contain three or fewer suits?

There are $C(39, 13)$ hands that contain only clubs, diamonds, and spades. In fact, for any three suits, there are $C(39, 13)$ hands that contain only those three suits. Since there are four 3-combinations of the suits, the answer is $4C(39, 13)$.

In Exercises 36–39, determine how many strings can be formed by ordering the letters *ABCDE* subject to the conditions given.

36.   *A* appears before *D*. (EXAMPLES: *BCAED, BCADE*.)

37H.   *A* and *D* are side by side. (EXAMPLES: *ADBCE, BCDAE*.)

38.   Neither the pattern *AB* nor the pattern *CD* appears.

39H.   Neither the pattern *AB* nor the pattern *BE* appears.

In Exercises 40–44, a coin is flipped 10 times.

40.   How many outcomes are possible?

41H.   How many outcomes have exactly three heads?

42.   How many outcomes have at most three heads?

43H.   How many outcomes have a head on the fifth toss?

44. How many outcomes have as many heads as tails?

45H. What is wrong with the following argument, which purports to show that there are $13^4 \cdot 48$ (unordered) five-card poker hands containing cards of all suits?

Pick one card of each suit. This can be done in $13 \cdot 13 \cdot 13 \cdot 13 = 13^4$ ways. Since the fifth card can be chosen in 48 ways, the answer is $13^4 \cdot 48$.

Exercises 46–49 refer to a shipment of 50 microprocessors, of which four are defective.

46. In how many ways can we select four microprocessors?

47H. In how many ways can we select four nondefective microprocessors?

48. In how many ways can we select four microprocessors containing exactly two defective microprocessors?

49H. In how many ways can we select four microprocessors containing at least one defective microprocessor?

50. In how many ways can six persons be seated around a circular table?

51H. In how many ways can six distinct keys be put on a ring? (Turning the ring over does not count as a different arrangement.)

52. In how many ways can five distinct Martians and five distinct Jovians wait in line?

53H. In how many ways can five distinct Martians and five distinct Jovians wait in line if no two Martians stand together?

54. In how many ways can five distinct Martians and five distinct Jovians be seated at a circular table?

55H. In how many ways can five distinct Martians and five distinct Jovians be seated at a circular table if no two Martians sit together?

56. In how many ways can five distinct Martians and eight distinct Jovians wait in line if no two Martians stand together?

57H. In how many ways can five distinct Martians and eight distinct Jovians be seated at a circular table if no two Martians sit together?

58. By interpreting $C(i, j)$ as the number of j-element subsets of an i-element set, show that

$$C(n, r) = C(n, n - r).$$

★59H. How many routes are there from A to B in the following figure if we are restricted to traveling only to the right or upward? (One such route is shown.)

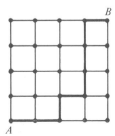

★**60.** Show that the number of n-bit strings having exactly k 0's, with no two 0's consecutive, is $C(n - k + 1, k)$.

★**61H.** Show that the product of any positive integer and its $k - 1$ successors is (evenly) divisible by $k!$.

★**62.** Show that there are $(2n - 1)(2n - 3) \cdots 3 \cdot 1$ ways to pick n pairs from $2n$ distinct items.

63. Write an algorithm that generates all 2-permutations of the distinct elements

$$S(1), \ldots, S(N).$$

64. Write an algorithm that generates all r-permutations of the distinct elements

$$S(1), \ldots, S(N).$$

65. Write an algorithm that generates all r-element subsets of a set having n elements.

66. Suppose that we have n objects, r distinct and $n - r$ identical. Give another derivation of the formula

$$P(n, r) = r!C(n, r)$$

by counting the number of orderings of the n objects in two ways:
(a) Count the orderings by first choosing positions for the r distinct objects.
(b) Count the orderings by first choosing positions for the $n - r$ identical objects.

2.3

Generalized Permutations and Combinations

In the preceding section, we dealt with orderings and selections without allowing repetitions. In this section we consider orderings of sequences containing repetitions and unordered selections in which repetitions are allowed.

EXAMPLE 2.3.1. In how many ways can we order the following letters?

MISSISSIPPI

Because of the duplication of letters, the answer is not 11!, but some number less than 11!.

Let us consider the problem to be filling 11 blanks,

— — — — — — — — — — —,

with the letters given. There are $C(11, 2)$ ways to choose positions for the two P's. Having selected the positions for the P's, there are $C(9, 4)$ ways to choose positions for the four S's. Having selected the positions for the S's, there are $C(5, 4)$ ways to choose positions for the four I's. Having made these selections, there is one position left to be filled by the M. By the First Counting Principle, the number of ways of ordering the letters is

$$C(11, 2)C(9, 4)C(5, 4) = \frac{11!}{2!9!} \frac{9!}{4!5!} \frac{5!}{4!1!}$$

$$= \frac{11!}{2!4!4!1!}$$

$$= 34{,}650.$$

The solution to Example 2.3.1 assumes a nice form. The number 11 that appears in the numerator is the total number of letters. The values in the denominator give the numbers of duplicates of each letter. The method can be used to establish a general formula.

THEOREM 2.3.2. *Suppose that a sequence S of n items has n_1 identical objects of type 1, n_2 identical objects of type 2, . . . , and n_t identical objects of type t. Then the number of orderings of S is*

$$\frac{n!}{n_1!n_2! \cdots n_t!}.$$

Proof. We assign positions to each of the n items to create an ordering of S. We may assign positions to the n_1 items of type 1 in $C(n, n_1)$ ways. Having made these assignments, we may assign positions to the n_2 items of type 2 in $C(n - n_1, n_2)$ ways, and so on. By the First Counting Principle, the number of orderings is

$$C(n, n_1)C(n - n_1, n_2)C(n - n_1 - n_2, n_3) \cdots C(n - n_1 - \cdots - n_{t-1}, n_t)$$

$$= \frac{n!}{n_1!(n - n_1)!} \frac{(n - n_1)!}{n_2!(n - n_1 - n_2)!} \cdots \frac{(n - n_1 - \cdots - n_{t-1})!}{n_t!0!}$$

$$= \frac{n!}{n_1!n_2! \cdots n_t!}.$$ ∎

EXAMPLE 2.3.3. In how many ways can eight distinct books be divided among three students if Bill gets four books and Shizuo and Marian each get two books?

Put the books in some fixed order. Now consider orderings of four B's, two S's, and two M's. An example is

BBBSMBMS

Each such ordering determines a distribution of books. For the example ordering, Bill gets books 1, 2, 3 and 6, Marian gets books 5 and 7, and Shizuo gets books 4 and 8. Thus the number of ways of ordering *BBBBSSMM* is the number of ways to distribute the books. By Theorem 2.3.2, this number is

$$\frac{8!}{4!2!2!} = 420.$$

Next, we turn to the problem of counting unordered selections where repetitions are allowed.

EXAMPLE 2.3.4. Consider three books: a computer science book, a physics book, and a history book. Suppose that the library has at least six copies of each of these books. In how many ways can we select six books?

The problem is to choose unordered, six-element selections from the set {computer science, physics, history}, repetitions allowed. A selection is uniquely determined by the number of each type of book selected. Let us denote a particular selection as

$$\text{CS} \quad \text{Physics} \quad \text{History}$$
$$\times \times \times \mid \times \times \mid \times$$

Here we have designated the selection consisting of three computer science books, two physics books, and one history book. Another example of a selection is

$$\text{CS} \quad \text{Physics} \quad \text{History}$$
$$\mid \times \times \times \times \mid \times \times$$

which denotes the selection consisting of no computer science books, four physics books, and two history books. We see that each ordering of six $\times$'s and two $\mid$'s denotes a selection. Thus our problem is to count the number of such orderings. But this is just the number of ways

$$C(8, 2) = 28$$

of selecting two positions for the $\mid$'s from eight possible positions. Thus there are 28 ways to select six books.

The method used in Example 2.3.4 can be used to derive a general result.

THEOREM 2.3.5. *If X is a set containing t elements, the number of unordered, k-element selections from X, repetitions allowed, is*

$$C(k + t - 1, t - 1) = C(k + t - 1, k).$$

Proof. Let $X = \{a_1, \ldots, a_t\}$. Consider the $k + t - 1$ slots

$$\underline{\quad} \ \underline{\quad} \ \underline{\quad} \ \cdots \ \underline{\quad} \ \underline{\quad}$$

and $k + t - 1$ symbols consisting of k $\times$'s and $t - 1$ $\mid$'s. Each placement of these symbols into the slots determines a selection. The number n_1 of $\times$'s up to the first $\mid$ represents the selection of n_1 a_1's; the number n_2 of $\times$'s between the first and second $\mid$'s represents the selection of n_2 a_2's; and so on. Since there are $C(k + t - 1, t - 1)$ ways to select the positions for the $\mid$'s, there are also $C(k + t - 1, t - 1)$ selections. This is the same as the number $C(k + t - 1, k)$ of ways to select the positions for the $\times$'s; hence there are

$$C(k + t - 1, t - 1) = C(k + t - 1, k)$$

unordered k-element selections from X, repetitions allowed. ∎

EXAMPLE 2.3.6. Suppose that there are piles of red, blue, and green balls and that each pile contains at least eight balls.
(a) In how many ways can we select eight balls?
(b) In how many ways can we select eight balls if we must have at least one ball of each color?
By Theorem 2.3.5, the number of ways of selecting eight balls is

$$C(8 + 3 - 1, 3 - 1) = C(10, 2) = 45.$$

We can also use Theorem 2.3.5 to solve part (b) if we first select one ball of each color. To complete the selection, we must choose five additional balls. This can be done in

$$C(5 + 3 - 1, 3 - 1) = C(7, 2) = 21$$

ways.

EXAMPLE 2.3.7. In how many ways can 12 identical mathematics books be distributed among the students Anna, Beth, Candy, and Dan?
We can use Theorem 2.3.5 to solve this problem if we consider the problem to be that of labeling each book with the student's name who receives it. This is the same as selecting 12 items (the names of the students) from the set {Anna, Beth, Candy, Dan}, repetitions allowed. By Theorem 2.3.5, the number of ways to do this is

$$C(12 + 4 - 1, 4 - 1) = C(15, 3) = 455.$$

EXAMPLE 2.3.8
(a) How many solutions in nonnegative integers are there to the equation

$$x_1 + x_2 + x_3 + x_4 = 29? \qquad (2.3.1)$$

(b) How many solutions in integers are there to (2.3.1) satisfying $x_1 > 0$, $x_2 > 1$, $x_3 > 2$, $x_4 \geq 0$?

(a) Each solution of (2.3.1) is equivalent to selecting 29 items, x_i of type i, $i = 1, 2, 3, 4$. According to Theorem 2.3.5, the number of selections is

$$C(29 + 4 - 1, 4 - 1) = C(32, 3) = 4960.$$

(b) Each solution of (2.3.1) satisfying the given conditions is equivalent to selecting 29 items, x_i of type i, $i = 1, 2, 3, 4$, where, in addition, we must have at least one item of type 1, at least two items of type 2, and at least three items of type 3. First, select one item of type 1, two items of type 2, and three items of type 3. Then, choose 23 additional items. By Theorem 2.3.5, this can be done in

$$C(23 + 4 - 1, 4 - 1) = C(26, 3) = 2600$$

ways.

EXAMPLE 2.3.9. How many times is the PRINT statement executed in the following program?

FOR $I_1 := 1$ to N
    FOR $I_2 := 1$ TO I_1
        FOR $I_3 := 1$ TO I_2
            .

            .

            .

                FOR $I_K := 1$ TO I_{K-1}
                    PRINT $I_1, I_2, \ldots, I_K$
                NEXT I_K

            .

            .

            .

        NEXT I_3
    NEXT I_2
NEXT I_1

Notice that each line of output consists of K integers

$$I_1 I_2 \cdots I_K, \qquad (2.3.2)$$

where

$$N \geq I_1 \geq I_2 \geq \cdots \geq I_K \geq 1 \qquad (2.3.3)$$

and that every sequence (2.3.2) satisfying (2.3.3) occurs. Thus the problem is to count the number of ways of choosing K integers, with repetitions allowed, from the set $\{1, 2, \ldots, N\}$. [Any such selection can be ordered to produce (2.3.3).] By Theorem 2.3.5, the total number of selections possible is

$$C(K + N - 1, K).$$

EXERCISES

In Exercises 1–3, determine the number of orderings of each set of letters.

1H.   *GUIDE*   **2.** *SCHOOL*   **3H.** *SALESPERSONS*

4.   In how many ways can 10 different books be divided among three students if the first student gets five books, the second three books, and the third two books?

Exercises 5–11 refer to piles of identical red, blue, and green balls where each pile contains at least 10 balls.

5H.   In how many ways can 10 balls be selected?

6. In how many ways can 10 balls be selected if at least one red ball must be selected?

7H. In how many ways can 10 balls be selected if at least one red ball, at least two blue balls, and at least three green balls must be selected?

8. In how many ways can 10 balls be selected if exactly one red ball must be selected?

9H. In how many ways can 10 balls be selected if exactly one red ball and at least one blue ball must be selected?

10. In how many ways can 10 balls be selected if at most one red ball is selected?

11H. In how many ways can 10 balls be selected if twice as many red balls as green balls must be selected?

In Exercises 12–17, find the number of integer solutions of

$$x_1 + x_2 + x_3 = 15$$

subject to the conditions given.

12. $x_1 \geq 0, x_2 \geq 0, x_3 \geq 0$ 13H. $x_1 \geq 1, x_2 \geq 1, x_3 \geq 1$

14. $x_1 = 1, x_2 \geq 0, x_3 \geq 0$ 15H. $x_1 \geq 0, x_2 > 0, x_3 = 1$

16. $0 \leq x_1 \leq 6, x_2 \geq 0, x_3 \geq 0$ ★17H. $0 \leq x_1 < 6, 1 \leq x_2 < 9, x_3 \geq 0$

★18. Find the number of solutions in integers to

$$x_1 + x_2 + x_3 + x_4 = 12$$

satisfying $0 \leq x_1 \leq 4, 0 \leq x_2 \leq 5, 0 \leq x_3 \leq 8$, and $0 \leq x_4 \leq 9$.

19H. How many integers between 1 and 1,000,000 have the sum of the digits equal to 15?

★20. How many integers between 1 and 1,000,000 have the sum of the digits equal to 20?

21H. How many bridge deals are there? (A deal consists of partitioning a 52-card deck into four hands each containing 13 cards.)

22. In how many ways can three teams containing four, two, and two persons be selected from a group of eight persons?

23H. A domino is a rectangle divided into two squares with each square numbered one of 0, 1, . . . , 6, repetitions allowed. How many distinct dominoes are there?

Exercises 24–30 refer to a bag containing 20 balls—six red, six green, and eight purple.

24. In how many ways can we select five balls if the balls are considered distinct?

25H. In how many ways can we select five balls if balls of the same color are considered identical?

26. In how many ways can we draw two red, three green, and two purple balls if the balls are considered distinct?

27H. We draw five balls, then replace the balls, and then draw five more balls. In how many ways can this be done if the balls are considered distinct?

28. We draw five balls without replacing them. We then draw five more balls. In how many ways can this be done if the balls are considered distinct?

29H. We draw five balls and at least one is red, then replace them. We then draw five balls and at most one is green. In how many ways can this be done if the balls are considered distinct?

★30. We draw five balls and at least one is red. Without replacing them, we then draw five balls and at most one is green. In how many ways can this be done if the balls are considered distinct?

31H. In how many ways can 15 identical mathematics books be distributed among six students?

32. In how many ways can 15 identical computer science books and 10 identical psychology books be distributed among five students?

33H. In how many ways can we place 10 identical balls in 12 boxes if each box can hold one ball?

34. In how many ways can we place 10 identical balls in 12 boxes if each box can hold 10 balls?

35H. Show that $(kn)!$ is (evenly) divisible by $(n!)^k$.

36. By considering the program

$$
\begin{aligned}
&\text{FOR } I_1 := 1 \text{ to } N \\
&\quad \text{FOR } I_2 := 1 \text{ to } I_1 \\
&\quad\quad \text{PRINT } I_1, I_2 \\
&\quad \text{NEXT } I_2 \\
&\text{NEXT } I_1
\end{aligned}
$$

and Example 2.3.9, deduce

$$
1 + 2 + \cdots + N = \frac{N(N + 1)}{2}.
$$

★37H. Prove the formula

$$
C(K - 1, K - 1) + C(K, K - 1) + \cdots + C(N + K - 2, K - 1) \\
= C(K + N - 1, K).
$$

38. Write an algorithm that lists all solutions in nonnegative integers to

$$
x_1 + x_2 + x_3 = N.
$$

★39H. How many legal bidding sequences are there in bridge?

2.4

Binomial Coefficients and Combinatorial Identities

The numbers $C(n, r)$ are called **binomial coefficients** because they appear in the expansion of the binomial $a + b$ raised to a power (see Theorem 2.4.1). This interplay between numbers that arise in counting problems and numbers that appear in algebraic expressions has important implications.. For example, in analyzing a problem involving counting, we may derive some algebraic relation, then solve the relation algebraically and thus produce a solution to the original problem. An identity that results from some counting process is called a **combinatorial identity** and the argument that leads to its formulation is called a **combinatorial argument**. In this section we briefly discuss these ideas.

Let us begin by discussing the binomial theorem from a combinatorial point of view. The binomial theorem gives a formula for the coefficients in the expansion of $(a + b)^n$. Since

$$(a + b)^n = \underbrace{(a + b)(a + b) \cdots (a + b),}_{n \text{ factors}} \tag{2.4.1}$$

the expansion results from selecting one term from each of the n factors, multiplying the selections together, and then summing all such products obtained. For example,

$$(a + b)^2 = (a + b)(a + b) = aa + ab + ba + bb = a^2 + 2ab + b^2.$$

In (2.4.1), a term of the form $a^{n-k}b^k$ arises from choosing a from $n - k$ factors and b from k factors. But this can be done in $C(n, k)$ ways, since $C(n, k)$ counts the number of ways of selecting k things from n items. Thus $a^{n-k}b^k$ appears $C(n, k)$ times. It follows that

$$(a + b)^n = C(n, 0)a^n b^0 + C(n, 1)a^{n-1}b^1 + C(n, 2)a^{n-2}b^2$$
$$+ \cdots + C(n, n - 1)a^1 b^{n-1} + C(n, n)a^0 b^n. \tag{2.4.2}$$

This result is known as the **Binomial Theorem**.

THEOREM 2.4.1   Binomial Theorem. *If a and b are real numbers and n is a positive integer, then*

$$(a + b)^n = \sum_{k=0}^{n} C(n, k)a^{n-k}b^k.$$

Proof.   The proof precedes the statement of the theorem.   ∎

The Binomial Theorem can also be proved using induction on n (see Exercise 16).

EXAMPLE 2.4.2.   Taking $n = 3$ in Theorem 2.4.1, we obtain

$$(a + b)^3 = C(3, 0)a^3 + C(3, 1)a^2b + C(3, 2)ab^2 + C(3, 3)b^3$$
$$= a^3 + 3a^2b + 3ab^2 + b^3.$$

If we set $a = b = 1$ in Theorem 2.4.1, we obtain the identity

$$2^n = (1 + 1)^n = \sum_{k=0}^{n} C(n, k). \tag{2.4.3}$$

This equation can also be proved by giving a combinatorial argument. Given an n-element set X, $C(n, k)$ counts the number of k-element subsets. Thus the right side of equation (2.4.3) counts the number of subsets of X. But in Theorem 1.7.5 we showed that the number of subsets of X is 2^n; we have reproved equation (2.4.3).

```
              1
           1     1
        1     2     1
     1     3     3     1
  1     4     6     4     1
1     5    10    10     5     1
```

FIGURE 2.4.1

Let us write the binomial coefficients in a triangular form known as **Pascal's triangle** (see Figure 2.4.1). The border consists of 1's, and any interior value is the sum of the two numbers above it. This relationship is stated formally in the next theorem.

THEOREM 2.4.3

$$C(n + 1, k) = C(n, k - 1) + C(n, k)$$

for $1 \le k \le n$.

Proof.   We will give a combinatorial proof.

Let X be a set with n elements. Choose $a \notin X$. Then $C(n + 1, k)$ is the number of k-element subsets of $Y = X \cup \{a\}$. Now the k-element subsets of Y can be divided into two disjoint classes:

1. Subsets of Y not containing a.
2. Subsets of Y containing a.

The subsets of class 1 are just k-element subsets of X and there are $C(n, k)$ of these. The subsets of class 2 consist of a $(k - 1)$-element subset of X together with a and there are $C(n, k - 1)$ of these. Therefore,

$$C(n + 1, k) = C(n, k - 1) + C(n, k). \qquad \blacksquare$$

Theorem 2.4.3 can also be proved using Theorem 2.2.7 (Exercise 17). Theorem 2.4.3 can be used to obtain other results.

THEOREM 2.4.4

$$\sum_{i=k}^{n} C(i, k) = C(n + 1, k + 1).$$

Proof. We use Theorem 2.4.3 in the form

$$C(i, k) = C(i + 1, k + 1) - C(i, k + 1)$$

to obtain

$$C(k, k) + C(k + 1, k) + C(k + 2, k) + \cdots + C(n, k) = 1$$
$$+ C(k + 2, k + 1) - C(k + 1, k + 1) + C(k + 3, k + 1)$$
$$- C(k + 2, k + 1) + \cdots + C(n + 1, k + 1)$$
$$- C(n, k + 1) = C(n + 1, k + 1). \qquad \blacksquare$$

Exercise 37, Section 2.3, shows another way to prove Theorem 2.4.4.

EXAMPLE 2.4.5. Use Theorem 2.4.4 to find the sum

$$1 + 2 + \cdots + n.$$

We may write

$$1 + 2 + \cdots + n = C(1, 1) + C(2, 1) + \cdots + C(n, 1)$$
$$= C(n + 1, 2) \qquad \text{by Theorem 2.4.4}$$
$$= \frac{(n + 1)n}{2}.$$

EXERCISES

1H. Expand $(x + y)^4$ using the Binomial Theorem.

2. Expand $(2c - 3d)^5$ using the Binomial Theorem.

3H. Find the eighth term in the expansion of $(x + y)^{11}$.

4. Find the seventh term in the expansion of $(2s - t)^{12}$.

In Exercises 5–9, find the coefficient of each term when the given expression is expanded.

5H. $x^2y^3z^5$; $(x + y + z)^{10}$ **6.** $w^2x^3y^2z^5$; $(2w + x + 3y + z)^{12}$

7H. a^2x^3; $(a + x + c)^2(a + x + d)^3$ **8.** a^2x^3; $(a + ax + x)(a + x)^4$

9H. a^3x^4; $(a + \sqrt{ax} + x)^2(a + x)^5$

In Exercises 10–12, find the number of terms in the expansion of each expression.

10. $(x + y + z)^{10}$

11H. $(w + x + y + z)^{12}$

★12. $(x + y + z)^{10}(w + x + y + z)^2$

13H. Find the eighth row of Pascal's triangle from the seventh row

$$1 \quad 7 \quad 21 \quad 35 \quad 35 \quad 21 \quad 7 \quad 1.$$

14. (a) Show that $C(n, k) < C(n, k + 1)$ if and only if $k < (n - 1)/2$.
(b) Use part (a) to deduce that the maximum of $C(n, k)$ for $k = 0, 1, \ldots, n$ is $C(n, n/2)$ if n is even and $C(n, (n - 1)/2)$ if n is odd.

15H. Show that

$$0 = \sum_{k=0}^{n} (-1)^k C(n, k).$$

16. Use induction on n to prove the Binomial Theorem.

17H. Prove Theorem 2.4.3 by using Theorem 2.2.7.

18. Give a combinatorial argument to show that

$$C(n, k) = C(n, n - k).$$

★19H. Prove Theorem 2.4.4 by giving a combinatorial argument.

20. Find the sum

$$1 \cdot 2 + 2 \cdot 3 + \cdots + (n - 1)n.$$

★21H. Use Theorem 2.4.4 to derive a formula for

$$1^2 + 2^2 + \cdots + n^2.$$

22. Use the Binomial Theorem to show that

$$\sum_{k=0}^{n} 2^k C(n, k) = 3^n.$$

23H. Suppose that n is even. Prove that

$$\sum_{k=0}^{n/2} C(n, 2k) = 2^{n-1} = \sum_{k=1}^{n/2} C(n, 2k - 1).$$

24H. Prove

$$(a + b + c)^n = \sum_{0 \le i+j \le n} \frac{n!}{i!j!(n - i - j)!} a^i b^j c^{n-i-j}.$$

25. Use Exercise 24 to write the expansion of $(x + y + z)^3$.

26. Prove

$$3^n = \sum_{0 \le i+j \le n} \frac{n!}{i!j!(n-i-j)!}$$

★27H. Give a combinatorial argument to prove that

$$\sum_{k=0}^{n} C(n, k)^2 = C(2n, n).$$

★28H. Given the polynomial

$$p(n) = a_m n^m + a_{m-1} n^{m-1} + \cdots + a_1 n + a_0,$$

show that there exist $b_0, \ldots, b_m$ such that

$$p(n) = b_m C(n, m) + b_{m-1} C(n, m-1) + \cdots + b_1 C(n, 1) + b_0 C(n, 0)$$

for $n = 1, 2, \ldots$. [Define $C(i, j) = 0$ if $i < j$.]

★29H. Show that if m is a positive integer, there exists a polynomial p of degree $m + 1$ such that

$$1^m + 2^m + \cdots + n^m = p(n)$$

for $n = 1, 2, \ldots$.

30. Use Exercise 29 to show that

$$1^m + 2^m + \cdots + n^m = O(n^{m+1}).$$

31. [Requires differentiation]   Make appropriate choices for a and b in the Binomial Theorem, and then differentiate to obtain the identity

$$n(1 + x)^{n-1} = \sum_{k=1}^{n} C(n, k)kx^{k-1}.$$

32. Use the result of Exercise 31 to show that

$$n2^{n-1} = \sum_{k=1}^{n} kC(n, k). \tag{2.4.4}$$

★33. Prove equation (2.4.4) by induction.

★34. Prove equation (2.4.4) by using a combinatorial argument.

2.5

Recurrence Relations

It is often possible to develop a relationship among the elements of a sequence. Such a relationship is called a **recurrence relation**. We will illustrate the concept with an example and then give a formal definition.

EXAMPLE 2.5.1. A person invests $1000 at 12 percent compounded annually. If A_n represents the amount at the end of n years, determine a relationship between A_n and A_{n-1}.

At the end of $n - 1$ years, the amount is A_{n-1}. After one more year, we will have the amount A_{n-1} plus the interest. Thus

$$A_n = A_{n-1} + (0.12)A_{n-1}$$

$$= (1.12)A_{n-1}. \qquad (2.5.1)$$

The initial value

$$A_0 = 1000 \qquad (2.5.2)$$

together with equation (2.5.1) allows us to compute the value of A_n for any n. For example,

$$A_3 = (1.12)A_2$$

$$= (1.12)(1.12)A_1$$

$$= (1.12)(1.12)(1.12)A_0$$

$$= (1.12)^3(1000) = 1404.93. \qquad (2.5.3)$$

Thus, at the end of the third year, the amount is $1404.93.

The computation (2.5.3) can be carried out for an arbitrary value of n to obtain

$$A_n = (1.12)A_{n-1}$$

$$\cdot$$
$$\cdot$$
$$\cdot$$

$$= (1.12)^n(1000). \qquad (2.5.4)$$

Equation (2.5.1) furnishes an example of a recurrence relation. A recurrence relation defines a sequence by giving the nth value in terms of certain of its predecessors. Explicitly given values of a sequence, such as (2.5.2), are called **initial conditions**. Example 2.5.1 illustrates the fact that a sequence can be defined by a recurrence relation together with certain initial conditions.

DEFINITION 2.5.2. A *recurrence relation* for the sequence $a_0, a_1, \ldots$ is an equation that relates a_n to certain of its predecessors $a_0, a_1, \ldots, a_{n-1}$.

A recurrence relation defines the nth term of a sequence indirectly; to compute a_n, we must first compute the terms $a_0, \ldots, a_{n-1}$. This is in contrast to an explicit formula for a_n, where we could compute a_n by just "plugging in n." Example 2.5.1 shows that sometimes an explicit formula can be derived from a recurrence relation and initial conditions. Equation (2.5.4) gives an explicit formula for the sequence defined by the recurrence relation (2.5.1) and the initial condition (2.5.2).

We can easily construct an algorithm based on a recurrence relation and initial conditions. The following algorithm, based on the recurrence relation (2.5.1) and initial condition (2.5.2), computes the sequence of Example 2.5.1. Sometimes an algorithm based on a recurrence relation is preferable to an algorithm based on an explicit formula.

ALGORITHM 2.5.3. This algorithm computes the sequence of Example 2.5.1. The algorithm is based on the recurrence relation (2.5.1) and the initial condition (2.5.2). The input is N, the term to be computed. The algorithm returns the value A, which is the Nth term of the sequence $A_0, A_1, \ldots$.
1. [Initialization.] $A := 1000$, $I := 0$ (I is the current term being computed.)
2. [Done?] If $I = N$, stop.
3. [Compute next term.] $A := (1.12) * A$.
4. [Update counter and then loop.] $I := I + 1$. Go to step 2.

A **recursive algorithm** is an algorithm that invokes itself. Many higher-level programming languages, such as Pascal, SNOBOL, ALGOL, PL/1, and C, directly support recursion by allowing routines to call themselves. Other languages, such as FORTRAN, do not allow routines to call themselves. As an example, we will rewrite Algorithm 2.5.3 as a recursive algorithm.

ALGORITHM 2.5.4. This recursive algorithm computes the sequence of Example 2.5.1. The input is N, the term to be computed. The algorithm returns the value A, which is the Nth term of the sequence $A_0, A_1, \ldots$.
1. [Trivial case?] If $N = 0$, set $A := 1000$ and return with the result in A.
2. [Recursive step.] Call Algorithm 2.5.4 to compute A_{N-1}. Set $A := (1.12) * A_{N-1}$ and return with the result in A.

Algorithm 2.5.4 is a direct translation of equations (2.5.1) and (2.5.2), which define the sequence $A_0, A_1, \ldots$. Step 1 corresponds to the initial condition (2.5.2) and step 2 corresponds to the recurrence relation (2.5.1).

The remainder of this section is devoted to giving additional examples of recurrence relations.

EXAMPLE 2.5.5. Let S_n denote the number of subsets of an n-element set. Since going from an $(n - 1)$-element set to an n-element set doubles the number of subsets, we obtain the recurrence relation

$$S_n = 2S_{n-1}$$

and the initial condition

$$S_0 = 1.$$

EXAMPLE 2.5.6 Fibonacci Sequence.   One of the oldest recurrence relations defines the **Fibonacci sequence**. The sequence originally arose in Fibonacci's book *Liber Abaci* (1202), where Fibonacci asked the question: "After one year, how many pairs of rabbits will there be, if at the beginning of the year there is one pair, and if every month each pair produces a new pair which becomes productive after one month?" It was further assumed that no deaths occur.

We let f_i denote the number of pairs of rabbits at the end of the ith month. Then

$$f_0 = 1. \tag{2.5.5}$$

After one month, there is still just one pair because a pair does not become productive until after one month. Therefore,

$$f_1 = 1. \tag{2.5.6}$$

Equations (2.5.5) and (2.5.6) are the initial conditions for the Fibonacci sequence. The increase in pairs of rabbits $f_n - f_{n-1}$ from month $n - 1$ to month n is due to each pair alive in month $n - 2$ producing an additional pair. That is,

$$f_n - f_{n-1} = f_{n-2}$$

or

$$f_n = f_{n-1} + f_{n-2}. \tag{2.5.7}$$

The recurrence relation (2.5.7), together with the initial conditions (2.5.5) and (2.5.6), defines the Fibonacci sequence. You should verify that the solution to Fibonacci's question is $f_{12} = 233$.

Although it is easy to obtain an explicit formula from the recurrence relation and initial condition for the sequence of Example 2.5.1, it is not immediately apparent how to obtain an explicit formula for the Fibonacci sequence. In the next section we will give a method that yields an explicit formula for the Fibonacci sequence. Of course, the lack of an explicit formula is no deterrent to writing an algorithm to compute the Fibonacci sequence.

EXAMPLE 2.5.7 Tower of Hanoi.   The Tower of Hanoi is a puzzle consisting of three pegs mounted on a board and n disks of various sizes

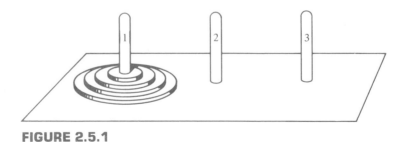

FIGURE 2.5.1

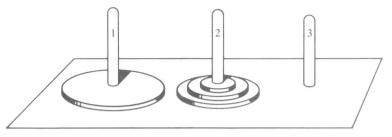

FIGURE 2.5.2

with holes in their centers (see Figure 2.5.1). It is assumed that if a disk is on a peg, only a disk of smaller diameter can be placed on top of the first disk. Given all the disks stacked on one peg as in Figure 2.5.1, the problem is to transfer the disks to another peg by moving one disk at a time.

If c_n denotes the number of moves in which the n-disk puzzle can be solved, find a recurrence relation and an initial condition for the sequence $c_1, c_2, \ldots$.

Suppose that we have n disks on peg 1 as in Figure 2.5.1. Then, in c_{n-1} moves, we can move the top $n - 1$ disks to peg 2 (see Figure 2.5.2). During these moves, the bottom disk on peg 1 stays fixed. Next, we move the remaining disk on peg 1 to peg 3. Finally, in c_{n-1} moves, we can move the $n - 1$ disks on peg 2 to peg 3. Therefore the desired recurrence relation is

$$c_n = 2c_{n-1} + 1. \tag{2.5.8}$$

The initial condition is

$$c_1 = 1.$$

EXAMPLE 2.5.8 Derangements. At a social gathering, n persons check their coats. When they leave, the coats are returned randomly and, unfortunately, no one receives the correct coat. Let D_n be the number of ways n persons can all receive the wrong coats. Show that the sequence D_1, $D_2, \ldots$ satisfies the recurrence relation

$$D_n = (n - 1)(D_{n-1} + D_{n-2}). \tag{2.5.9}$$

We see that D_n is the number of permutations

$$m_1, \ldots, m_n$$

of $1, \ldots, n$, where $m_i \neq i$ for $i = 1, \ldots, n$. Such permutations are called **derangements**.

Suppose that there are C derangements of $1, \ldots, n$ of the form

$$2, m_2, \ldots, m_n.$$

By interchanging 2 and 3, we see that there are also C derangements of $1, \ldots, n$ of the form

$$3, m_2, \ldots, m_n.$$

It follows that there are C derangements of $1, \ldots, n$ of the form

$$k, m_2, \ldots, m_n$$

where k is a fixed integer between 2 and n. Since there were $n - 1$ possibilities, $2, \ldots, n$, for the first term,

$$D_n = (n - 1)C. \qquad (2.5.10)$$

We partition the derangements of $1, \ldots, n$

$$2, m_2, \ldots, m_n$$

into those of the form

$$2, 1, m_3, \ldots, m_n \qquad (2.5.11)$$

and

$$2, m_2, m_3, \ldots, m_n \qquad (2.5.12)$$

where $m_2 \neq 1$. In (2.5.11), the $n - 2$ numbers $m_3, \ldots, m_n$ are all out of position, so that (2.5.11) consists of 2, 1 followed by a derangement of $3, \ldots, n$. There are D_{n-2} of these. In (2.5.12), the $n - 1$ numbers $m_2, \ldots, m_n$ are all out of position so that (2.5.12) consists of 2 followed by a derangement of $1, 3, 4, \ldots, n$. There are D_{n-1} of these. It follows that

$$C = D_{n-1} + D_{n-2}.$$

Combining this equation with (2.5.10), we obtain the recurrence relation

$$D_n = (n - 1)(D_{n-1} + D_{n-2}).$$

It is possible to extend the definition of recurrence relation to include functions indexed over n-tuples of positive integers. Our last example is of this form.

EXAMPLE 2.5.9 Ackermann's Function.   Ackermann's function can be defined by the recurrence relations

$$A(m, 0) = A(m - 1, 1), \qquad\qquad m = 1, 2, \ldots, \quad (2.5.13)$$

$$A(m, n) = A(m - 1, A(m, n - 1)), \qquad m = 1, 2, \ldots,$$

$$n = 1, 2, \ldots, \quad (2.5.14)$$

and the initial conditions

$$A(0, n) = n + 1, \qquad n = 0, 1, \ldots. \qquad (2.5.15)$$

Ackermann's function is of theoretical importance because of its rapid rate of growth.

The computation

$$A(1, 1) = A(0, A(1, 0)) \qquad \text{by } (2.5.14)$$
$$= A(0, A(0, 1)) \qquad \text{by } (2.5.13)$$
$$= A(0, 2) \qquad \text{by } (2.5.15)$$
$$= 3 \qquad \text{by } (2.5.15)$$

illustrates the use of equations (2.5.13)–(2.5.15).

EXERCISES

1H.   (a) A person invests \$2000 at 14 percent compounded annually. If A_n represents the amount at the end of n years, find a recurrence relation and an initial condition for the sequence $A_0, A_1, \ldots$.
(b) Give an explicit formula for A_n.

2.   (a) A person invests \$3000 at 12.5 percent compounded quarterly. If A_n represents the amount at the end of n years, find a recurrence relation and an initial condition for the sequence $A_0, A_1, \ldots$.
(b) Give an explicit formula for A_n.

3H.   If P_n denotes the number of permutations of n distinct objects, find a recurrence relation and an initial condition for the sequence $P_1, P_2, \ldots$.

★4.   Let D_n be the number of derangements of n objects. Show that

$$D_n = n!\left(1 - \frac{1}{1!} + \frac{1}{2!} - \cdots + \frac{(-1)^n}{n!}\right).$$

5H.   Suppose that we have n dollars and that each day we buy either orange juice (\$1), milk (\$2), or beer (\$2). If C_n is the number of ways of spending all the money, show that

$$C_n = C_{n-1} + 2C_{n-2}.$$

6.   Suppose that we have n dollars and that each day we buy either tape (\$1), paper (\$1), pens (\$2), pencils (\$2), or binders (\$3). If C_n is the number of ways of spending all the money, derive a recurrence relation for the sequence $C_1, C_2, \ldots$.

7H.   Let C_n denote the number of regions into which the plane is divided by n lines. Assume that each pair of lines meets in a point, but that no three lines meet in a point. Derive a recurrence relation for the sequence $C_1, C_2, \ldots$.

Exercises 8 and 9 refer to the sequence S_n defined by

$$S_1 = 0, \qquad S_2 = 1$$

$$S_n = \frac{S_{n-1} + S_{n-2}}{2}, \qquad n = 3, 4, \ldots .$$

8.   Compute S_1, S_2, S_3, and S_4.

★9H.   Guess a formula for S_n and show that it is correct by using induction.

         Exercises 10–14 refer to the Fibonacci sequence f_0, f_1, ...

10.   Show that $f_{n+2}^2 - f_{n+1}^2 = f_n f_{n+3}$, $n = 0, 1, \ldots$.

11H.   Show that $f_{n+1}^2 = f_n f_{n+2} + (-1)^{n+1}$, $n = 0, 1, \ldots$.

12.   A robot can move forward in steps of size 1 meter or 2 meters. Let C_n denote the number of ways the robot can walk n meters. Show that $C_n = f_n$, $n = 1$, 2, ...

13H.   Let C_n denote the number of bit strings of length n having no two 0's consecutive. Show that $C_n = f_{n+1}$, $n = 1, 2, \ldots$.

14.   Show that

$$f_{n+1} = \sum_{i=0}^{n} C(n + 1 - i, i), \qquad n = 1, 2, \ldots ,$$

where $C(j, k)$ is defined to be 0 if $j < k$.

★15H.   If α is a bit string, let $C(\alpha)$ be the maximum number of consecutive 0's in α. (EXAMPLES: $C(10010) = 2$, $C(00110001) = 3$.) Let S_n be the number of n-bit strings α with $C(\alpha) \le 2$. Develop a recurrence relation for S_1, S_2, ...

★16H.   Let S_n denote the number of n-bit strings that do not contain the pattern 010. Develop a recurrence relation for S_1, S_2, ... How many eight-bit strings contain the pattern 010?

17.   Write an algorithm that generates all eight-bit strings that do not contain the pattern 010.

18.   Let P_n be the number of partitions of an n-element set. Show that the sequence P_0, P_1, . . . satisfies the recurrence relation

$$P_n = \sum_{k=0}^{n-1} C(n - 1, k)P_k.$$

★19H.   Let F_n denote the number of functions f from $X = \{1, \ldots, n\}$ into X having the property that if i is in the range of f, then $1, 2, \ldots, i$ are also in the range of f. (Set $F_0 = 1$.) Show that the sequence F_0, F_1, . . . satisfies the recurrence relation

$$F_n = \sum_{j=0}^{n-1} C(n, j)F_j.$$

20.   Derive a recurrence relation for $C(n, k)$, the number of k-element subsets of an n-element set. Specifically, write $C(n + 1, k)$ in terms of $C(n, i)$ for appropriate i.

21H.   Derive a recurrence relation for $S(k, n)$, the number of ways of choosing k items, allowing repetitions, from n available types. Specifically, write $S(k, n)$ in terms of $S(k - 1, i)$ for appropriate i.

22. Let $S(n, k)$ denote the number of functions from $\{1, \ldots, n\}$ onto $\{1, \ldots, k\}$. Show that $S(n, k)$ satisfies the recurrence relation

$$S(n, k) = k^n - \sum_{i=1}^{k-1} C(k, i)S(n, i).$$

23. Write explicit solutions for the cases $n = 3, 4, 5$ of the Tower of Hanoi puzzle.

24H. Write a recursive algorithm for solving the Tower of Hanoi puzzle.

Exercises 25–29 refer to Ackermann's function $A(m, n)$.

25H. Compute $A(2, 2)$, $A(2, 3)$, and $A(3, 2)$.

26. Use induction to show that

$$A(1, n) = n + 2, \qquad n = 0, 1, \ldots .$$

27H. Use induction to show that

$$A(2, n) = 3 + 2n, \qquad n = 0, 1, \ldots .$$

28H. Guess a formula for $A(3, n)$ and prove it by using induction.

★29. Write a recursive algorithm to compute Ackermann's function.

What we and others have called Ackermann's function is actually derived from Ackermann's original function defined by

$$AO(0, y, z) = z + 1, \qquad\qquad y, z \ \geq 0$$
$$AO(1, y, z) = y + z, \qquad\qquad y, z \ \geq 0$$
$$AO(2, y, z) = yz, \qquad\qquad y, z \ \geq 0$$
$$AO(x + 3, y, 0) = 1, \qquad\qquad x, y \ \geq 0$$
$$AO(x + 3, y, z + 1) = AO(x + 2, y, AO(x + 3, y, z)), \qquad x, y, z \geq 0.$$

Exercises 30–33 refer to this function and to Ackermann's function A.

30. Show that $A(x, y) = AO(x, 2, y + 3) - 3$ for $y \geq 0$ and $x = 0, 1, 2, 3$.

31H. Show that $AO(x, 2, 1) = 2$ for $x \geq 2$.

32. Show that $AO(x, 2, 2) = 4$ for $x \geq 2$.

★33H. Show that $A(x, y) = AO(x, 2, y + 3) - 3$ for $x, y \geq 0$.

34. Write nonrecursive algorithms to compute the sequences of Exercises 1–3, 5–10, 15, 16, and 18–22.

35. Write recursive algorithms to compute the sequences of Exercises 1–3, 5–10, 15, 16, and 18–22.

36H. [Requires infinite series] If there are n possible outcomes, all equally likely, and an event E occurs in k of these outcomes, we say that the **probability** of E is k/n.

(a) Show that under the conditions of Example 2.5.8, the probability P_n that no one receives the correct coat is

$$P_n = 1 - \frac{1}{1!} + \frac{1}{2!} - \cdots + \frac{(-1)^n}{n!}.$$

(b) Show that if $n \geq 5$, $P_n \doteq 1/e = 0.3678\ldots$, where the error in the approximation is no more than 10^{-3}.

This result shows that if $n \geq 5$, P_n is almost constant.

⋆37H. Let S_n denote the number of permutations of n distinct objects that fix exactly k objects. Show that S_k, S_{k+1}, $\ldots$ satisfies the recurrence relation

$$S_n = \frac{n(n - k - 1)S_{n-1} + n(n - 1)S_{n-2}}{n - k}.$$

2.6

Solving Recurrence Relations

To solve a recurrence relation involving the sequence $a_0, a_1, \ldots$ is to find an explicit formula for the general term a_n. In this section we discuss two methods of solving recurrence relations: iteration and a special method that applies to linear homogeneous recurrence relations with constant coefficients. For more powerful methods, such as methods that make use of generating functions, consult [Brualdi].

To solve a recurrence relation involving the sequence $a_0, a_1, \ldots$ by **iteration**, we use the recurrence relation to write the nth term a_n in terms of certain of its predecessors $a_{n-1}, \ldots, a_0$. We then successively use the recurrence relation to replace each of $a_{n-1}, \ldots$ by certain of their predecessors. We continue until an explicit formula is obtained. The iterative method was used to solve the recurrence relation of Example 2.5.1 [see equations (2.5.4)].

EXAMPLE 2.6.1. We can solve the recurrence relation

$$S_n = 2S_{n-1}$$

of Example 2.5.5, subject to the initial condition

$$S_0 = 1,$$

by iteration:

$$S_n = 2S_{n-1} = 2(2S_{n-1})_{-1} = 2(2S_{n-2})$$
$$= 2(2S_{n-2})$$

$$\cdot$$
$$\cdot$$
$$\cdot$$

$$= 2^n S_0$$
$$= 2^n.$$

EXAMPLE 2.6.2. Find an explicit formula for c_n, the number of moves in which the n-disk Tower of Hanoi puzzle can be solved (see Example 2.5.7).

In Example 2.5.7, we obtained the recurrence relation

$$c_n = 2c_{n-1} + 1 \qquad (2.6.1)$$

and the initial condition

$$c_1 = 1.$$

Applying the iterative method to (2.6.1), we obtain

$$c_n = 2c_{n-1} + 1$$
$$= 2(2c_{n-2} + 1) + 1$$
$$= 2^2 c_{n-2} + 2 + 1$$
$$= 2^2(2c_{n-3} + 1) + 2 + 1$$
$$= 2^3 c_{n-3} + 2^2 + 2 + 1$$

$$\cdot$$
$$\cdot$$
$$\cdot$$

$$= 2^{n-1}c_1 + 2^{n-2} + 2^{n-3} + \cdots + 2 + 1$$
$$= 2^{n-1} + 2^{n-2} + 2^{n-3} + \cdots + 2 + 1$$
$$= 2^n - 1.$$

The last step results from the formula for the sum of a geometric series.

EXAMPLE 2.6.3.   Let D_n be the number of derangements of n objects (see Example 2.5.8). Show that

$$D_n - nD_{n-1} = -[D_{n-1} - (n-1)D_{n-2}]. \qquad (2.6.2)$$

Use iteration to transform (2.6.2) to

$$D_n = nD_{n-1} + (-1)^n; \qquad (2.6.3)$$

then apply iteration to (2.6.3) to derive an explicit formula for D_n.

Equation (2.6.2) follows immediately from (2.5.9). Applying iteration to (2.6.2), we obtain

$$
\begin{aligned}
D_n - nD_{n-1} &= -[D_{n-1} - (n-1)D_{n-2}] \\
&= -\{-[D_{n-2} - (n-2)D_{n-3}]\} \\
&= D_{n-2} - (n-2)D_{n-3} \\
& \cdot \\
& \cdot \\
& \cdot \\
&= (-1)^n(D_2 - 2D_1). \qquad (2.6.4)
\end{aligned}
$$

Since $D_2 = 1$ and $D_1 = 0$, (2.6.4) yields (2.6.3).

Applying iteration to (2.6.3), we obtain

$$
\begin{aligned}
D_n &= nD_{n-1} + (-1)^n \\
&= n[(n-1)D_{n-2} + (-1)^{n-1}] + (-1)^n \\
&= n(n-1)D_{n-2} + n(-1)^{n-1} + (-1)^n \\
&= n(n-1)[(n-2)D_{n-3} + (-1)^{n-2}] + n(-1)^{n-1} + (-1)^n \\
&= n(n-1)(n-2)D_{n-3} + n(n-1)(-1)^{n-2} \\
& + n(-1)^{n-1} + (-1)^n \\
& \cdot \\
& \cdot \\
& \cdot \\
&= n(n-1)\cdots 2D_1 + [n(n-1)\cdots 3] - [n(n-1)\cdots 4] \\
& + \cdots + n(-1)^{n-1} + (-1)^n \\
&= [n(n-1)\cdots 3] - [n(n-1)\cdots 4] + \cdots + n(-1)^{n-1} \\
& + (-1)^n, \qquad n > 2.
\end{aligned}
$$

For example,

$$D_5 = 5 \cdot 4 \cdot 3 - 5 \cdot 4 + 5 - 1 = 44.$$

We turn next to a special class of recurrence relations.

DEFINITION 2.6.4. A *linear homogeneous recurrence relation of order k with constant coefficients* is a recurrence relation of the form

$$a_n = c_1 a_{n-1} + c_2 a_{n-2} + \cdots + c_k a_{n-k}. \tag{2.6.5}$$

Notice that a linear homogeneous recurrence relation of order k with constant coefficients (2.6.5), together with the k initial conditions

$$a_0 = C_0, \quad a_1 = C_1, \ldots, a_{k-1} = C_{k-1},$$

uniquely defines a sequence $a_0, a_1, \ldots$.

EXAMPLE 2.6.5. The recurrence relations

$$A_n = (1.12)A_{n-1} \tag{2.6.6}$$

of Example 2.5.1 and

$$f_n = f_{n-1} + f_{n-2} \tag{2.6.7}$$

of Example 2.5.6 (Fibonacci sequence) are both linear homogeneous recurrence relations with constant coefficients. The relation (2.6.6) is of order 1 and (2.6.7) is of order 2.

We will illustrate the general method of solving linear homogeneous recurrence relations with constant coefficients by finding an explicit formula for the Fibonacci sequence.

The solution of (2.6.6) was of the form

$$A_n = Ct^n.$$

Let us search for a solution of the Fibonacci recurrence relation (2.6.7) of the form t^n. We must have

$$t^n = t^{n-1} + t^{n-2}$$

or

$$t^n - t^{n-1} - t^{n-2} = 0$$

or

$$t^2 - t - 1 = 0. \tag{2.6.8}$$

The solutions of (2.6.8) are

$$r = \frac{1 \pm \sqrt{5}}{2}.$$

At this point we have the two solutions of (2.6.7),

$$S_n = \left(\frac{1 + \sqrt{5}}{2}\right)^n, \qquad T_n = \left(\frac{1 - \sqrt{5}}{2}\right)^n. \tag{2.6.9}$$

We can verify (see Theorem 2.6.6) that if s_n and t_n are solutions of (2.6.7), then $bs_n + dt_n$ is also a solution of (2.6.7). Therefore,

$$U_n = bS_n + dT_n$$

$$= b\left(\frac{1 + \sqrt{5}}{2}\right)^n + d\left(\frac{1 - \sqrt{5}}{2}\right)^n$$

is a solution of (2.6.7).

To satisfy the initial conditions

$$f_0 = 1 = f_1$$

of the Fibonacci sequence, we must have

$$U_0 = 1 = U_1$$

or

$$bS_0 + dT_0 = b + d = 1$$

$$bS_1 + dT_1 = \frac{b(1 + \sqrt{5})}{2} + \frac{d(1 - \sqrt{5})}{2} = 1.$$

Solving these equations for b and d, we obtain

$$b = \frac{1}{\sqrt{5}}\left(\frac{1 + \sqrt{5}}{2}\right), \qquad d = -\frac{1}{\sqrt{5}}\left(\frac{1 - \sqrt{5}}{2}\right).$$

Therefore,

$$f_n = \frac{1}{\sqrt{5}}\left(\frac{1 + \sqrt{5}}{2}\right)^{n+1} - \frac{1}{\sqrt{5}}\left(\frac{1 - \sqrt{5}}{2}\right)^{n+1}.$$

In the case of the Fibonacci sequence, it is simpler to compute the nth term f_n using the recurrence relation than the explicit formula.

At this point we will summarize and justify the techniques used to find the explicit formula for the Fibonacci sequence.

THEOREM 2.6.6. *Let*

$$a_n = c_1 a_{n-1} + c_2 a_{n-2} \tag{2.6.10}$$

be a second-order linear homogeneous recurrence relation with constant coefficients.

If S_n and T_n are solutions of (2.6.10), then $U_n = bS_n + dT_n$ is also a solution of (2.6.10).

If r is a root of

$$t^2 - c_1 t - c_2 = 0, \qquad (2.6.11)$$

then r^n is a solution of (2.6.10).

   Let a_n be a solution of (2.6.10) satisfying

$$a_0 = C_0, \qquad a_1 = C_1. \qquad (2.6.12)$$

If r_1 and r_2 are roots of (2.6.11) and $r_1 \neq r_2$, there exist constants b and d such that

$$a_n = br_1^n + dr_2^n, \qquad n = 0, 1, \ldots .$$

   Proof.   Since S_n and T_n are solutions of (2.6.10),

$$S_n = c_1 S_{n-1} + c_2 S_{n-2}$$

$$T_n = c_1 T_{n-1} + c_2 T_{n-2}.$$

If we multiply the first equation by b and the second by d and add, we obtain

$$U_n = bS_n + dT_n = c_1(bS_{n-1} + dT_{n-1}) + c_2(bS_{n-2} + dT_{n-2})$$

$$= c_1 U_{n-1} + c_2 U_{n-2}.$$

Therefore, U_n is a solution of (2.6.10).

   Since r is a root of (2.6.11),

$$r^2 = c_1 r + c_2.$$

Now,

$$c_1 r^{n-1} + c_2 r^{n-2} = r^{n-2}(c_1 r + c_2)$$

$$= r^{n-2} r^2 = r^n;$$

thus r^n is a solution of (2.6.10).

   If we set $U_n = br_1^n + dr_2^n$, then U_n satisfies (2.6.10). To meet the initial conditions (2.6.12), we must have

$$U_0 = b + d = C_0$$

$$U_1 = br_1 + dr_2 = C_1.$$

If we multiply the first equation by r_1 and subtract, we obtain

$$d(r_1 - r_2) = r_1 C_0 - C_1.$$

Since $r_1 - r_2 \neq 0$, we can solve for d. Similarly, we can solve for b. With these choices for b and d, we have

$$U_0 = C_0, \qquad U_1 = C_1.$$

Since U_n satisfies (2.6.10), it follows that $a_n = U_n$, $n = 0, 1, \ldots .$ ■

Theorem 2.6.6 states that any solution of (2.6.10) may be given in terms of two basic solutions r_1^n and r_2^n. However, in case (2.6.11) has two equal roots r, we obtain only one basic solution r^n. The next theorem shows that in this case, nr^n furnishes the other basic solution.

THEOREM 2.6.7. *Let*

$$a_n = c_1 a_{n-1} + c_2 a_{n-2} \qquad (2.6.13)$$

be a second-order linear homogeneous recurrence relation with constant coefficients.

Let a_n be a solution of (2.6.13) satisfying

$$a_0 = C_0, \qquad a_1 = C_1.$$

If both roots of

$$t^2 - c_1 t - c_2 = 0 \qquad (2.6.14)$$

are equal to r, there exist constants b and d such that

$$a_n = br^n + dnr^n, \qquad n = 0, 1, \ldots.$$

Proof. The proof of Theorem 2.6.6 shows that r^n is a solution of (2.6.13). We show that nr^n is also a solution of (2.6.13).

Since r is the only solution of (2.6.14), we must have

$$t^2 - c_1 t - c_2 = (t - r)^2.$$

It follows that

$$c_1 = 2r, c_2 = -r^2.$$

Now

$$c_1[(n - 1)r^{n-1}] + c_2[(n - 2)r^{n-2}] = 2r(n - 1)r^{n-1} - r^2(n - 2)r^{n-2}$$

$$= r^n[2(n - 1) - (n - 2)] = nr^n.$$

Therefore, nr^n is a solution of (2.6.13).

By Theorem 2.6.6, $U_n = br^n + dnr^n$ is a solution of (2.6.13).

The proof that there are constants b and d such that $U_0 = C_0$ and $U_1 = C_1$ is similar to the argument given in Theorem 2.6.6 and is left as an exercise (Exercise 32). It follows that $a_n = U_n$, $n = 0, 1, \ldots$. ∎

EXAMPLE 2.6.8. Solve the recurrence relation

$$d_n = 4(d_{n-1} - d_{n-2}) \qquad (2.6.15)$$

subject to the initial conditions

$$d_0 = 1 = d_1.$$

According to Theorem 2.6.6, $S_n = r^n$ is a solution of (2.6.15) where r is a solution of

$$t^2 - 4t + 4 = 0. \qquad (2.6.16)$$

Thus we obtain the solution

$$S_n = 2^n$$

of (2.6.15). Since 2 is the only solution of (2.6.16), by Theorem 2.6.7,

$$T_n = n2^n$$

is also a solution of (2.6.15). Thus the general solution of (2.6.15) is of the form

$$U_n = aS_n + bT_n.$$

We must have

$$U_0 = 1 = U_1.$$

These last equations become

$$aS_0 + bT_0 = a + 0b = 1$$

$$aS_1 + bT_1 = 2a + 2b = 1.$$

Solving for a and b, we obtain

$$a = 1, \qquad b = -\frac{1}{2}.$$

Therefore, the solution of (2.6.15) is

$$d_n = 2^n - n2^{n-1}.$$

For the general linear homogeneous recurrence relation of order k with constant coefficients (2.6.5), if r is a root of

$$t^k - c_1 t^{k-1} - c_2 t^{k-2} - \cdots - c_k = 0$$

of multiplicity m, it can be shown that

$$r^n, nr^n, \ldots, n^{m-1}r^n$$

are solutions of (2.6.5). This fact can be used, just as in the previous examples for recurrence relations of order 2, to solve a linear homogeneous recurrence relation of order k with constant coefficients. For a precise statement and a proof of the general result, see [Brualdi].

EXERCISES

Tell whether or not each recurrence relation in Exercises 1–10 is a linear homogeneous recurrence relation with constant coefficients. Give the order of each linear homogeneous recurrence relation with constant coefficients.

1H. $a_n = -3a_{n-1}$

2. $a_n = 2na_{n-1}$

3H. $a_n = 2na_{n-2} - a_{n-1}$

4. $a_n = a_{n-1} + n$

5H. $a_n = 7a_{n-2} - 6a_{n-3}$

6. $a_n = a_{n-1} + 1 + 2^{n-1}$

7H. $a_n = (\lg 2n)a_{n-1} - [\lg(n-1)]a_{n-2}$

8. $a_n = 6a_{n-1} - 9a_{n-2}$

9H. $a_n = -a_{n-1} - a_{n-2}$

10. $a_n = -a_{n-1} + 5a_{n-2} - 3a_{n-3}$

In Exercises 11–24, solve each recurrence relation for the initial conditions given.

11H.   Exercise 1; $a_0 = 2$

12.   Exercise 2; $a_0 = 1$

13H.   Exercise 4; $a_0 = 0$

14. $a_n = 6a_{n-1} - 8a_{n-2}$; $a_0 = 1, a_1 = 0$

15H. $2a_n = 7a_{n-1} - 3a_{n-2}$; $a_0 = a_1 = 1$

16.   Exercise 5; $a_0 = 0, a_1 = 1, a_2 = 0$

17H.   Exercise 6; $a_0 = 0$

18.   Exercise 8; $a_0 = a_1 = 1$

19H.   Exercise 9; $a_0 = 1, a_1 = 2$

20.   Exercise 10; $a_0 = 0, a_1 = 1, a_2 = 2$

21H.   Exercise 5, Section 2.5

22.   Exercise 7, Section 2.5

23H.   The recurrence relation preceding Exercise 8, Section 2.5

24.   Exercise 37, Section 2.5

25H.   Solve the recurrence relation

$$\sqrt{a_n} = \sqrt{a_{n-1}} + 2\sqrt{a_{n-2}}$$

with initial conditions $a_0 = a_1 = 1$.

★26.   Solve the recurrence relation

$$a_n = \frac{\sqrt{a_{n-1}}}{a_{n-2}^2}$$

with initial conditions $a_0 = 1, a_1 = 2$.

★27H.   Solve the recurrence relation

$$a_n = -2na_{n-1} + 3n(n-1)a_{n-2}$$

with initial conditions $a_0 = 1, a_1 = 2$.

28.   The equation

$$a_n = c_1a_{n-1} + c_2a_{n-2} + f(n) \qquad (2.6.17)$$

is called a **second-order linear inhomogeneous recurrence relation with constant coefficients**.

Let $g(n)$ be a solution of (2.6.17). Show that any solution U_n of (2.6.17) is of the form

$$U_n = V_n + g(n),$$

where V_n is a solution of the homogeneous equation (2.6.10).

29H. The equation

$$a_n = f(n)a_{n-1} + g(n)a_{n-2} \qquad (2.6.18)$$

is called a **second-order linear homogeneous recurrence relation**. The coefficients $f(n)$ and $g(n)$ are not necessarily constant. Show that if S_n and T_n are solutions of (2.6.18), then $bS_n + dT_n$ is also a solution of (2.6.18).

30. Show that if S_n and T_n are solutions of (2.6.5), then $bS_n + dT_n$ is also a solution of (2.6.5).

31H. Show that if r is a solution of

$$t^k - c_1 t^{k-1} - \cdots - c_k = 0,$$

then r^n satisfies (2.6.5).

32. Suppose that both roots of

$$t^2 - c_1 t - c_2 = 0$$

are equal to r and suppose that a_n satisfies

$$a_n = c_1 a_{n-1} + c_2 a_{n-2}$$

$$a_0 = C_0, \qquad a_1 = C_1.$$

Show that there exist constants b and d such that

$$a_n = br^n + dnr^n, \qquad n = 0, 1, \ldots,$$

thus completing the proof of Theorem 2.6.7.

2.7

An Application to the Analysis of Algorithms

Recurrence relations often arise in the analysis of algorithms. Consider, for example, using binary search to locate an item in an ordered sequence (Algorithm 2.7.2). Here the idea is to divide the sequence into two nearly equal parts. We first check whether the item is located at the dividing point (step 4). If the item is not found, an additional comparison (step 5) will locate the half of the sequence in which the item might appear. We continue to halve the sequence until either we locate the item or we find that it is not in the sequence.

The **floor** of x, denoted $\lfloor x \rfloor$, used in Algorithm 2.7.2, is the greatest integer less than or equal to x. The **ceiling** of x, denoted $\lceil x \rceil$, is the least integer greater than or equal to x.

EXAMPLE 2.7.1

$$\lfloor 8.3 \rfloor = 8, \quad \lceil 9.1 \rceil = 10, \quad \lfloor -8.7 \rfloor = -9, \quad \lceil -11.3 \rceil = -11,$$
$$\lfloor 6 \rfloor = 6, \quad \lceil -8 \rceil = -8$$

ALGORITHM 2.7.2   Binary Search.   The input is the sequence

$$S(1), S(2), \ldots, S(N)$$

sorted in increasing order. This algorithm searches this sequence for the value KEY. The algorithm returns the position of KEY in M if it is found; otherwise, it returns the value 0 in M.

1. [Initialization.] $I := 1$; $J := N$. (I and J mark the boundaries of the sequence currently being searched.)

2. [Not found?] If $I > J$, then (not found) set $M := 0$ and stop.

3. [Subdivide.] $M := \lfloor (I + J)/2 \rfloor$.

4. [Found?] If KEY $= S(M)$, then (found) stop.

5. [Reset boundaries.] If KEY $< S(M)$, then $J := M - 1$; otherwise, set $I := M + 1$.

6. [Loop.] Go to step 2.

EXAMPLE 2.7.3.   We will illustrate Algorithm 2.7.2 for the array

$$S(1) = `B', \quad S(2) = `D', \quad S(3) = `F', \quad S(4) = `S'.$$

First, suppose that KEY $= `S'$. At step 1, $I = 1$ and $J = 4$. At step 2, since $I = 1 \leq 4 = J$, we proceed to step 3. At step 3, $M = 2$. At step 4, since KEY $= `S' \neq `D' = S(2)$, we proceed to step 5. At step 5, KEY $= `S' \geq `D' = S(2)$; so we set $I = 3$. At step 2, since $I = 3 \leq 4 = J$, we proceed to step 3. At step 3, $M = 3$. At step 4, since KEY $= `S' \neq `F' = S(3)$, we proceed to step 5. At step 5, KEY $= `S' \geq `F' = S(3)$; so we set $I = 4$. At step 2, since $I = 4 \leq 4 = J$, we proceed to step 3. At step 3, $M = 4$. At step 4, KEY $= `S' = S(4)$, so the algorithm terminates with KEY found at location 4.

Now, suppose that KEY $= `C'$ and $S$ is as before. At step 1, $I = 1$ and $J = 4$. At step 2, since $I = 1 \leq 4 = J$, we proceed to step 3. At step 3, $M = 2$. At step 4, since KEY $= `C' \neq `D' = S(2)$, we proceed to step 5. At step 5, KEY $= `C' < `D' = S(2)$; so we set $J = 1$. At step 2, since $I = 1 \leq 1 = J$, we proceed to step 3. At step 3, $M = 1$. At step 4, since KEY $= `C' \neq `B' = S(1)$, we proceed to step 5. At step 5, KEY $= `C' \geq `B' = S(1)$; so we set $I = 2$. At step 2, $I = 2 > 1 = J$, so we set $M = 0$ and terminate the algorithm. In this case, KEY is not found.

We want to examine the worst-case time of this algorithm. We will measure the time of this algorithm by counting the number of iterations of the loop 2–3–4–5–6. We let a_N be the number of iterations of the loop in the worst case for an input of size N.

Suppose that $N > 1$. Initially, $I = 1$ and $J = N$, so at step 2, $I < J$ and we proceed to step 3. At step 3, M is set to $\lfloor (I + J)/2 \rfloor = \lfloor (1 + N)/2 \rfloor$. In the worst case, KEY $\neq S(M)$ and we proceed to step 5. At step 5, if KEY $< S(M)$, we execute the loop again with input the array

$$S(1), \ldots, S(M - 1). \tag{2.7.1}$$

If KEY $\geq S(M)$, we execute the loop again with input the array

$$S(M + 1), \ldots, S(N). \tag{2.7.2}$$

We will show that in the worst case, the size of the array input is $\lfloor N/2 \rfloor$. By assumption, $a_{\lfloor N/2 \rfloor}$ additional iterations of the loop are required. Since we have already executed the loop once, we obtain the recurrence relation

$$a_N = 1 + a_{\lfloor N/2 \rfloor}. \tag{2.7.3}$$

The initial condition is

$$a_1 = 1. \tag{2.7.4}$$

To show that, in the worst case, the size of the array input to the loop is $\lfloor N/2 \rfloor$, we must consider separately the possibilities: N is even, N is odd. We will give the argument for N even and leave the case N odd as an exercise (Exercise 16).

If N is even, $\lfloor N/2 \rfloor = N/2$. Since $1 + N$ is odd,

$$M = \left\lfloor \frac{1 + N}{2} \right\rfloor = \frac{N}{2}.$$

The size of the array (2.7.1) is

$$M - 1 = \frac{N}{2} - 1.$$

The size of the array (2.7.2) is

$$N - (M + 1) + 1 = \frac{N}{2}.$$

In the worst case, the array size input to the loop is $N/2 = \lfloor N/2 \rfloor$ as claimed.

The recurrence relation (2.7.3) is not readily solved by the methods of the preceding section. However, in case N is a power of 2, say

$$N = 2^k,$$

(2.7.3) becomes

$$a_{2^k} = 1 + a_{2^{k-1}}, \quad k = 1, 2, \ldots.$$

If we let $b_k = a_{2^k}$, we obtain the recurrence relation

$$b_k = 1 + b_{k-1}, \quad k = 1, 2, \ldots, \tag{2.7.5}$$

and the initial condition

$$b_0 = 1.$$

The recurrence relation (2.7.5) can be solved by the iterative method:

$$b_k = 1 + b_{k-1}$$

$$= 2 + b_{k-2}$$

$$\cdot$$

$$\cdot$$

$$\cdot$$

$$= k + b_0$$

$$= k + 1.$$

Thus, if $N = 2^k$,

$$a_N = a_{2^k} = b_k = k + 1 = 1 + \lg N. \tag{2.7.6}$$

Equation (2.7.6) suggests the general formula that can be verified using induction.

THEOREM 2.7.4. *The solution of the recurrence relation (2.7.3) with initial condition (2.7.4) is*

$$a_N = \lfloor \lg N \rfloor + 1. \tag{2.7.7}$$

Proof. We use induction on N.
If $N = 1$,

$$\lfloor \lg 1 \rfloor + 1 = 1 = a_1,$$

so the basis step is verified.
Suppose that $N > 1$ and

$$a_K = \lfloor \lg K \rfloor + 1 \tag{2.7.8}$$

for all $K < N$.
If N is even,

$$a_N = 1 + a_{\lfloor N/2 \rfloor} \qquad\qquad \text{by (2.7.3)}$$

$$= 1 + a_{N/2}$$

$$= 1 + \left\lfloor \lg \frac{N}{2} \right\rfloor + 1 \qquad\qquad \text{by (2.7.8)}$$

$$= 2 + \lfloor \lg N - 1 \rfloor$$

$$= 2 + \lfloor \lg N \rfloor - 1 \qquad \text{since } \lfloor x - 1 \rfloor = \lfloor x \rfloor - 1$$

$$= 1 + \lfloor \lg N \rfloor.$$

If N is odd,

$$a_N = 1 + a_{\lfloor N/2 \rfloor}$$

$$= 1 + a_{(N-1)/2}$$

$$= 1 + \left\lfloor \lg \frac{N-1}{2} \right\rfloor + 1$$

$$= 2 + \lfloor \lg (N-1) - 1 \rfloor$$

$$= 2 + \lfloor \lg (N-1) \rfloor - 1$$

$$= 1 + \lfloor \lg (N-1) \rfloor$$

$$= 1 + \lfloor \lg N \rfloor \qquad \text{if } N \text{ is odd,} \quad \lfloor \lg N \rfloor = \lfloor \lg (N-1) \rfloor.$$

In either case, $a_N = 1 + \lfloor \lg N \rfloor$, so the inductive step is complete. ∎

COROLLARY 2.7.5. *For an input of size N, in the worst case, binary search (Algorithm 2.7.2) requires* $1 + \lfloor \lg N \rfloor$ *iterations.*

Algorithm 2.7.2 furnishes an example of a **divide-and-conquer algorithm**. In a divide-and-conquer algorithm, the original problem is divided into subproblems. In Algorithm 2.7.2, the input array

$$S(1), \ldots, S(N)$$

is partitioned into two almost evenly divided arrays

$$S(1), \ldots, S\left(\left\lfloor \frac{1+N}{2} \right\rfloor - 1 \right) \qquad S\left(\left\lfloor \frac{1+N}{2} \right\rfloor + 1 \right), \ldots, S(N).$$

The original problem is then reduced to solving the problem in one of these smaller arrays.

EXERCISES

In Exercises 1–6, compute the quantity indicated.

1H. $\lfloor 13.5 \rfloor$ **2.** $\lceil 13.5 \rceil$ **3H.** $\lfloor -13.5 \rfloor$ **4.** $\lceil -13.5 \rceil$

5H. $\lfloor 16 \rfloor$ **6.** $\lceil 16 \rceil$

7H. Show that for any real number x, $\lfloor x - 1 \rfloor = \lfloor x \rfloor - 1$ and $\lceil x - 1 \rceil = \lceil x \rceil - 1$.

8. Show that if n is an odd integer, $\lfloor n/2 \rfloor = (n-1)/2$ and $\lceil n/2 \rceil = (n+1)/2$.

9H. Show that if n is an odd integer greater than 1, then $\lfloor \lg n \rfloor = \lfloor \lg (n - 1) \rfloor$.

10. State and prove a result analogous to Exercise 9 for the ceiling function.

Exercises 11–14 refer to the array

$$S(1) = \text{`}C\text{'}, \quad S(2) = \text{`}G\text{'}, \quad S(3) = \text{`}J\text{'}, \quad S(4) = \text{`}M\text{'}, \quad S(5) = \text{`}X\text{'}.$$

11. Show how Algorithm 2.7.2 executes in case KEY = 'G'.

12. Show how Algorithm 2.7.2 executes in case KEY = 'P'.

13. Show how Algorithm 2.7.2 executes in case KEY = 'C'.

14. Show how Algorithm 2.7.2 executes in case KEY = 'Z'.

15H. In the best case of Algorithm 2.7.2, how many loop iterations are required?

16. Show that for N odd, the size of the array input to the loop in Algorithm 2.7.2 is $\lfloor N/2 \rfloor$.

17H. **(a)** Show that Algorithm 2.7.2 works correctly if step 3 is changed to $M :=$ $\lceil (I + J)/2 \rceil$.
(b) For the change of part (a), determine the size of the array input in the worst case to the loop of Algorithm 2.7.2.

Exercises 18–22 refer to the following algorithm.

ALGORITHM 2.7.6   Finding the Largest and Smallest Elements in an Array.   This recursive algorithm finds the largest and smallest elements in the array

$$S(1), \ldots, S(N).$$

The largest element is returned in LARGE and the smallest in SMALL.
    1. [Trivial?] If $N = 1$, set LARGE $:= S(1)$, SMALL $:= S(1)$, and return.
    2. [Divide array.] Set $M := \lfloor (1 + N)/2 \rfloor$.
    3. [Solve subproblems.] Call Algorithm 2.7.6 with input the array $S(1), \ldots, S(M)$. Set LARGELEFT $:=$ LARGE and SMALLLEFT $:=$ SMALL. Call Algorithm 2.7.6 with input the array $S(M + 1), \ldots, S(N)$. Set LARGERIGHT $:=$ LARGE and SMALLRIGHT $:=$ SMALL.
    4. [Find largest element.] If LARGELEFT $>$ LARGERIGHT, then LARGE $:=$ LARGELEFT; otherwise, set LARGE $:=$ LARGERIGHT.
    5. [Find smallest element.] If SMALLLEFT $<$ SMALLRIGHT, then SMALL $:=$ SMALLLEFT; otherwise, set SMALL $:=$ SMALLRIGHT.
    6. [Return.] Return.

Let a_N be the number of comparisons (steps 4 and 5) required for an input of size N.

18. Show that $a_1 = 0$ and $a_2 = 2$.

19H. Find a_3.

20. Establish the recurrence relation

$$a_{\lfloor N/2 \rfloor} + a_{\lfloor (N+1)/2 \rfloor} + 2 = a_N \qquad (2.7.9)$$

for $N > 1$.

21H. Solve the recurrence relation (2.7.9) for the case that N is power of 2 to obtain

$$a_N = 2N - 2, \qquad N = 1, 2, 4, \ldots .$$

★22. Use induction to show that

$$a_N = 2N - 2, \qquad N = 1, 2, 3, \ldots .$$

Exercises 23–27 refer to Algorithm 2.7.6 with step 1 replaced by

    1a. [First trivial case?] If $N = 1$, set LARGE $:=$ $S(1)$, SMALL $:=$ $S(1)$, and return.
    1b. [Second trivial case?] If $N > 2$, go to step 2.
    1c. [Case $N = 2$.] If $S(1) > S(2)$, set LARGE $:=$ $S(1)$, SMALL $:=$ $S(2)$, and return; otherwise, set LARGE $:=$ $S(2)$, SMALL $:=$ $S(1)$, and return.

Let a_N be the number of comparisons (steps 1c, 4, and 5) for an input of size N.

23H. Show that $a_1 = 0$ and $a_2 = 1$.

24. Compute a_3 and a_4.

25H. Show that the recurrence relation (2.7.9) holds for $N > 2$.

26. Solve the recurrence relation (2.7.9) for the case that N is a power of 2 to obtain

$$a_N = \tfrac{3}{2}N - 2, \qquad N = 2, 4, 8, \ldots .$$

★27. Find a formula for a_N, $N = 1, 2, \ldots$, and prove it by using induction.

★28H. Modify Algorithm 2.7.6 by replacing step 1 with steps 1a, 1b, and 1c above. Replace step 2 by

    2. [Divide array.] If N is even and $N/2$ is odd, set $M := (N/2) - 1$; otherwise, set $M := \lfloor (1 + N)/2 \rfloor$.

Show that in the worst case, this modified algorithm requires at most $\lceil (3N/2) - 2 \rceil$ comparisons to find the largest and smallest elements in an array of size N.

Exercises 29–36 refer to a divide-and-conquer algorithm that accepts as input the array

$$S(1), \ldots, S(N).$$

If $N > 1$, the subproblems

$$S(1), \ldots, S\left(\left\lfloor \frac{1+N}{2} \right\rfloor\right) \quad \text{and} \quad S\left(\left\lfloor \frac{1+N}{2} \right\rfloor + 1\right), \ldots, S(N)$$

are solved recursively. Solutions to subproblems $S(1), \ldots, S(M)$ and $T(1), \ldots, T(K)$ can be combined in time $b_{M,K}$ to solve the original problem. Let a_N be the time required by the algorithm for an input of size N.

29H. Write a recurrence relation for a_N assuming that $b_{M,K} = 3$.

30. Write a recurrence relation for a_N assuming that $b_{M,K} = M + K$.

31H. Solve the recurrence relation of Exercise 29 assuming that $a_1 = 0$ for the case when N is a power of 2.

32. Solve the recurrence relation of Exercise 29 assuming that $a_1 = 1$ for the case when N is a power of 2.

33H. Solve the recurrence relation of Exercise 30 assuming that $a_1 = 0$ for the case when N is a power of 2.

34. Solve the recurrence relation of Exercise 30 assuming that $a_1 = 1$ for the case when N is a power of 2.

★35H. Assume that if $M_1 \geq M_2$ and $K_1 \geq K_2$, then $b_{M_1,K_1} \geq b_{M_2,K_2}$. Use induction to show that the sequence $a_1, a_2, \ldots$ is increasing.

★36H. Assuming that $b_{M,K} = M + K$ and $a_1 = 0$, show that $a_N \leq 4N \lg N$.

2.8

Notes

An elementary book on counting methods is [Niven]. References on combinatorics are [Bogart; Brualdi; Even, 1973; Liu, 1968; and Riordan]. [Vilenkin] contains many worked-out combinatorial examples. The following general references on discrete mathematics contain chapters on counting methods and recurrence relations: [Liu, 1985; Sahni; Stanat; and Tucker]. [Even, 1973; Hu, 1982; and Reingold] treat combinatorial algorithms. [Gardner, 1979] has a chapter on the Fibonacci sequence.

COMPUTER EXERCISES

1. Write a program that generates all 2-permutations of the elements *ABCDE*.

2. Write a program that generates all *r*-permutations, for arbitrary *r*, of the elements *ABCDEF*.

3. Write a program that lists all *r*-element subsets, for arbitrary *r*, of the set $\{A, B, C, D, E\}$.

4. Write a program that lists all eight-bit strings having exactly three 0's in a row.

5. Write a program that lists all permutations of *ABCDE* in which *A* appears before *D*.

6. Write a program that lists all permutations of *ABCDE* in which *C* and *E* are side by side in either order.

7. Write a program that lists all ways of selecting six books from three available types. (At least six copies of each type are available.)

8. Write a program that lists all solutions in nonnegative integers of

$$x_1 + x_2 + x_3 = 10.$$

9. Write a program that lists all solutions in integers of

$$x_1 + x_2 + x_3 = 12$$

satisfying

$$0 < x_1, \quad 2 < x_2, \quad 1 < x_3 < 6.$$

10. Write a program that generates Pascal's triangle to level N for arbitrary N.

11. Write a nonrecursive program to compute the nth Fibonacci number.

12. Write a recursive program to compute the nth Fibonacci number. When you run the program, restrict yourself to $n < 5$. Explain why this recursive program is grossly inefficient.

13. Write a recursive program that solves the Tower of Hanoi puzzle. Your program should print out, or display graphically, the sequence of moves constituting the solution.

14. Write a program to compute the number of derangements of n objects.

15. Write a program to compute Ackermann's function.

16. Write nonrecursive programs to compute the sequences of Exercises 1–3, 5–10, 15, 16, and 18–22, Section 2.5.

17. Write recursive programs to compute the sequences of Exercises 1–3, 5–10, 15, 16, and 18–22, Section 2.5.

18. (a) Implement Algorithm 2.7.2 as a program. Allow the program to optionally output the number of iterations of the loop.
(b) Generate some random values for KEY and calculate the average number of loop iterations required by Algorithm 2.7.2. If all positions in the array, including gaps, are equally likely, it can be shown (see [Baase]) that Algorithm 2.7.2 requires about $\lg n + \frac{1}{2}$ loop iterations. Compare your experimental results with this theoretical value.

19. Implement Algorithm 2.7.6 as a program.

20. Implement Algorithm 2.7.6, with the changes given before Exercise 23, Section 2.7, as a program.

21. Implement Algorithm 2.7.6, with the changes given in Exercise 28, Section 2.7, as a program.

CHAPTER 3

Graph Theory

Although the first paper in graph theory goes back to 1736 (see Example 3.1.4) and several important results in graph theory were obtained in the nineteenth century, it is only since the 1920s that there has been a sustained, widespread, intense interest in graph theory. Indeed, the first text on graph theory ([König]) appeared in 1936. Undoubtedly, one of the reasons for the recent interest in graph theory is its applicability in many diverse fields, including computer science, chemistry, operations research, electrical engineering, linguistics, and economics.

This chapter begins with several examples illustrating how graphs arise in a variety of problems. We then study certain important properties of graphs, including paths and circuits. A shortest-path algorithm is presented which efficiently finds a shortest path between two given points. Finally, we study the questions of when two graphs are essentially the same (i.e., when two graphs are isomorphic) and when a graph can be drawn in the plane without having any of its edges cross.

3.1

Examples

In Figure 3.1.1 we see a map of eight cities and the connecting roads. Such a diagram provides an example of a **graph**. A graph consists of **vertices** and

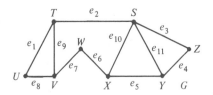

FIGURE 3.1.1

edges. In Figure 3.1.1 the vertices represent cities and the edges represent roads. The formal definition of a graph follows.

DEFINITION 3.1.1. A *graph* (or *undirected graph*) G consists of a set V of *vertices* (or *nodes*) and a set E of *edges* (or *arcs*) such that each edge $e \in E$ is associated with an unordered pair of vertices. If an edge e is associated with a unique pair of vertices v and w, we write $e = (v, w)$ or $e = (w, v)$. In this context, (v, w) denotes an edge in an undirected graph and *not* an ordered pair.

A *directed graph* (or *digraph*) G consists of a set V of *vertices* (or *nodes*) and a set E of *edges* (or *arcs*) such that each edge $e \in E$ is associated with an ordered pair of vertices. If an edge e is associated with a unique ordered pair (v, w) of vertices, we write $e = (v, w)$.

An edge $e = (v, w)$ in a graph (undirected or directed) is said to be *incident on* v and w. The vertices v and w are said to be *incident on e* and to be *adjacent vertices*.

If G is a graph (undirected or directed) with vertices V and edges E, we write $G = (V, E)$.

Unless specified otherwise, the sets E and V are assumed finite.

EXAMPLE 3.1.2. In Figure 3.1.1 the (undirected) graph G consists of the set

$$\{S, T, U, V, W, X, Y, Z\}$$

of vertices and the set

$$\{e_1, e_2, \ldots, e_{11}\}$$

of edges. Edge e_1 is associated with the unordered pair $\{U, T\}$ of vertices and edge e_{10} is associated with the unordered pair $\{S, X\}$ of vertices. Edge e_1 is denoted (U, T) or (T, U) and edge e_{10} is denoted (S, X) or (X, S). Edge e_4 is incident on Y and Z and the vertices Y and Z are adjacent.

EXAMPLE 3.1.3. A directed graph is shown in Figure 3.1.2. The directed edges are indicated by arrows. Edge e_1 is associated with the ordered pair (v_2, v_1) of vertices and edge e_7 is associated with the ordered pair (v_6, v_6) of vertices. Edge e_1 is denoted (v_2, v_1) and edge e_7 is denoted (v_6, v_6).

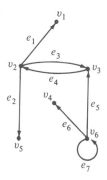

FIGURE 3.1.2

In this chapter we will be concerned primarily with undirected graphs. In Chapter 5 we will discuss some applications of directed graphs.

Definition 3.1.1 allows distinct edges to be associated with the same pair of vertices. For example, in Figure 3.1.3 edges e_1 and e_2 are both associated with the vertex pair $\{v_1, v_2\}$. Such edges are called **parallel edges**. An edge of the form (v, v) is called a **loop**. For example, in Figure 3.1.3 edge $e_3 = (v_2, v_2)$ is a loop. Notice that in Figure 3.1.3, no edge is incident on vertex v_4. A graph with neither loops nor parallel edges is called a **simple graph**.

Some definitions of graph do not permit loops and parallel edges. One would expect that if agreement has not been reached on the definition of graph, most other terms in graph theory would also not have standard definitions. This is indeed the case. In reading articles and books about graphs, it is necessary to check on the definitions being used.

If a real problem can be modeled as a question about graphs, one would hope to be able to apply general graph theory to provide an answer to the graph question. Interpreting this answer in the real world solves the original problem. Of course, the situation is usually not this neat. There may be no applicable general theory. Moreover, sometimes we do not want or need a general theory; the graph model is used simply to aid our thinking or to communicate the problem to others. In this section we illustrate a variety of problems that make use of graphs.

EXAMPLE 3.1.4 Königsberg Bridge Problem.   The first paper in graph theory was Leonhard Euler's in 1736. The paper presented a general theory that included a solution to what is now called the Königsberg Bridge Problem.

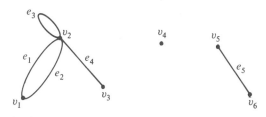

FIGURE 3.1.3

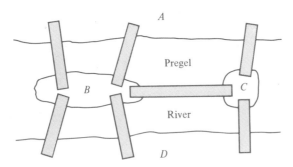

FIGURE 3.1.4

Two islands lying in the Pregel River in Königsberg (now Kaliningrad in the Soviet Union) were connected to each other and the river banks by bridges as shown in Figure 3.1.4. The problem is to start at any location—A, B, C, or D; walk over each bridge exactly once; then return to the starting location. Such a route is called an **Euler circuit**.

The bridge configuration can be modeled as a graph as shown in Figure 3.1.5. The vertices represent the locations and the edges represent the bridges.

The solution is easily obtained by using the concept of the degree of a vertex. The **degree of a vertex** v, $\delta(v)$, is the number of edges incident on v. (By definition, each loop on v contributes 2 to the degree of v.) In Figure 3.1.5 we find that

$$\delta(A) = 3$$
$$\delta(B) = 5$$
$$\delta(C) = 3$$
$$\delta(D) = 3.$$

Now, suppose that we can start at any vertex, move along each edge exactly once, and return to the starting position. Let v be a vertex other than the initial vertex. Each time we visit v, we also leave v (see Figure 3.1.6). Moreover, each edge incident on v is traversed once. Thus the edges incident on v occur in pairs, and therefore v has even degree. The edges incident on the initial vertex v also occur in pairs. The edge on which we leave v for the first time is paired with the edge on which we return to v for the last time. The remaining edges also occur in pairs just as for a noninitial vertex. Therefore, if we can start at any vertex, move along each edge exactly once, and

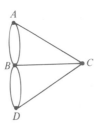

FIGURE 3.1.5

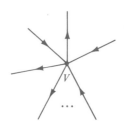

FIGURE 3.1.6

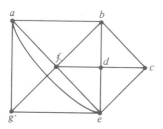

FIGURE 3.1.7

return to the starting position, all vertices must have even degree. Thus the graph of Figure 3.1.5 does not have an Euler circuit.

In Section 3.3 we will discuss Euler circuits in more detail. We will show that, if G is a graph in which every vertex has even degree and if it is possible to go from any vertex to any other vertex on some path, then G has an Euler circuit.

A problem similar in its statement to the problem of finding an Euler circuit in a graph is finding a Hamiltonian circuit in a graph. A **Hamiltonian circuit** in a graph G is a route that begins and ends at the same vertex and which passes through each vertex of G exactly once. For example, the route (a, b, c, d, e, f, g, a) is a Hamiltonian circuit for the graph of Figure 3.1.7. An Euler circuit visits each edge once, whereas a Hamiltonian circuit visits each vertex once. Notice that the graph of Figure 3.1.7 does not have an Euler circuit.

There are apparently profound differences between the problems of finding Euler circuits and finding Hamiltonian circuits. Unlike the situation for Euler circuits, no easily verified necessary and sufficient conditions are known for the existence of a Hamiltonian circuit in a graph. Although there are algorithms (see, e.g., [Even, 1979]) for finding an Euler circuit, if there is one, in time $O(n)$ for a graph having n edges, every known algorithm for finding Hamiltonian circuits requires either exponential or factorial time in the worst case.

The Hamiltonian circuit is named for Sir William Rowan Hamilton, who marketed a puzzle in the mid-1800s in the form of a dodecahedron (see Figure 3.1.8). Each corner bore the name of a city and the problem was to start at any city, travel along the edges, visit each city once, and return to the initial city. The graph of the edges of the dodecahedron is given in Figure 3.1.9. We leave as an exercise (Exercise 15) the problem of solving Hamilton's puzzle.

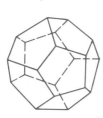

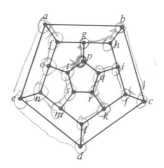

FIGURE 3.1.8   **FIGURE 3.1.9**

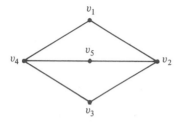

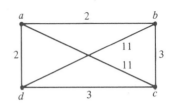

FIGURE 3.1.10   **FIGURE 3.1.11**

EXAMPLE 3.1.5.   Show that the graph of Figure 3.1.10 does not contain a Hamiltonian circuit.

  Since there are five vertices, a Hamiltonian circuit must have five edges. Suppose that we could eliminate edges from the graph to produce a Hamiltonian circuit. We would have to eliminate one edge incident at v_2 and one edge incident at v_4, since each vertex in a Hamiltonian circuit has degree 2. But this leaves four edges—not enough for a Hamiltonian circuit. Therefore, the graph of Figure 3.1.10 does not contain a Hamiltonian circuit.

A problem related to that of finding a Hamiltonian circuit in a graph is the **traveling salesperson problem** (formerly known as the traveling salesman problem). The problem is: Given a set of cities and the distance between each pair, find the shortest route that visits each city once and returns to the initial city. For example, given the cities and distances of Figure 3.1.11, the circuit (a, b, c, d, a) solves the traveling salesperson problem.

EXAMPLE 3.1.6 Similarity Graphs.   This example deals with the problem of grouping "like" objects into classes based on properties of the objects. For example, suppose that a particular algorithm is implemented in BASIC by a number of persons and we want to group "like" programs into classes based on certain properties of the programs (see Table 3.1.1). Suppose that we select as properties

1. The number of lines in the program.
2. The number of GOTO statements in the program.
3. The number of subroutine calls in the program.

TABLE 3.1.1

Program	Number of Program Lines	Number of GOTOs	Number of Subroutine Calls
1	66	20	1
2	41	10	2
3	68	5	8
4	90	34	5
5	75	12	14

A **similarity graph** G is constructed as follows. The vertices correspond to programs. A vertex is denoted (v_1, v_2, v_3) where v_i is the value of property i above. We define a function s as follows. For each pair of vertices $v = (v_1, v_2, v_3)$ and $w = (w_1, w_2, w_3)$, we set

$$s(v, w) = |v_1 - w_1| + |v_2 - w_2| + |v_3 - w_3|.$$

If we let P_i be the vertex corresponding to program i, we obtain

$$s(P_1, P_2) = 36, \qquad s(P_1, P_3) = 24, \qquad s(P_1, P_4) = 42,$$

$$s(P_1, P_5) = 30, \qquad s(P_2, P_3) = 38, \qquad s(P_2, P_4) = 76,$$

$$s(P_2, P_5) = 48, \qquad s(P_3, P_4) = 54, \qquad s(P_3, P_5) = 20,$$

$$s(P_4, P_5) = 46.$$

If v and w are vertices corresponding to two programs, $s(v, w)$ is a measure of how dissimilar the programs are. A large value of $s(v, w)$ indicates dissimilarity, while a small value indicates similarity.

Choose a number S. An edge exists between distinct vertices v and w if $s(v, w) \leq S$. We say that v and w are **in the same class** if $v = w$ or there is a sequence of edges of the form

$$(v, v_1), (v_1, v_2), \ldots, (v_n, w).$$

In Figure 3.1.12 we show the graph corresponding to the programs of Table 3.1.1 with $S = 25$. In this graph the programs are grouped into three classes $\{1, 3, 5\}$, $\{2\}$, and $\{4\}$. In a real problem, an appropriate value for S might be selected by trial and error or the value of S might be selected automatically according to some predetermined criteria.

Example 3.1.6 belongs to the subject called pattern recognition. **Pattern recognition** is concerned with grouping data into classes based on properties of the data. Pattern recognition by computer has much practical significance. For example, computers have been programmed to detect cancer from X-rays, to select tax returns to be audited, to analyze satellite pictures, to recognize text, and to forecast weather.

EXAMPLE 3.1.7 Instant Insanity. Instant Insanity is a puzzle consisting of four cubes each of whose faces are painted one of the four colors,

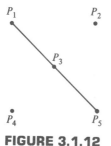

FIGURE 3.1.12

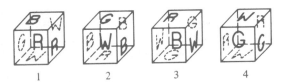

FIGURE 3.1.13

red, white, blue, or green (see Figure 3.1.13). (There are different versions of the puzzle depending on which faces are painted which colors.) The problem is to stack the cubes, one on top of the other, so that whether viewed from front, back, left, or right, one sees all four colors (see Figure 3.1.14). Since 331,776 different stacks are possible (see Exercise 42), a solution by hand by trial and error is impractical. A solution, using graphs, makes it possible to discover a solution, if there is one, in a few minutes!

First, notice that any particular stacking can be represented by two graphs, one representing the front/back colors and the other representing the left/right colors. For example, in Figure 3.1.15 we have represented the stacking of Figure 3.1.14. The vertices represent the colors and an edge connects two vertices if the opposite faces have those colors. For example, in the front/back graph, the edge labeled 1 connects R and W, since the front and back faces of cube 1 are red and white. As another example, in the left/right graph, W has a loop, since both the left and right faces of cube 3 are white.

We can also construct a stacking from a pair of graphs like those in Figure 3.1.15 which represent a solution of the Instant Insanity puzzle. Begin with the front/back graph. Cube 1 is to have red and white opposing faces. Assign one of these colors, say red, arbitrarily to the front. Then cube 1 has a white back face. The other edge incident on W is 2, so make 2's front face white.

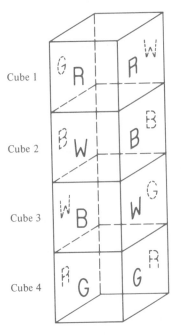

FIGURE 3.1.14

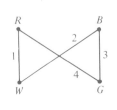

(a) Front/back

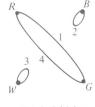

(b) Left/right

FIGURE 3.1.15

This gives 2 a blue back face. The other edge incident on B is 3, so make 3's front face blue. This gives 3 a green back face. The other edge incident on G is 4. Cube 4 then gets a green front face and a red back face. The front and back faces are now properly aligned. At this point, the left and right faces are randomly arranged; however, we will show how to correctly orient the left and right faces without altering the colors of the front and back faces.

Cube 1 is to have red and green opposing left and right faces. Assign one of these colors, say green, arbitrarily to the left. Then cube 1 has a red right face. Notice that we can obtain this left/right orientation without changing the colors of the front and back by rotating the cube (see Figure 3.1.16). We can similarly orient cubes 2, 3, and 4. Notice that cubes 2 and 3 have the same colors on opposing sides. The stacking of Figure 3.1.14 has been reconstructed.

It is apparent from the discussion above that a solution to the Instant Insanity puzzle can be obtained if we can find two graphs like those of Figure 3.1.15. The properties needed are

Each vertex should have degree 2. (3.1.1)

Each cube should be represented by an edge once in each graph. (3.1.2)

The two graphs should not have any edges in common. (3.1.3)

Property (3.1.1) assures us that each color can be used twice, once on the front (or left) and once on the back (or right). Property (3.1.2) assures us that each cube can be properly aligned and property (3.1.3) assures us that, after orienting the front and back sides, we can successfully orient the left and right sides.

To obtain a solution, we first draw a graph that represents all of the faces of all of the cubes. In Figure 3.1.17, we have drawn the graph that represents the cubes of Figure 3.1.13. We then extract two graphs satisfying properties (3.1.1)–(3.1.3). Try your hand at the method by finding another solution to the puzzle represented by Figure 3.1.17.

If $G = (V, E)$ and $G' = (V', E')$ are graphs with $V' \subseteq V$ and $E' \subseteq E$, we call G' a **subgraph** of G. Thus the solution of Instant Insanity described above

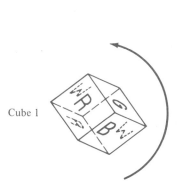

Cube 1

FIGURE 3.1.16

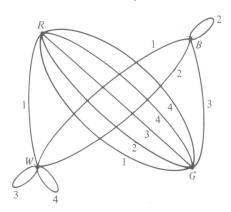

FIGURE 3.1.17

involves finding certain subgraphs of the graph representing the puzzle. Notice that if G' is a subgraph of a graph G and e is an edge in G' incident on vertices v and w, then v and w must be included in the vertex set for G'.

EXERCISES

1H.   Answer the following questions for the graph of Figure 3.1.1.
   (a) Is there a route from V to Y that passes through each city exactly once? If none exists, what is the maximum number of cities one can visit once in going from V to Y?
   (b) Answer part (a) for cities U and V.

2.   Find the degree of each vertex for the graph shown.

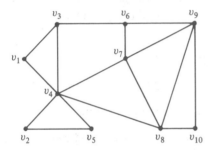

3H.   Is it possible to trace the graph of Exercise 2, beginning and ending at the same point, without lifting your pencil from the paper and without retracing lines? Justify your answer.

4.   Is it possible to trace the graph of Exercise 2 without lifting your pencil from the paper if it is not required that you begin and end at the same point? Retracing lines is not permitted. Justify your answer.

5H.   Repeat Exercises 2–4 for the graph shown.

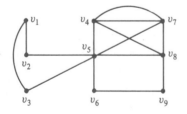

★6.   Repeat Exercises 2–4 for the graph shown.

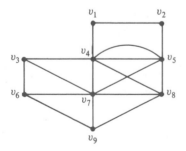

7H.   Repeat Exercise 4 for the graph of Figure 3.1.5.

8.   The following graph is continued to an arbitrary, finite depth. Can the figure be drawn without lifting your pencil from the paper and without retracing lines? If the answer is yes, develop an algorithm for drawing it.

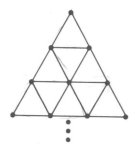

The graph K_n, called the **complete graph on n vertices**, has n vertices and every vertex is joined to every other vertex by an edge. (There are no loops or parallel edges.)

9H.   Draw K_3, K_4, and K_5.

10.   Find a formula for the number of edges in K_n.

11H.   When does K_n have an Euler circuit?

A graph in which the vertices can be partitioned into disjoint sets V_1 and V_2 with every edge incident on one vertex in V_1 and one vertex in V_2 is called a **bipartite graph**.

The graph $K_{m,n}$, called the **complete bipartite graph on m and n vertices**, has disjoint sets V_1 of m vertices and V_2 of n vertices. Every vertex in V_1 is joined to every vertex in V_2 by an edge. (There are no parallel edges.)

12.   Draw $K_{2,3}$, $K_{2,4}$, and $K_{3,3}$.

13H.   Find a formula for the number of edges in $K_{m,n}$.

14.   When does $K_{m,n}$ have an Euler circuit?

15H.   Solve Hamilton's puzzle; that is, find a Hamiltonian circuit for the graph of Figure 3.1.9.

Find Hamiltonian circuits in each graph.

16.

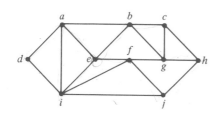

17H.

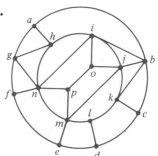

Show that none of the graphs contains a Hamiltonian circuit.

18.   **19H.**   **20.**

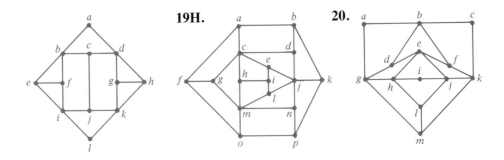

Determine whether or not each graph contains a Hamiltonian circuit. If there is a Hamiltonian circuit, exhibit it; otherwise, give an argument which shows that there is no Hamiltonian circuit.

21H.   **22.**   **23H.**

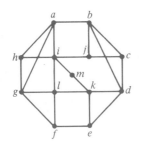 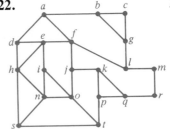 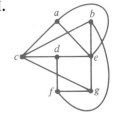

24. Give an example of a graph that has an Euler circuit, but contains no Hamiltonian circuit.

25H. Give an example of a graph that has an Euler circuit which is also a Hamiltonian circuit.

26. Give an example of a graph that has an Euler circuit and a Hamiltonian circuit which are not identical.

★27H. For which values of *m* and *n* does the graph contain a Hamiltonian circuit?

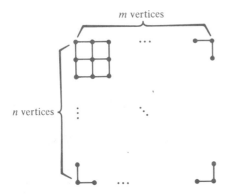

28. Show that if $n \geq 3$, the complete graph on n vertices K_n contains a Hamiltonian circuit.

29H. When does the complete bipartite graph $K_{m,n}$ contain a Hamiltonian circuit?

30. Show that the circuit (a, b, c, d, a) solves the traveling salesperson problem for the graph of Figure 3.1.11.

31. Show that the circuit (e, b, a, c, d, e) provides a solution to the traveling salesperson problem for the graph shown.

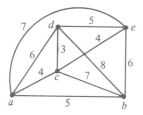

32. Solve the traveling salesperson problem for the graph.

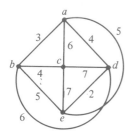

33H. Let G be a bipartite graph with disjoint vertex sets V_1 and V_2. Show that if G has a Hamiltonian circuit, V_1 and V_2 have the same number of elements.

34. Let the vertices of a graph be the squares of an ordinary chessboard. Describe an edge set appropriate for modeling the problem of determining whether a knight can move around the chessboard so that it visits every square exactly once. What does a solution correspond to in graph terms?

35H. Draw the similarity graph that results from setting $S = 40$ in Example 3.1.6. How many classes are there?

36. Draw the similarity graph that results from setting $S = 50$ in Example 3.1.6. How many classes are there?

37H. Let G be a similarity graph. Suppose that we select subgraphs G' of G so that each G' is a complete graph, each pair of subgraphs has no vertices in common, and every vertex in G is in some G'. What properties, in "similarity" terms, would the classes defined by the subgraphs have?

38. In general, is "is similar to" an equivalence relation?

39. Suggest additional properties for Example 3.1.6 that might be useful in comparing programs.

40. How might one automate the selection of S to group data into classes using a similarity graph?

41H. Show that there are 24 orientations of a cube.

42. Number the cubes of an Instant Insanity puzzle 1, 2, 3, and 4. Show that the number of stackings in which the cubes are stacked 1, 2, 3, and 4, reading from bottom to top, is 331,776.

★43H. How many Instant Insanity graphs are there; that is, how many graphs are there with four vertices and 12 edges—three of each of four types?

44. (a) Find all subgraphs of Figure 3.1.17 satisfying properties (3.1.1) and (3.1.2) as listed in Example 3.1.7.
(b) Find all the solutions to the Instant Insanity puzzle of Example 3.1.7.

45H. (a) Represent the following Instant Insanity puzzle by a graph.

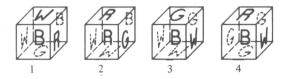

(b) Find a solution to the puzzle.
(c) Find all subgraphs of your graph of part (a) satisfying properties (3.1.1) and (3.1.2) as listed in Example 3.1.7.
(d) Use part (c) to show that the puzzle has a unique solution.

Find solutions to the following Instant Insanity puzzles.

46.

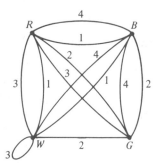

47H.

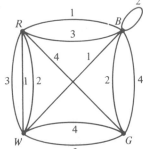

★48. Give an example of an Instant Insanity puzzle with no solution yet for which each cube contains all four colors. Show that your example has the required properties.

★49. Design an Instant Insanity puzzle that has exactly four solutions.

Exercises 50–54 refer to a modified version of Instant Insanity, where a solution is defined to be a stacking that when viewed from the front, back, left, or right shows one color. (The front, back, left, and right are of different colors.)

50.   Develop a graph model for this puzzle and explain how to obtain a solution from your model.

51.   If Instant Insanity, as given in the text, has a solution, must this modified version also have a solution? If the answer is yes, prove it; otherwise, give a counterexample.

52.   If this modified version of Instant Insanity has a solution, must the version given in the text also have a solution? If the answer is yes, prove it; otherwise, give a counterexample.

53.   Is it possible for both versions of Instant Insanity to have solutions? If the answer is yes, prove it; otherwise, give a counterexample.

54.   Is it possible for neither version of Instant Insanity to have a solution even if each cube contains all four colors? If the answer is yes, prove it; otherwise, give a counterexample.

55.   Find all subgraphs having at least one vertex of the graph shown. (There are 17.)

Exercises 56–58 refer to the following graph. The vertices represent offices. An edge connects two offices if there is a communication link between the two. Notice that any office can communicate with any other either directly through a communication link or by having others relay the message.

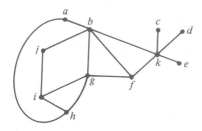

56.   Show that communication among all offices is still possible even if some communication links are broken.

57H.   What is the maximum number of communication links that can be broken with communication among all offices still possible?

58.   Show a configuration in which the maximum number of communication links are broken with communication among all offices still possible.

59H. In the graph, the vertices represent cities and the numbers on the edges represent the costs of building the indicated roads. Find a least-expensive road system that connects all the cities. Is the solution unique?

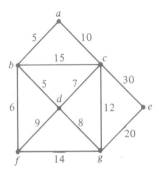

Representations of Graphs

In the preceding section we represented a graph by drawing it. Sometimes, as for example in using a computer to analyze a graph, we need a more formal representation. Our first method of representing a graph uses the **adjacency matrix**.

Consider the graph of Figure 3.2.1. To obtain the adjacency matrix of this graph, we first select an arbitrary ordering of the vertices, say a, b, c, d, e. Next, we label the rows and columns of a matrix with the ordered vertices. The entry in this matrix is 1 if the row and column vertices are connected by an edge and 0 otherwise. The adjacency matrix for this graph is

$$
\begin{array}{c}
 \\ a \\ b \\ c \\ d \\ e
\end{array}
\begin{array}{c}
\begin{array}{ccccc} a & b & c & d & e \end{array} \\
\left(\begin{array}{ccccc}
0 & 1 & 0 & 0 & 1 \\
1 & 0 & 1 & 0 & 1 \\
0 & 1 & 1 & 0 & 1 \\
0 & 0 & 0 & 0 & 1 \\
1 & 1 & 1 & 1 & 0
\end{array}\right)
\end{array}.
$$

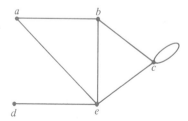

FIGURE 3.2.1

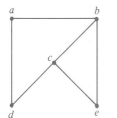

$$
A = \begin{array}{c@{\quad}ccccc}
 & a & b & c & d & e \\
a & \begin{bmatrix} 0 \\ 1 \\ 0 \\ 1 \\ 0 \end{bmatrix} & \begin{matrix} 1 \\ 0 \\ 1 \\ 0 \\ 1 \end{matrix} & \begin{matrix} 0 \\ 1 \\ 0 \\ 1 \\ 1 \end{matrix} & \begin{matrix} 1 \\ 0 \\ 1 \\ 0 \\ 0 \end{matrix} & \begin{matrix} 0 \\ 1 \\ 1 \\ 0 \\ 0 \end{bmatrix}
\end{array}
$$

FIGURE 3.2.2

Notice that while the adjacency matrix allows us to represent loops, it does not allow us to represent parallel edges. Also, note that if the graph has no loops, we can obtain the degree of a vertex by summing its row or column.

The adjacency matrix is not a very efficient way to represent a graph. Since the matrix is symmetric about the main diagonal, the information, except that on the main diagonal, appears twice.

An example of a simple graph and its adjacency matrix A is given in Figure 3.2.2. Suppose that we square the matrix A:

$$
A^2 = \begin{pmatrix} 0 & 1 & 0 & 1 & 0 \\ 1 & 0 & 1 & 0 & 1 \\ 0 & 1 & 0 & 1 & 1 \\ 1 & 0 & 1 & 0 & 0 \\ 0 & 1 & 1 & 0 & 0 \end{pmatrix} \begin{pmatrix} 0 & 1 & 0 & 1 & 0 \\ 1 & 0 & 1 & 0 & 1 \\ 0 & 1 & 0 & 1 & 1 \\ 1 & 0 & 1 & 0 & 0 \\ 0 & 1 & 1 & 0 & 0 \end{pmatrix} = \begin{array}{c} \\ a \\ b \\ c \\ d \\ e \end{array}\begin{array}{ccccc} a & b & c & d & e \\ \begin{pmatrix} 2 & 0 & 2 & 0 & 1 \\ 0 & 3 & 1 & 2 & 1 \\ 2 & 1 & 3 & 0 & 1 \\ 0 & 2 & 0 & 2 & 1 \\ 1 & 1 & 1 & 1 & 2 \end{pmatrix} \end{array}.
$$

Consider the entry for row a, column c in A^2, obtained by multiplying pairwise the entries in row a by the entries in column c of the A matrix and summing

$$
\begin{array}{c} b \quad d \\ a(0 \quad 1 \quad 0 \quad 1 \quad 0) \end{array} \begin{array}{c} c \\ \begin{pmatrix} 0 \\ 1 \\ 0 \\ 1 \\ 1 \end{pmatrix} \begin{array}{c} b \\ \\ d \end{array} \end{array} = 0 \cdot 0 + 1 \cdot 1 + 0 \cdot 0 + 1 \cdot 1 + 0 \cdot 1 = 2.
$$

The only way a nonzero product appears in this sum is if both entries to be multiplied are 1. This happens if there is a vertex v whose entry in row a is 1 and whose entry in column c is 1. In other words, there must be edges of the form (a, v) and (v, c). In this case the sum is increased by 1. In this example the sum 2 represents the edge pairs

$$(a, b) \qquad (b, c)$$

and

$$(a, d) \qquad (d, c).$$

Either pair is called an **edge sequence** of length 2 from a to c. Consequently, the entry in row x and column y of the matrix A^2 is the number of edge sequences of length 2 from vertex x to vertex y. Before continuing the discussion, we will state the formal definition of edge sequence.

DEFINITION 3.2.1. Let G be a graph with vertices V and edges E. An *edge sequence of length n* from vertex x to vertex y is a set of edges S which can be arranged so that

$$S = \{(v_0, v_1), (v_1, v_2), \ldots, (v_{n-1}, v_n)\}$$

where $v_0 = x$ and $v_n = y$.

The definition of edge sequence specifically allows repetition of edges and/or vertices. In particular, the v_i are allowed to equal x or y.

It is easy to see why the entries on the main diagonal of A^2 give the degrees of the vertices. Consider, for example, vertex c. The degree of c is 3, since c is incident on the three edges (c, b), (c, d), and (c, e). But each of these edges can be converted to an edge sequence of length 2 from c to c:

$$(c, b) \quad (b, c)$$

$$(c, d) \quad (d, c)$$

$$(c, e) \quad (e, c).$$

Similarly, an edge sequence of length 2 from c to c defines an edge incident on c. Thus the number of edge sequences of length 2 from c to c is 3, the degree of c.

If we take the nth power of an adjacency matrix, the entries will give the number of edge sequences of length n. This theorem can be verified using induction.

THEOREM 3.2.2. *If A is the adjacency matrix of a simple graph, the ijth entry of A^n is the number of edge sequences of length n from vertex i to vertex j.*

Proof. We will use induction on n.

In case $n = 1$, A^1 is simply A. The ijth entry is 1 if there is an edge from i to j, which is an edge sequence of length 1, and 0 otherwise. Thus the theorem is true in case $n = 1$. The Basis Step has been verified.

Assume that the theorem is true for n. Now

$$A^{n+1} = A^n A$$

so that the ikth entry in A^{n+1} is obtained by multiplying pairwise the elements in the ith row of A^n by the elements in the kth column of A and summing

kth column of A

$$i\text{th row of } A^n \ (s_1 \quad s_2 \cdots s_j \cdots s_m) \begin{pmatrix} t_1 \\ t_2 \\ \cdot \\ \cdot \\ \cdot \\ t_j \\ \cdot \\ \cdot \\ \cdot \\ t_m \end{pmatrix}$$

$$= s_1t_1 + s_2t_2 + \cdots + s_jt_j + \cdots + s_mt_m$$

$$= ik\text{th entry in } A^{n+1}.$$

By induction, s_j gives the number of edge sequences of length n from i to j in the graph G. Now t_j is either 0 or 1. If t_j is 0 there is no edge from j to k, so there are $s_jt_j = 0$ edge sequences of length $n + 1$ from i to k where the last edge is (j, k). If t_j is 1, there is an edge from vertex j to vertex k (see Figure 3.2.3). Since there are s_j edge sequences of length n from vertex i to vertex j, there are $s_j \, t_j = s_j$ edge sequences of length $n + 1$ from i to k where the last edge is (j, k) (see Figure 3.2.3). Summing over all j, we will count all edge sequences of length $n + 1$ from i to k. Thus the ikth entry in A^{n+1} gives the number of edge sequences of length $n + 1$ from i to k and the Inductive Step is verified.

By the Principle of Mathematical Induction, the theorem is established. ∎

EXAMPLE 3.2.3. For the matrix A of Figure 3.2.2, we find that

$$A^4 = \begin{array}{c} \\ a \\ b \\ c \\ d \\ e \end{array} \begin{array}{c} \begin{array}{ccccc} a & b & c & d & e \end{array} \\ \begin{pmatrix} 9 & 3 & 11 & 1 & 6 \\ 3 & 15 & 7 & 11 & 8 \\ 11 & 7 & 15 & 3 & 8 \\ 1 & 11 & 3 & 9 & 6 \\ 6 & 8 & 8 & 6 & 8 \end{pmatrix} \end{array}.$$

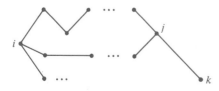

s_j edge sequences of length n from i to j

FIGURE 3.2.3

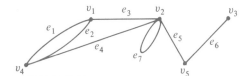

FIGURE 3.2.4

The entry from row d, column e is 6, which means there are six edge sequences of length 4 from d to e. By inspection, we find them to be

$$(d, a), \quad (a, d), \quad (d, c), \quad (c, e)$$
$$(d, c), \quad (c, d), \quad (d, c), \quad (c, e)$$
$$(d, a), \quad (a, b), \quad (b, c), \quad (c, e)$$
$$(d, c), \quad (c, e), \quad (e, c), \quad (c, e)$$
$$(d, c), \quad (c, e), \quad (e, b), \quad (b, e)$$
$$(d, c), \quad (c, b), \quad (b, c), \quad (c, e).$$

Another useful matrix representation of a graph is known as the **incidence matrix**. For example, to obtain the incidence matrix of the graph in Figure 3.2.4, we label the rows with the vertices and the columns with the edges. The entry for row v and column e is 1 if e is incident on v and 0 otherwise. Thus the incidence matrix for the graph of Figure 3.2.4 is

$$\begin{array}{c} \\ v_1 \\ v_2 \\ v_3 \\ v_4 \\ v_5 \end{array} \begin{array}{cccccccc} e_1 & e_2 & e_3 & e_4 & e_5 & e_6 & e_7 \\ \left(\begin{array}{ccccccc} 1 & 1 & 1 & 0 & 0 & 0 & 0 \\ 0 & 0 & 1 & 1 & 1 & 0 & 1 \\ 0 & 0 & 0 & 0 & 0 & 1 & 0 \\ 1 & 1 & 0 & 1 & 0 & 0 & 0 \\ 0 & 0 & 0 & 0 & 1 & 1 & 0 \end{array} \right). \end{array}$$

A column like e_7 is understood to represent a loop.

The incidence matrix allows us to represent both parallel edges and loops. Notice that in a graph without loops each column has two 1's and that the sum of a row gives the degree of the vertex identified with that row.

EXERCISES

In Exercises 1–5, write the adjacency matrix of each graph.

1H.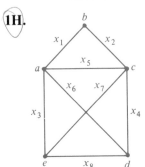

2.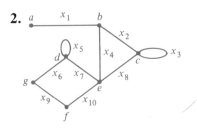

3H.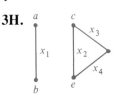

4. The complete graph on five vertices K_5. (The graph K_n is defined before Exercise 9, Section 3.1.)

5H. The complete bipartite graph $K_{2,3}$. (The graph $K_{m,n}$ is defined before Exercise 12, Section 3.1.)

In Exercises 6–10, write the incidence matrix of each graph.

6. The graph of Exercise 1

7H. The graph of Exercise 2

8. The graph of Exercise 3

9H. The complete graph on five vertices K_5

10. The complete bipartite graph $K_{2,3}$

In Exercises 11–13, draw the graph represented by each adjacency matrix.

11H.
$$
\begin{array}{c c c c c c}
 & a & b & c & d & e \\
a & 1 & 0 & 0 & 1 & 0 \\
b & 0 & 0 & 1 & 0 & 1 \\
c & 0 & 1 & 1 & 1 & 1 \\
d & 1 & 0 & 1 & 0 & 0 \\
e & 0 & 1 & 1 & 0 & 0
\end{array}.
$$

12.
$$
\begin{array}{c c c c c c}
 & a & b & c & d & e \\
a & 0 & 1 & 0 & 0 & 0 \\
b & 1 & 0 & 0 & 0 & 0 \\
c & 0 & 0 & 0 & 1 & 1 \\
d & 0 & 0 & 1 & 0 & 1 \\
e & 0 & 0 & 1 & 1 & 1
\end{array}.
$$

13H. The 7×7 matrix whose ijth entry is 1 if $i + 1$ divides $j + 1$ (evenly) or $j + 1$ divides $i + 1$ and whose ijth entry is 0 otherwise.

14. For the graphs in Exercises 1–3, list all edge sequences of length 3 from c to e.

15H. Compute the squares of the adjacency matrices of K_5 and the graphs of Exercises 1 and 3.

16. Let A be the adjacency matrix for the graph of Exercise 1. What is the entry in row a, column d, of A^5?

17H. Suppose that a graph has an adjacency matrix of the form

$$
A = \left(\begin{array}{c|c} & A' \\ \hline A'' & \end{array} \right),
$$

where all entries of the submatrices A' and A'' are 0. What must the graph look like?

18. Repeat Exercise 17 with "adjacency" replaced by "incidence."

19H. How might the definition of adjacency matrix be changed to allow for the representation of parallel edges?

20. Let A be an adjacency matrix of a graph. Why is A^n symmetric about the main diagonal for every positive integer n?

In Exercises 21 and 22, draw the graphs represented by the incidence matrices.

21H.
$$\begin{array}{c} a \\ b \\ c \\ d \\ e \end{array} \begin{pmatrix} 1 & 0 & 0 & 0 & 0 & 1 \\ 0 & 1 & 1 & 0 & 1 & 0 \\ 1 & 0 & 0 & 1 & 0 & 0 \\ 0 & 1 & 0 & 1 & 0 & 0 \\ 0 & 0 & 1 & 0 & 1 & 1 \end{pmatrix}.$$

22.
$$\begin{array}{c} a \\ b \\ c \\ d \\ e \end{array} \begin{pmatrix} 0 & 1 & 0 & 0 & 1 & 1 \\ 0 & 1 & 1 & 0 & 1 & 0 \\ 0 & 0 & 0 & 0 & 0 & 1 \\ 1 & 0 & 0 & 1 & 0 & 0 \\ 1 & 0 & 0 & 1 & 0 & 0 \end{pmatrix}.$$

23H. What must a graph look like if some row of its incidence matrix consists only of 0's?

24. Let A be the adjacency matrix of a graph G with n vertices. Let
$$Y = A + A^2 + \cdots + A^{n-1}.$$

If some off-diagonal entry in the matrix Y is zero, what can you say about the graph G?

Exercises 25–27 refer to the adjacency matrix A of K_5.

25. Let n be a positive integer. Show that all the diagonal elements of A^n coincide and that all the off-diagonal elements of A^n coincide.

Let d_n be the common value of the diagonal elements of A^n and let a_n be the common value of the off-diagonal elements of A^n.

26H. Show that
$$d_{n+1} = 4a_n; \quad a_{n+1} = d_n + 3a_n; \quad a_{n+1} = 3a_n + 4a_{n-1}.$$

27H. Show that
$$a_n = \tfrac{1}{5}[4^n + (-1)^{n+1}]$$
$$d_n = \tfrac{4}{5}[4^{n-1} + (-1)^n].$$

★28. Derive results similar to Exercises 26 and 27 for the adjacency matrix A of the graph K_m.

★29. Let A be the adjacency matrix of the graph $K_{m,n}$. Find a formula for A^j.

Paths and Circuits

In this section we give formal definitions of paths and circuits and explore these concepts in some detail. We will often abbreviate the edge sequence

$$\{(v_0, v_1), (v_1, v_2), \ldots, (v_{n-1}, v_n)\}$$

to

$$(v_0, v_1, v_2, \ldots, v_n).$$

DEFINITION 3.3.1. Let G be a graph and let v and w be vertices in G.

A *path* from v to w of length n is an edge sequence from v to w of length n in which the edges are distinct.

A *simple path* from v to w of length n is a path of the form

$$(v_0, v_1, v_2, \ldots, v_n), \quad \text{vertices are distinct}$$

where $v_0 = v$ and $v_n = w$ and $v_0, v_1, \ldots, v_n$ are distinct.

A *circuit* (or *cycle*) is a path from v to v.

A *simple circuit* is a circuit of the form

$$(v_0, v_1, v_2, \ldots, v_n)$$

where $v_0 = v_n$ and $v_0, v_1, \ldots, v_{n-1}$ are distinct. vertices are distinct

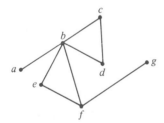

FIGURE 3.3.1

EXAMPLE 3.3.2. For the graph of Figure 3.3.1, we have

Edge Sequence	Path?	Simple Path?	Circuit?	Simple Circuit?
(a, b, d, c, b, a)	No	No	No	No
(f, e, b, d, c, b, a)	Yes	No	No	No
(f, e, b, d)	Yes	Yes	No	No
(b, f, e, b, d, c, b)	Yes	No	Yes	No
(e, f, b, e)	Yes	No	Yes	Yes

An Euler circuit is a circuit that includes all the edges and vertices of a given graph. In Example 3.1.4 we showed that if a graph G has an Euler circuit, every

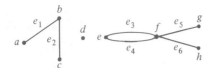

FIGURE 3.3.2

vertex in G has even degree. Before discussing the converse, we need one more definition.

DEFINITION 3.3.3. We say that a graph G is *connected* if, given any distinct vertices v and w, there is a path from v to w.

EXAMPLE 3.3.4. The graph of Figure 3.3.1 is connected, but the graph of Figure 3.3.2 is not connected.

If a graph has an Euler circuit, it must be connected, since we can always choose a portion of the Euler circuit to serve as a path between any two distinct given vertices. The following example illustrates a general method of constructing an Euler circuit in a connected graph in which all vertices have even degree.

EXAMPLE 3.3.5. Find an Euler circuit in the graph of Figure 3.3.3.
    To begin, we choose a vertex arbitrarily and tour the graph, being careful not to reuse an edge. After arriving at a vertex, we arbitrarily choose an edge on which to leave. We can always leave a vertex after reaching it, since the degree of every vertex is even. For example, we might start at v_6 and travel along the path

$$(v_6, v_4, v_7, v_5, v_1, v_3, v_6). \tag{3.3.1}$$

Eventually, we will return to the initial vertex. At this point we may have traversed all the edges in which case we have found an Euler circuit. Otherwise, we remove the edges traversed and any isolated vertices. In our case we are left with the subgraph shown in Figure 3.3.4. Since we removed an even number of edges from each vertex and the degree of every vertex in the original graph is even, the degree of every vertex in the resulting subgraph is also even. We then repeat the procedure using the subgraph.
    Since the original graph was connected, the path (3.3.1) has a vertex in common with the subgraph of Figure 3.3.4. Choose such a vertex, say v_3, and take a tour. Suppose that this time we obtain

$$(v_3, v_4, v_1, v_2, v_5, v_4, v_2, v_3). \tag{3.3.2}$$

Now all the edges have been traversed. If some edges remained, we would repeat the procedure again. By our construction, paths (3.3.1) and (3.3.2) share the vertex v_3. To obtain an Euler circuit, we begin with (3.3.1). When we encounter v_3, we detour to travel the circuit (3.3.2) and then finish with (3.3.1). In this way we obtain the Euler circuit

$$(v_6, v_4, v_7, v_5, v_1, v_3, v_4, v_1, v_2, v_5, v_4, v_2, v_3, v_6).$$

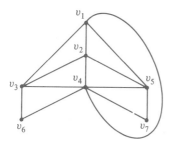

FIGURE 3.3.3

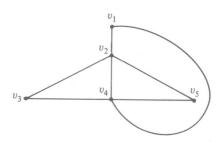

FIGURE 3.3.4

Since this method applies to any connected graph in which all the vertices have even degree, the method, together with Example 3.1.4, gives the following theorem.

THEOREM 3.3.6. *A graph has an Euler circuit if and only if it is connected and all the vertices have even degree.*

  Proof. Suppose that a graph has an Euler circuit. It is obviously connected. The argument which shows that all vertices have even degree was given in Example 3.1.4.
  The proof of the converse is illustrated in Example 3.3.5. ■

What can be said about a connected graph in which not all the vertices have even degree? The first observation (Theorem 3.3.8) is that there are an even number of vertices of odd degree. This follows from the fact (Theorem 3.3.7) that the sum of all of the degrees in a graph is an even number.

THEOREM 3.3.7. *If G is a graph with vertices $\{v_1, v_2, \ldots, v_n\}$, the sum*

$$\delta(v_1) + \delta(v_2) + \cdots + \delta(v_n)$$

is even.

  Proof. Each edge contributes 1 to the degree of each of the vertices on which it is incident. Therefore, if there are N edges in G, we must have

$$2N = \delta(v_1) + \delta(v_2) + \cdots + \delta(v_n).$$ ■

THEOREM 3.3.8. *In any graph, there are an even number of vertices of odd degree.*

  Proof. Let us divide the vertices into two groups: those with even degree $x_1, \ldots, x_m$ and those with odd degree $y_1, \ldots, y_n$. Let

$$S = \delta(x_1) + \delta(x_2) + \cdots + \delta(x_m)$$

$$T = \delta(y_1) + \delta(y_2) + \cdots + \delta(y_n).$$

By Theorem 3.3.7, $S + T$ is even. Since S is the sum of even numbers, S is even. Thus T is even. But T is the sum of n odd numbers, and therefore n is even. ∎

Suppose that a connected graph G has exactly two vertices v and w of odd degree. Let us temporarily insert an edge e from v to w. The resulting graph G' is connected and every vertex has even degree. By Theorem 3.3.6, G' has an Euler circuit. If we delete e from this Euler circuit, we obtain a path from v to w containing all the edges and vertices of G. We have shown that if a graph has exactly two vertices v and w of odd degree, there is a path containing all the edges and vertices from v to w. The converse can be proved similarly.

THEOREM 3.3.9. *A graph has a path from v to w ($v \neq w$) containing all the edges and vertices if and only if it is connected and v and w are the only vertices having odd degree.*

Proof. Suppose that a graph has a path P from v to w containing all the edges and vertices. The graph is surely connected. If we add an edge from v to w, the resulting graph has an Euler circuit, namely, the path P together with the added edge. By Theorem 3.3.6, every vertex has even degree. Removing the added edge affects only the degrees of v and w that are each reduced by 1. Thus in the original graph, v and w have odd degree and all other vertices have even degree.

The converse was discussed just before the statement of the theorem. ∎

Generalizations of Theorem 3.3.9 are given as Exercises 26 and 28.

We conclude this section by showing that any graph G can be partitioned into connected subgraphs called **components**. For example, the graph of Figure 3.3.2 can be partitioned into three connected subgraphs $G_i = (V_i, E_i)$, $i = 1, 2, 3$, where

$$V_1 = \{a, b, c\}, \qquad E_1 = \{e_1, e_2\},$$

$$V_2 = \{d\}, \qquad E_2 = \varnothing,$$

$$V_3 = \{e, f, g, h\}, \qquad E_3 = \{e_3, e_4, e_5, e_6\}.$$

DEFINITION 3.3.10. A *component* of a graph G is a connected subgraph G' of G that has the following property: If G'' is a connected subgraph of G and G' is a subgraph of G'', then $G' = G''$.

In other words, a component of a graph G is a maximal connected subgraph of G.

THEOREM 3.3.11. *If G is a graph, the components of G form a partition of G. Specifically, if $G_i = (V_i, E_i)$, $i = 1, \ldots, n$, is the set of components of G, then $G = (\bigcup_{i=1}^{n} V_i, \bigcup_{i=1}^{n} E_i)$ and if $i \neq j$, then $V_i \cap V_j = \varnothing = E_i \cap E_j$.*

Proof. In this proof, if $G_i = (V_i, E_i)$, $i = 1, 2$, we define

1. G_1 *meets* G_2 to mean that G_1 and G_2 share at least one vertex.
2. $G_1 \cup G_2$ to be the graph $(V_1 \cup V_2, E_1 \cup E_2)$.
3. $G_1 \subseteq G_2$ to mean that G_1 is a subgraph of G_2.

First, we note that if $G_1, \ldots, G_n$ are connected subgraphs of G and G_i meets G_j for $i, j = 1, \ldots, n$, then $\cup_{i=1}^n G_i$ is a connected subgraph of G.

Let v be a vertex in G. Let $\mathcal{S}$ be the set of connected subgraphs of G containing v. By our remark above, $G' = \cup \mathcal{S}$ is a connected subgraph of G. In fact, it is a component. For if G'' is a connected subgraph of G and $G'' \supseteq G'$, then v is in G''. Therefore, $G'' \in \mathcal{S}$ and $G'' \subseteq \cup \mathcal{S} = G'$. Hence $G' = G''$.

We must show that the family of components $\mathcal{C}$ of G partitions G.

Let $C_1, C_2 \in \mathcal{C}$. Suppose that C_1 meets C_2. By our remark above, $C_1 \cup C_2$ is connected. Since $C_1 \cup C_2 \supseteq C_1$ and C_1 is a component, $C_1 \cup C_2 = C_1$. Thus $C_2 \subseteq C_1$. Similarly, $C_1 \subseteq C_2$. Therefore, $C_1 = C_2$. Thus the family $\mathcal{C}$ is pairwise disjoint.

Since a single vertex or a single edge forms a connected subgraph of G, by our remark above, it is contained in a component. In other words, $\cup \mathcal{C} = G$. Therefore, $\mathcal{C}$ is a partition of G. ∎

EXERCISES

In Exercises 1–7, tell whether the given edge sequence in the graph is

(a) A path
(b) A simple path
(c) A circuit
(d) A simple circuit

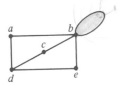

1H. (b, b)

2. (e, d, c, b)

3H. (a, d, c, d, e)

4. (d, c, b, e, d)

5H. $(b, c, d, a, b, e, d, c, b)$

6. (b, c, d, e, b, b)

7H. (a, d, c, b, e)

In Exercises 8–16, draw a graph having the given properties or explain why no such graph exists.

144   **GRAPH THEORY**

8.   Six vertices each of degree 3

9H.   Five vertices each of degree 3

10.   Four vertices each of degree 1

11H.   Six vertices; four edges

12.   Four edges; four vertices having degrees 1, 2, 3, 4

13H.   Four vertices having degrees 1, 2, 3, 4

14.   Simple graph; six vertices having degrees 1, 2, 3, 4, 5, 5

15H.   Simple graph; five vertices having degrees 2, 3, 3, 4, 4

16.   Simple graph; five vertices having degrees 2, 2, 4, 4, 4

17H.   Find all the simple circuits in the graph.

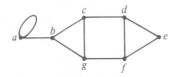

18.   Find all simple paths from *a* to *e* in the graph of Exercise 17.

19H.   Find all connected subgraphs of the graph containing all of the vertices of the
           original graph and having as few edges as possible. Which are paths? Which
           are simple paths? Which are circuits? Which are simple circuits?

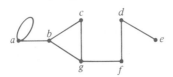

           Find Euler circuits for each graph.

20.

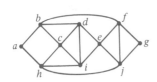

21H.

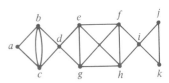

22.   For which values of *m* and *n* does the graph of Exercise 27, Section 3.1, have
           an Euler circuit?

           Verify that there are an even number of vertices of odd degree in each graph.

23H.

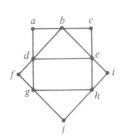

24.

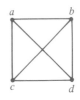

25H. For the graph of Exercise 23, find a path from d to e containing all the edges.

26. Let G be a connected graph with four vertices v_1, v_2, v_3, and v_4 of odd degree. Show that there are paths from v_1 to v_2 and from v_3 to v_4 such that every edge in G is in exactly one of the paths.

27H. Illustrate Exercise 26 using the graph.

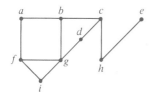

28. State and prove a generalization of Exercise 26 where there are an arbitrary number of vertices of odd degree.

In Exercises 29–31, tell whether each assertion is true or false. If false, give a counterexample and if true, explain.

29H. Let G be a graph and let v and w be distinct vertices. If there is a path from v to w, there is a simple path from v to w.

30. Let G be a graph and let v be a vertex. If v is contained in a circuit, v is contained in a simple circuit.

31H. If a graph contains a circuit that includes all the edges, the circuit is an Euler circuit.

32. Let G be a connected graph. Suppose that an edge e is in a circuit. Show that G with e removed is still connected.

33H. Give an example of a connected graph such that the removal of any edge results in a graph that is not connected. (Assume that removing an edge does not remove any vertices.)

★34. Can a knight move around a chessboard so that it makes every move exactly once? (A move is considered to be made when the move is made in either direction.)

35H. Show that a path which is not a simple path must contain a circuit.

36. Show that if $G_1, \ldots, G_n$ are connected subgraphs of a graph G and G_i meets G_j, for $i, j = 1, \ldots, n$, then $\bigcup_{i=1}^{n} G_i$ is a connected subgraph of G. (See the proof of Theorem 3.3.11 for the definition of union and meet.)

37H. Show that if G' is a connected subgraph of a graph G, then G' is contained in a component.

In Exercises 38–40, write the adjacency matrix of each component of the graph with the given adjacency matrix.

38.

$$\begin{array}{c} \\ a \\ b \\ c \\ d \\ e \\ f \end{array} \begin{array}{cccccc} a & b & c & d & e & f \\ \begin{pmatrix} 0 & 0 & 1 & 0 & 0 & 1 \\ 0 & 1 & 0 & 1 & 1 & 0 \\ 1 & 0 & 0 & 0 & 0 & 1 \\ 0 & 1 & 0 & 0 & 1 & 0 \\ 0 & 1 & 0 & 1 & 0 & 0 \\ 1 & 0 & 1 & 0 & 0 & 0 \end{pmatrix} \end{array}.$$

39H.

$$\begin{array}{c} \\ a \\ b \\ c \\ d \\ e \\ f \end{array} \begin{array}{cccccc} a & b & c & d & e & f \\ \begin{pmatrix} 1 & 1 & 1 & 1 & 0 & 1 \\ 1 & 0 & 1 & 1 & 1 & 0 \\ 1 & 1 & 0 & 1 & 1 & 1 \\ 1 & 1 & 1 & 0 & 1 & 1 \\ 0 & 1 & 1 & 1 & 0 & 1 \\ 1 & 0 & 1 & 1 & 1 & 0 \end{pmatrix} \end{array}.$$

40.

$$\begin{array}{c} \\ a \\ b \\ c \\ d \\ e \\ f \\ g \\ h \\ i \\ j \end{array} \begin{array}{cccccccccc} a & b & c & d & e & f & g & h & i & j \\ \begin{pmatrix} 0 & 0 & 0 & 0 & 0 & 0 & 0 & 0 & 0 & 0 \\ 0 & 1 & 0 & 0 & 0 & 1 & 0 & 0 & 0 & 1 \\ 0 & 0 & 0 & 0 & 0 & 0 & 0 & 1 & 0 & 0 \\ 0 & 0 & 0 & 0 & 1 & 0 & 0 & 0 & 1 & 0 \\ 0 & 0 & 0 & 1 & 0 & 0 & 1 & 0 & 1 & 0 \\ 0 & 1 & 0 & 0 & 0 & 0 & 0 & 0 & 0 & 1 \\ 0 & 0 & 0 & 0 & 1 & 0 & 0 & 0 & 1 & 0 \\ 0 & 0 & 1 & 0 & 0 & 0 & 0 & 0 & 0 & 0 \\ 0 & 0 & 0 & 1 & 1 & 0 & 1 & 0 & 0 & 0 \\ 0 & 1 & 0 & 0 & 0 & 1 & 0 & 0 & 0 & 0 \end{pmatrix} \end{array}.$$

41. Show that the decomposition given in Theorem 3.3.11 is unique; that is, if a graph G is partitioned into connected subgraphs, the connected subgraphs are components.

42. Let G be a graph. Define a relation R on the vertices V of G as vRw if there is a path from v to w or if $v = w$.
(a) Show that R is an equivalence relation on V.
(b) What are the equivalence classes?

43H. Is it possible in a department of 25 persons racked by dissension for each person to get along with exactly five others?

44. Can the distinct dominoes (see Exercise 23, Section 2.3) be arranged in a circle so that touching dominoes have adjacent squares with identical numbers?

★45H. Show that the maximum number of edges in a simple, disconnected graph with n vertices is $(n - 1)(n - 2)/2$.

★46. Show that the maximum number of edges in a simple, bipartite graph with n vertices is $\lfloor n^2/4 \rfloor$.

A vertex v in a connected graph G is an **articulation point** if the removal of v and all edges incident on v disconnects G.

47H. Give an example of a graph with six vertices that has exactly two articulation points.

48. Give an example of a graph with six vertices that has no articulation points.

49. Show that a vertex v in a connected graph G is an articulation point if and only if there are vertices w and x in G having the property that every path from w to x passes through v.

50. Let v be a vertex in a directed graph. The **indegree** of v, in(v), is the number of edges of the form (w, v). The **outdegree** of v, out(v), is the number of edges of the form (v, w).

Prove that a directed graph G contains a (directed) Euler circuit,

$$(v_0, v_1), (v_1, v_2), \ldots, (v_{n-1}, v_n),$$

$v_0 = v_n$, if and only if the graph obtained by ignoring the directions of the edges of G is connected and in(v) = out(v) for every vertex v in G.

A **de Bruijn sequence** for n (in 0's and 1's) is a sequence

$$a_1, \ldots, a_{2^n}$$

of 2^n bits having the property that if s is a bit string of length n, for some m,

$$s = a_m a_{m+1} \cdots a_{m+n-1}. \tag{3.3.3}$$

In (3.3.3), we define $a_{2^n+i} = a_i$, for $i = 1, \ldots, 2^n - 1$.

51. Verify that 00011101 is a de Bruijn sequence for $n = 3$.
52. Let G be a directed graph with vertices corresponding to all bit strings of length $n - 1$. A directed edge exists from vertex $x_1 \cdots x_{n-1}$ to $x_2 \cdots x_n$. Show that an Euler circuit in G corresponds to a de Bruijn sequence.
★53H. Show that there is a de Bruijn sequence for every $n = 1, 2, \ldots$.

3.4

A Shortest-Path Algorithm

A **weighted graph** is a graph in which data are associated with the edges. The value $w(i, j)$ associated with the edge (i, j) is called the **weight** of (i, j). For example, if we interpret cities as vertices and the roads between them as edges and if we assign to each road its length, we obtain a weighted graph (see Figure 3.4.1). Often weights are used to represent costs. For example, if vertices rep-

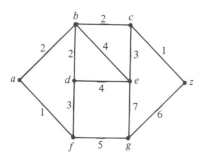

FIGURE 3.4.1

resent cities and edges represent projected roads, the weight of an edge might represent the cost of building that road. The **weight of a graph** is the sum of the weights of its edges. The weight of a path is usually referred to as the **length of the path**. In weighted graphs, we often want to find a **shortest path** (i.e., a path having minimum length) between two given vertices. Algorithm 3.4.1, due to Dijkstra, which efficiently solves this problem, is the topic of this section.

Throughout this section, G denotes a connected, weighted graph. We assume that the weights are positive numbers and that we want to find a shortest path from vertex a to vertex z. The assumption that G is connected can be dropped (see Exercise 12).

Dijkstra's algorithm involves assigning labels to vertices. We let $L(v)$ denote the label of vertex v. At any point, some vertices have temporary labels and the rest have permanent labels. We let T denote the set of vertices having temporary labels. In illustrating the algorithm, we will circle vertices having permanent labels. We will show later that if $L(v)$ is the permanent label of vertex v, then $L(v)$ is the length of a shortest path from a to v. Initially, all vertices have temporary labels. Each iteration of the algorithm changes the status of one label from temporary to permanent; thus we may terminate the algorithm when z receives a permanent label. At this point $L(z)$ gives the length of a shortest path from a to z.

ALGORITHM 3.4.1 Dijkstra's Shortest-Path Algorithm.   This algorithm finds the length of a shortest path from vertex a to vertex z in a connected, weighted graph. The weight of edge (i, j) is $w(i, j) > 0$ and the label of vertex x is $L(x)$. At termination, $L(z)$ is the length of a shortest path from a to z.

1. [Initialization.] Set $L(a) := 0$. For all vertices $x \neq a$, set $L(x) := \infty$. Let T be the set of vertices.

2. [Done?] If $z \notin T$, stop. ($L(z)$ is the length of a shortest path from a to z.)

3. [Get next vertex.] Choose $v \in T$ with the smallest value of $L(v)$. Set $T := T - \{v\}$.

4. [Revise labels.] For each vertex $x \in T$ adjacent to v, set

$$L(x) := \min \{L(x), L(v) + w(v, x)\}.$$

Go to step 2.

We will apply Algorithm 3.4.1 to the graph of Figure 3.4.1. (The vertices in T are uncircled and have temporary labels. The circled vertices have permanent labels.) Figure 3.4.2 shows the result of executing step 1. At step 2, z is not circled. We proceed to step 3, where we select vertex a, the uncircled vertex with the smallest label, and circle it (see Figure 3.4.3). At step 4, we update each of the uncircled vertices, b and f, adjacent to a. We obtain the new labels

$$L(b) = \min \{\infty, 0 + 2\} = 2$$

$$L(f) = \min \{\infty, 0 + 1\} = 1$$

(see Figure 3.4.3.) At this point, we return to step 2.

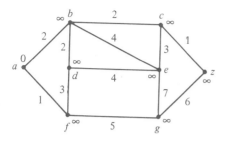

FIGURE 3.4.2

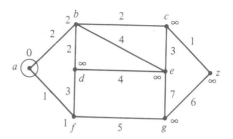

FIGURE 3.4.3

Since z is not circled, we proceed to step 3, where we select vertex f, the uncircled vertex with the smallest label, and circle it (see Figure 3.4.4). At step 4 we update each label of the uncircled vertices, d and g, adjacent to f. We obtain the labels shown in Figure 3.4.4.

You should verify that the next iteration of the algorithm produces the labeling shown in Figure 3.4.5 and that at the termination of the algorithm, z is labeled 5 indicating that the length of a shortest path from a to z is 5. A shortest path is given by (a, b, c, z).

Next, we show that Algorithm 3.4.1 is correct. We will use induction on the number of iterations of the loop consisting of steps 2–4 to show that just before step 2 is executed:

> The label on a circled vertex v gives the length of a shortest path from a to v. (3.4.1)

> There is a shortest path from a to a circled vertex v consisting only of circled vertices. (3.4.2)

> The label on an uncircled vertex v gives the length of a shortest path of the form
>
> $$(a, v_1, \ldots, v_n),$$
>
> where $v = v_n$ and $a, v_1, \ldots, v_{n-1}$ are circled. If there is no path of the form $(a, v_1, \ldots, v_n)$ where $v = v_n$ and $a, v_1, \ldots, v_{n-1}$ are circled, the label on v is ∞. (3.4.3)

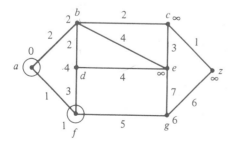

FIGURE 3.4.4

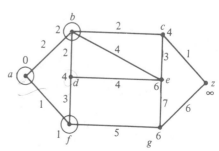

FIGURE 3.4.5

Because Algorithm 3.4.1 terminates when z is circled, it follows from (3.4.1) that $L(z)$ gives the length of a shortest path from a to z as desired.

Initially, a is labeled 0 and all other vertices are labeled ∞. At this point, (3.4.1)–(3.4.3) hold. The Basis Step has been verified.

Assume that (3.4.1)–(3.4.3) hold the ith time we arrive at step 2. Consider an additional iteration of the algorithm. First, at step 3, we circle the uncircled vertex v having the minimum label. We first show that (3.4.1) holds for v; that is, that the label $L(v)$ now gives the length of a shortest path from a to v. Suppose that there is a path

$$(a, v_1, \ldots, v_n), \tag{3.4.4}$$

where $v = v_n$ and whose length is less than $L(v)$. In view of (3.4.3), we must have some vertex v_j, with $1 \le j < n$, uncircled. Let k be the least index with v_k uncircled. Since the weight of every edge is positive, the length L' of the path

$$(a, v_1, \ldots, v_k) \tag{3.4.5}$$

is less than the length of path (3.4.4). Since v_k is the only uncircled vertex in path (3.4.5), using (3.4.3) we see that the label on v_k is at most L'. Since $L' < L(v)$, we have contradicted the choice of v. Therefore, (3.4.1) holds for v. The same kind of argument shows that (3.4.2) holds for v.

Finally, we must show that after updating the labels of the uncircled vertices adjacent to v at step 4, (3.4.3) holds. First, we will show that if x is an uncircled vertex that is not adjacent to v, $L(x)$ is the length of a shortest path of the form

$$(a, v_1, \ldots, v_n), \tag{3.4.6}$$

where $x = v_n$ and only $a, v_1, \ldots, v_{n-1}$ are circled. Consider a shortest path of the form (3.4.6). Because of (3.4.2),

$$(a, v_1, \ldots, v_{n-1}) \tag{3.4.7}$$

is a shortest path from a to v_{n-1}. Moreover, since $v_{n-1} \ne v$, there is a path consisting of circled vertices, none of which is v, from a to v_{n-1} whose length is the same as the length of (3.4.7). Therefore, $L(x)$ is the correct label for x.

Now suppose that x is adjacent to v. There are two kinds of paths of the form (3.4.6): those with $v_{n-1} = v$ and those with $v_{n-1} \ne v$. The argument in the preceding paragraph shows that the length of a shortest path (3.4.6) with $v_{n-1} \ne v$ is $L(x)$. A shortest path (3.4.6) with $v_{n-1} = v$ is obtained by taking a shortest path from a to v [whose length is $L(v)$] and appending the edge (v, x). This path has length $L(v) + w(v, x)$. Thus the length of a shortest path (3.4.6) is

$$\min \{L(x), L(v) + w(v, x)\}.$$

Therefore, after relabeling the vertices adjacent to v, (3.4.3) holds. We have verified that (3.4.1)–(3.4.3) hold prior to the $(i + 1)$st execution of step 2. The Inductive Step is complete. Therefore, Algorithm 3.4.1 is correct.

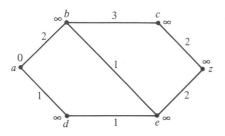

FIGURE 3.4.6

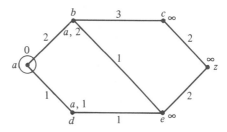

FIGURE 3.4.7

Algorithm 3.4.1 finds the length of a shortest path from a to z. In most applications, we would also want to identify a shortest path. A slight modification of Algorithm 3.4.1 allows us to find a shortest path.

EXAMPLE 3.4.2.   Find a shortest path from a to z and its length for the graph of Figure 3.4.6.

We will apply Algorithm 3.4.1 with a slight modification. In addition to circling a vertex, we will also label it with the name of the vertex from which it was labeled.

Figure 3.4.6 shows the result of executing step 1 of Algorithm 3.4.1. First, we circle a (see Figure 3.4.7). Next, we label the vertices b and d adjacent to a. Vertex b is labeled a, 2 to indicate its value and the fact that it was labeled from a. Similarly, vertex d is labeled a, 1.

Next, we circle vertex d and update the label of the vertex e adjacent to d (see Figure 3.4.8). Then we circle vertex b and update the labels of vertices c and e (see Figure 3.4.9). Next, we circle vertex e and update the label of vertex z (see Figure 3.4.10). At this point, we may circle z, so the algorithm terminates. The length of a shortest path from a to z is 4. Starting at z, we can retrace the labels to find the shortest path

$$(a, d, e, z).$$

If a shortest-path algorithm receives as input a simple graph having n vertices, we will define the time required by the algorithm to be the number of times it examines an edge of the graph. Our next theorem shows that Algorithm 3.4.1 is $O(n^2)$.

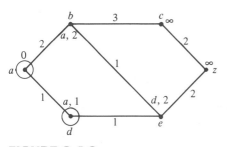

FIGURE 3.4.8

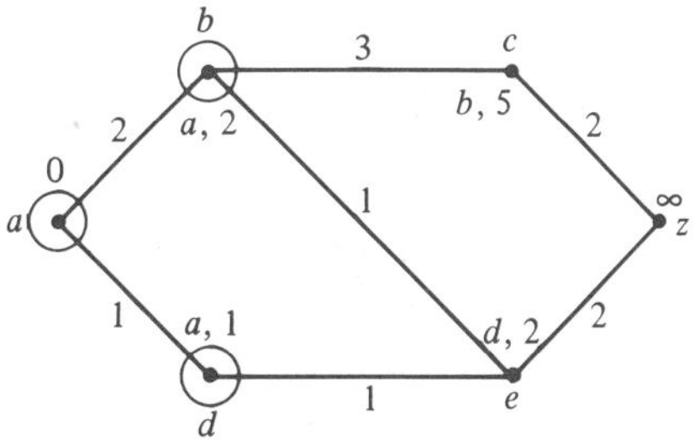

FIGURE 3.4.9

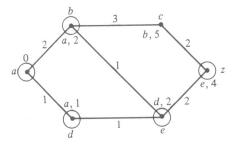

FIGURE 3.4.10

THEOREM 3.4.3. *Let G be a simple, connected, weighted graph having n vertices. If f(n) is the number of times Algorithm* 3.4.1 *examines an edge of G, then f(n) = O(n²).*

Proof. Algorithm 3.4.1 examines edges at step 4. Since G has n vertices, the maximum number of edges examined at step 4 is $n - 1$. Step 4 is in a loop consisting of steps 2–4. This loop executes as long as $z \in T$. Since every iteration of the loop removes one element from T and initially T has n elements, the loop executes at most n times. Therefore, the number of times Algorithm 3.4.1 examines an edge of G is at most $n(n - 1)$. We have

$$f(n) \leq n(n - 1) \leq n^2$$

as desired.   ■

Suppose that weights are assigned to K_n, the complete graph on n vertices. Any algorithm that finds a shortest path from one vertex to another must examine each edge at least once. Since K_n has $n(n - 1)/2$ edges (see Exercise 10, Section 3.1), it must examine $O(n^2)$ edges of K_n. It follows from Theorem 3.4.3 that Algorithm 3.4.1 is optimal.

EXERCISES

In Exercises 1–5, find the length of a shortest path and a shortest path between each pair of vertices in the weighted graph.

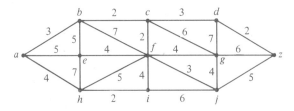

1H. a, f **2.** a, g **3H.** a, z **4.** b, j **5H.** h, d

6.   Write an algorithm that finds the length of a shortest path between two given vertices in a connected, weighted graph and also finds a shortest path.

7H.   Write an algorithm that finds the lengths of the shortest paths from a given vertex to every other vertex in a connected, weighted graph G.

8.   If $f(n)$ is the number of times your algorithm in Exercise 7 examines an edge of a simple graph G with n vertices, determine $g(n)$ so that $f(n) = O(g(n))$.

★9H.   Write an algorithm that finds the lengths of the shortest paths between every pair of vertices in a connected, weighted graph G.

10.   If $f(n)$ is the number of times your algorithm in Exercise 9 examines an edge of a simple graph G with n vertices, determine $g(n)$ so that $f(n) = O(g(n))$.

★11H.   Write an algorithm that finds the lengths of the shortest paths between all vertex pairs in a simple, connected, weighted graph having n vertices in time $O(n^3)$.

12.   Modify Algorithm 3.4.1 so that it accepts a weighted graph that is not necessarily connected. At termination, what is $L(z)$ if there is no path from a to z?

13H.   True or false? When a connected, weighted graph and vertices a and z are input to the following algorithm, at termination LENGTH gives the length of a shortest path from a to z. If the algorithm is correct, prove it; otherwise, give an example of a connected, weighted graph and vertices a and z for which it fails.

> ### ALGORITHM 3.4.4
> 1. Set LENGTH $:= 0$ and $v := a$. Let T be the set of vertices.
> 2. If $v = z$, stop; otherwise, set $T := T - \{v\}$, select $x \in T$ with $w(v, x)$ minimal, set LENGTH $:=$ LENGTH $+ w(v, x)$, set $v := x$, and go to step 2.

14H.   Write an algorithm that finds the length of a longest path between a given pair of vertices in a connected, circuit-free, weighted graph G.

15.   Does your algorithm of Exercise 14 work if the graph is not circuit-free?

16H.   If $f(n)$ is the number of times your algorithm in Exercise 14 examines an edge of a simple graph G with n vertices, determine $g(n)$ so that $f(n) = O(g(n))$.

17.   Give a real-world example in which one might be interested in finding a longest path between two points.

18H.   Write an algorithm that finds the lengths of the longest paths from a given vertex to every other vertex in a connected, circuit-free, weighted graph G.

19.   If $f(n)$ is the number of times your algorithm in Exercise 18 examines an edge of a simple graph G with n vertices, determine $g(n)$ so that $f(n) = O(g(n))$.

★20H.   Write an algorithm that finds the lengths of the longest paths between every pair of vertices in a connected, circuit-free, weighted graph G.

21.   If $f(n)$ is the number of times your algorithm in Exercise 20 examines an edge of a simple graph G with n vertices, determine $g(n)$ so that $f(n) = O(g(n))$.

22.   Write an algorithm that finds the length of a shortest (directed) path from a given vertex to another given vertex in a connected, weighted digraph.

23.   True or false? Algorithm 3.4.1 finds the length of a shortest path in a connected, weighted graph even if some weights are negative. If true, prove it; otherwise, provide a counterexample.

Isomorphisms of Graphs

The following instructions are given to two persons who cannot see each other's paper: "Draw and label five vertices a, b, c, d, and e. Connect a and b, b and c, c and d, d and e, and a and e." The graphs produced are shown in Figure 3.5.1. Surely these figures define the same graph even though they appear dissimilar. Such graphs are said to be **isomorphic**.

DEFINITION 3.5.1. Graphs G_1 and G_2 are *isomorphic* if there is a one-to-one, onto function f from the vertices of G_1 to the vertices of G_2 and a one-to-one, onto function g from the edges of G_1 to the edges of G_2, so that an edge e is incident on v and w in G_1 if and only if the edge $g(e)$ is incident on $f(v)$ and $f(w)$ in G_2. The pair of functions f and g is called an *isomorphism*.

EXAMPLE 3.5.2. An isomorphism for the graphs G_1 and G_2 of Figure 3.5.1 is defined by

$$f(a) = A, \quad f(b) = B, \quad f(c) = C, \quad f(d) = D, \quad f(e) = E$$

$$g(x_i) = y_i, \quad i = 1, \ldots, 5.$$

According to Definition 3.5.1, if graphs G_1 and G_2 are isomorphic, the function f pairs the vertices in G_1 with the vertices in G_2 and the function g pairs the edges in G_1 with the edges in G_2 in such a way that incidence is preserved. It follows that if graphs are isomorphic, for appropriate orderings of the vertices and edges, they have the same incidence matrix. The converse is also true.

THEOREM 3.5.3. *Graphs G_1 and G_2 are isomorphic if and only if, for some ordering of the vertices and edges, their incidence matrices are identical.*

Proof. Suppose that G_1 and G_2 are isomorphic. Then there is a one-to-one, onto function f from the vertices

$$V = \{v_1, \ldots, v_m\}$$

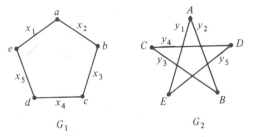

G_1

G_2

FIGURE 3.5.1

of G_1 to the vertices of G_2 and a one-to-one, onto function g from the edges

$$E = \{e_1, \ldots, e_n\}$$

of G_1 to the edges of G_2 having the property specified in Definition 3.5.1.

Let M_1 be the incidence matrix of G_1 relative to the orderings $v_1, \ldots,$ v_m and $e_1, \ldots, e_n$ and let M_2 be the incidence matrix of G_2 relative to the orderings $f(v_1), \ldots, f(v_m)$ and $g(e_1), \ldots, g(e_n)$. By Definition 3.5.1, an edge e is incident on v in G_1 if and only if the edge $g(e)$ is incident on $f(v)$ in G_2. It follows that M_1 and M_2 are identical.

Similarly, identical incidence matrices define isomorphic graphs. ∎

EXAMPLE 3.5.4. The incidence matrix of the graph G_1 of Figure 3.5.1 relative to the orderings a, b, c, d, e and x_1, x_2, x_3, x_4, x_5 is

$$
\begin{array}{c}
\\ a \\ b \\ c \\ d \\ e
\end{array}
\begin{array}{c}
\begin{array}{ccccc} x_1 & x_2 & x_3 & x_4 & x_5 \end{array} \\
\left(\begin{array}{ccccc}
1 & 1 & 0 & 0 & 0 \\
0 & 1 & 1 & 0 & 0 \\
0 & 0 & 1 & 1 & 0 \\
0 & 0 & 0 & 1 & 1 \\
1 & 0 & 0 & 0 & 1
\end{array}\right)
\end{array}.
$$

The incidence matrix of the graph G_2 of Figure 3.5.1 relative to the orderings A, B, C, D, E and y_1, y_2, y_3, y_4, y_5 is

$$
\begin{array}{c}
\\ A \\ B \\ C \\ D \\ E
\end{array}
\begin{array}{c}
\begin{array}{ccccc} y_1 & y_2 & y_3 & y_4 & y_5 \end{array} \\
\left(\begin{array}{ccccc}
1 & 1 & 0 & 0 & 0 \\
0 & 1 & 1 & 0 & 0 \\
0 & 0 & 1 & 1 & 0 \\
0 & 0 & 0 & 1 & 1 \\
1 & 0 & 0 & 0 & 1
\end{array}\right)
\end{array}.
$$

By Theorem 3.5.3, G_1 and G_2 are isomorphic.

An interesting problem is to determine whether two graphs are isomorphic. Although every known algorithm to test whether two graphs are isomorphic requires exponential or factorial time in the worst case, there are algorithms that can determine whether "most" pairs of graphs are isomorphic in linear time (see [Read] and [Babai]).

One way to show that two graphs G_1 and G_2 are *not* isomorphic is to find a property that isomorphic graphs must share, but which G_1 and G_2 do *not* share. A property that is preserved under a graph isomorphism is called an **invariant**. By Definition 3.5.1, if graphs G_1 and G_2 are isomorphic, there are one-to-one, onto functions from the edges (respectively, vertices) of G_1 to the edges (respectively, vertices) of G_2. Thus, if G_1 and G_2 are isomorphic, then G_1 and G_2 have the same number of edges and the same number of vertices. Therefore, if e and n are nonnegative integers, the properties, "has e edges" and "has n vertices" are invariants.

G_1 G_2

FIGURE 3.5.2

EXAMPLE 3.5.5. The graphs G_1 and G_2 in Figure 3.5.2 are not isomorphic, since G_1 has seven edges and G_2 has six edges and "has seven edges" is an invariant.

EXAMPLE 3.5.6. Show that if k is a positive integer, "has a vertex of degree k" is an invariant.

Suppose G_1 and G_2 are isomorphic graphs and f (respectively, g) is a one-to-one, onto function from the vertices (respectively, edges) of G_1 onto the vertices (respectively, edges) of G_2. Suppose that G_1 has a vertex v of degree k. Then there are k edges $e_1, \ldots, e_k$ incident on v. By Definition 3.5.1, $g(e_1), \ldots, g(e_k)$ are incident on $f(v)$. Because g is one-to-one, $\delta(f(v)) \geq k$.

Let E be an edge that is incident on $f(v)$ in G_2. Since g is onto, there is an edge e in G_1 with $g(e) = E$. Since $g(e)$ is incident on $f(v)$ in G_2, by Definition 3.5.1, e is incident on v in G_1. Since $e_1, \ldots, e_k$ are the only edges in G_1 incident on v, $e = e_i$ for some $i \in \{1, \ldots, k\}$. Now $g(e_i) = g(e) = E$. Thus $\delta(f(v)) = k$, so G_2 has a vertex, namely $f(v)$, of degree k.

EXAMPLE 3.5.7. Since "has a vertex of degree 3" is an invariant, the graphs G_1 and G_2 of Figure 3.5.3 are not isomorphic; G_1 has vertices (a and f) of degree 3, but G_2 does not have a vertex of degree 3. Notice that G_1 and G_2 have the same numbers of edges and vertices.

Another invariant that is sometimes useful is "has a simple circuit of length k." We leave the proof that this property is an invariant to the exercises (Exercise 11).

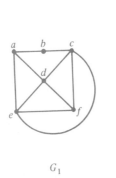 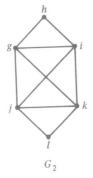

G_1 G_2

FIGURE 3.5.3

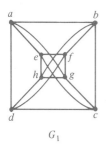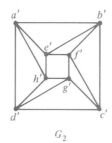

FIGURE 3.5.4

EXAMPLE 3.5.8. Since "has a simple circuit of length 3" is an invariant, the graphs G_1 and G_2 of Figure 3.5.4 are not isomorphic; the graph G_2 has a simple circuit of length 3, but all simple circuits in G_2 have length at least 4. Notice that G_1 and G_2 have the same numbers of edges and vertices and that every vertex in G_1 or G_2 has degree 4.

EXERCISES

In Exercises 1–10, determine whether the graphs G_1 and G_2 are isomorphic. If the graphs are isomorphic, find a function f for Definition 3.5.1; otherwise, give an invariant that the graphs do not share.

1H.

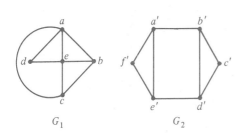

2.

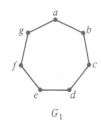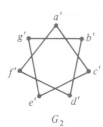

3H.

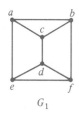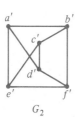

4.

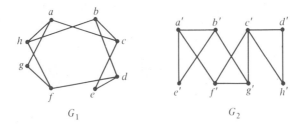

G_1 G_2

5H.

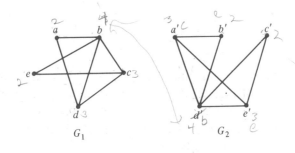

G_1 G_2

6.

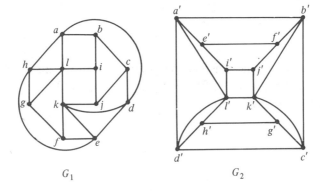

G_1 G_2

7H.

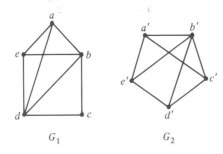

G_1 G_2

★8.

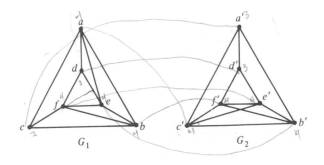

G_1 G_2

★9H.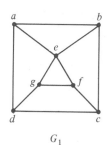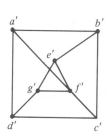

★10.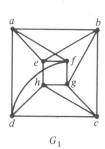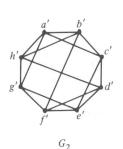

In Exercises 11–15, show that the property given is an invariant.

11H. Has a simple circuit of length k

12. Has n vertices of degree k

13H. Is connected

14. Has n simple circuits of length k

15H. Has an edge (v, w), where $\delta(v) = i$ and $\delta(w) = j$

16. Find an invariant not given in this section or in Exercises 11–15. Prove that your property is an invariant.

In Exercises 17–19, tell whether each property is an invariant or not. If the property is an invariant, prove that it is; otherwise, give a counterexample.

17H. Has an Euler circuit

18. Has a vertex inside some simple circuit

19H. Is bipartite

20. Draw all nonisomorphic simple graphs having three vertices.

21H. Draw all nonisomorphic simple graphs having four vertices.

22. Draw all nonisomorphic, circuit-free, connected simple graphs having five vertices.

23H. Draw all nonisomorphic, circuit-free, connected simple graphs having six vertices.

24. Show that simple graphs G_1 and G_2 are isomorphic if and only if their vertices may be ordered so that their adjacency matrices are identical.

The **complement** of a simple graph G is a simple graph $\overline{G}$ with the same vertices as G. An edge exists in $\overline{G}$ if and only if it does not exist in G.

★25H. Show that if G is a simple graph, either G or $\overline{G}$ is connected.

26. A simple graph G is **self-complementary** if G and $\overline{G}$ are isomorphic. Find a self-complementary graph having five vertices.

27H. Let G_1 and G_2 be simple graphs. Show that G_1 and G_2 are isomorphic if and only if $\overline{G}_1$ and $\overline{G}_2$ are isomorphic.

28. Given two graphs G_1 and G_2, suppose that there is a one-to-one, onto function f from the vertices of G_1 to the vertices of G_2 and a one-to-one, onto function g from the edges of G_1 to the edges of G_2, so that if an edge e is incident on v and w in G_1, the edge $g(e)$ is incident on $f(v)$ and $f(w)$ in G_2. Are G_1 and G_2 isomorphic?

A **homomorphism** from a graph G_1 to a graph G_2 is a function f from the vertex set of G_1 to the vertex set of G_2 with the property that if v and w are adjacent in G_1, then $f(v)$ and $f(w)$ are adjacent in G_2.

29H. Suppose that G_1 and G_2 are simple graphs. Show that if f is a homomorphism of G_1 to G_2 and f is one-to-one and onto, G_1 and G_2 are isomorphic.

In Exercises 30–34, for each pair of graphs, give an example of a homomorphism from G_1 to G_2.

30.

31H.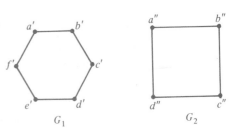

32. $G_1 = G_1$ of Exercise 31; $G_2 = G_1$ of Exercise 30

33H. $G_1 = G_1$ of Exercise 30

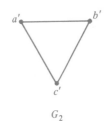

34.

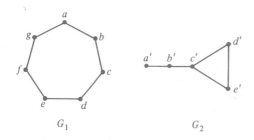

G_1 G_2

★35H. [Hell] Show that the only homomorphism from the graph to itself is the identity function.

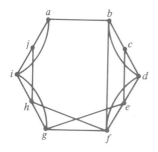

3.6

Planar Graphs

Three cities, C_1, C_2, and C_3, are to be directly connected by expressways to each of three other cities, C_4, C_5, and C_6. Can this road system be designed so that the expressways do not cross? A system in which the roads do cross is illustrated in Figure 3.6.1. If you try drawing a system in which the roads do not cross, you will soon be convinced that it cannot be done. Later in this section we explain carefully why it cannot be done.

> **DEFINITION 3.6.1.** A graph is *planar* if it can be drawn in the plane without its edges crossing.

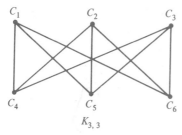

$K_{3,3}$

FIGURE 3.6.1

In designing printed circuits it is desirable to have as few lines cross as possible; thus the designer of printed circuits faces the problem of planarity.

If a connected, planar graph is drawn in the plane, the plane is divided into contiguous regions called **faces**. A face is characterized by the circuit that forms its boundary. For example, in the graph of Figure 3.6.2, face A is bounded by the circuit $(5, 2, 3, 4, 5)$ and face C is bounded by the circuit $(1, 2, 5, 1)$. The outer face D is considered to be bounded by the circuit $(1, 2, 3, 4, 6, 1)$. The graph of Figure 3.6.2 has $f = 4$ faces, $e = 8$ edges, and $v = 6$ vertices. Notice that f, e, and v satisfy the equation

$$f = e - v + 2. \tag{3.6.1}$$

In 1752, Euler proved that equation (3.6.1) holds for any connected, planar graph. At the end of this section we will show how to verify (3.6.1), but for now let us show how (3.6.1) can be used to show that certain graphs are not planar.

EXAMPLE 3.6.2.   Show that the graph $K_{3,3}$ of Figure 3.6.1 is not planar.

Suppose that $K_{3,3}$ is planar. Since every circuit has at least four edges, each face is bounded by at least four edges. Thus the number of edges that bound faces is at least $4f$. In a planar graph, each edge belongs to at most two bounding circuits. Therefore,

$$2e \geq 4f.$$

Using (3.6.1), we find that

$$2e \geq 4(e - v + 2). \tag{3.6.2}$$

For the graph of Figure 3.6.1, $e = 9$ and $v = 6$, so (3.6.2) becomes

$$18 = 2 \cdot 9 \geq 4(9 - 6 + 2) = 20,$$

which is a contradiction. Therefore, $K_{3,3}$ is not planar.

By a similar kind of argument (see Exercise 15), we can show that the graph K_5 of Figure 3.6.3 is not planar.

Obviously, if a graph contains $K_{3,3}$ or K_5 as a subgraph, it cannot be planar. The converse is almost true. To state the situation precisely, we must introduce some new terms.

DEFINITION 3.6.3.   If a graph G has a vertex v of degree 2 and edges (v, v_1) and (v, v_2), with $v_1 \neq v_2$, we say that the edges (v, v_1) and (v, v_2) are

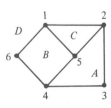

FIGURE 3.6.2

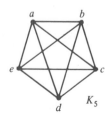

FIGURE 3.6.3

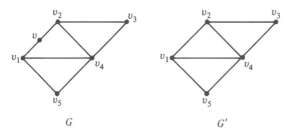

FIGURE 3.6.4

in *series*. A *series reduction* consists of deleting the vertex v from the graph G and replacing the edges (v, v_1) and (v, v_2) by the edge (v_1, v_2). The resulting graph G' is said to be *obtained from G by a series reduction*. By convention, G is said to be obtainable from itself by a series reduction.

EXAMPLE 3.6.4. In the graph G of Figure 3.6.4, the edges (v, v_1) and (v, v_2) are in series. The graph G' of Figure 3.6.4 is obtained from G by a series reduction.

DEFINITION 3.6.5. Graphs G_1 and G_2 are *homeomorphic* if G_1 and G_2 can be reduced to isomorphic graphs by performing a sequence of series reductions.

According to Definitions 3.6.3 and 3.6.5, G is homeomorphic to itself. Also, G_1 and G_2 are homeomorphic if G_1 (respectively, G_2) can be reduced to G_2 (respectively, G_1) by performing a sequence of series reductions.

EXAMPLE 3.6.6. The graphs G_1 and G_2 of Figure 3.6.5 are homeomorphic, since they can both be reduced to the graph G' of Figure 3.6.5 by a sequence of series reductions.

We can now state a necessary and sufficient condition for a graph to be planar. The theorem was first stated and proved by Kuratowski in 1930. The proof appears in [Even, 1979].

THEOREM 3.6.7 Kuratowski's Theorem. *A graph G is planar if and only if G does not contain a subgraph homeomorphic to K_5 or $K_{3,3}$.*

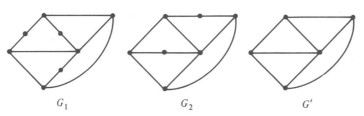

FIGURE 3.6.5

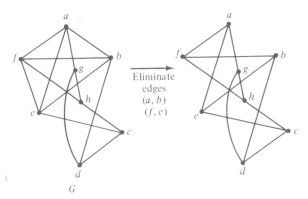

FIGURE 3.6.6

EXAMPLE 3.6.8. Show that the graph G of Figure 3.6.6 is not planar by using Kuratowski's Theorem.

Let us try to find $K_{3,3}$ in the graph G of Figure 3.6.6. We first note that the vertices a, b, f, and e each have degree 4. In $K_{3,3}$ each vertex has degree 3, so let us eliminate the edges (a, b) and (f, e) so that all vertices have degree 3 (see Figure 3.6.6). We note that if we eliminate one more edge, we will obtain two vertices of degree 2 and we can then carry out two series reductions. The resulting graph will have nine edges and since $K_{3,3}$ has nine edges, this approach looks promising. Using trial and error, we finally see that if we eliminate edge (g, h) and carry out the series reductions, we obtain an isomorphic copy of $K_{3,3}$ (see Figure 3.6.7). Therefore, the graph G of Figure 3.6.6 is not planar, since it contains a subgraph homeomorphic to $K_{3,3}$.

Although Theorem 3.6.7 does give an elegant characterization of planar graphs, it does not lead to an efficient algorithm for recognizing planar graphs. However, algorithms are known that can determine whether a graph having n vertices is planar in time $O(n)$ (see [Even, 1979]).

We will conclude this section by proving Euler's formula.

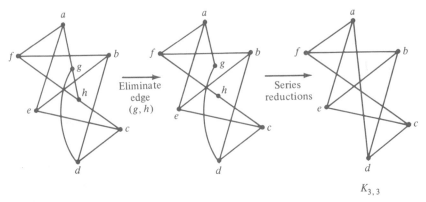

FIGURE 3.6.7

$f = 2, e = 1, v = 1$ $f = 1, e = 1, v = 2$

FIGURE 3.6.8

THEOREM 3.6.9   Euler's Formula for Graphs.   *If G is a connected, planar graph with e edges, v vertices, and f faces, then*

$$f = e - v + 2. \tag{3.6.3}$$

Proof.   We will use induction on the number of edges.

Suppose that $e = 1$. Then G is one of the two graphs shown in Figure 3.6.8. In either case, the formula holds. We have verified the Basis of Induction step.

Suppose that the formula holds for connected, planar graphs with n edges. Let G be a graph with $n + 1$ edges. First, suppose that G contains no circuits. Pick a vertex v and trace a path starting at v. Since G is circuit-free, every time we trace an edge, we arrive at a new vertex. Eventually, we will reach a vertex a, with degree 1, which we cannot leave (see Figure 3.6.9). We delete a and the edge x incident on a from the graph G. The resulting graph G' has n edges; hence, by the inductive assumption, (3.6.3) holds for G'. Since G has one more edge than G', one more vertex than G', and the same number of faces as G', it follows that (3.6.3) also holds for G.

Now suppose that G contains a circuit. Let x be an edge in a circuit (see Figure 3.6.10). Now x is part of a boundary for two faces. This time we delete the edge x but no vertices to obtain the graph G' (see Figure 3.6.10). Again G' has n edges; hence, by the inductive assumption, (3.6.3) holds for G'. Since G has one more face than G', one more edge than G', and the same number of vertices as G', it follows that (3.6.3) also holds for G.

Since we have verified the Inductive Step, by the Principle of Mathematical Induction, the theorem is proved.   ■

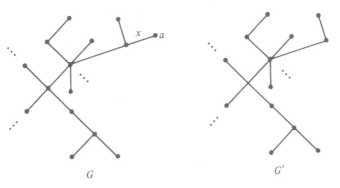

FIGURE 3.6.9

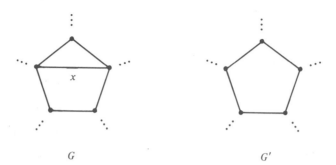

FIGURE 3.6.10

EXERCISES

In Exercises 1–3, show that each graph is planar by redrawing it so that no edges cross.

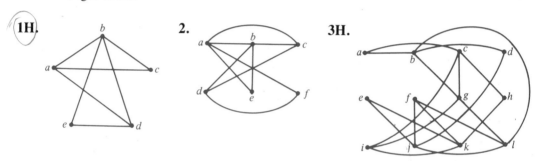

In Exercises 4 and 5, show that each graph is not planar by finding a subgraph homeomorphic to either K_5 or $K_{3,3}$.

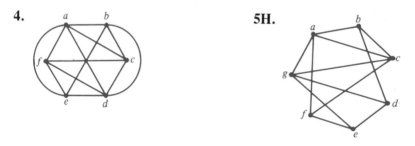

In Exercises 6–8, determine whether each graph is planar. If the graph is planar, redraw it so that no edges cross; otherwise, find a subgraph homeomorphic to either K_5 or $K_{3,3}$.

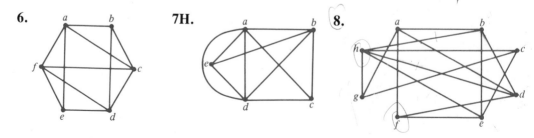

9H. A connected, planar graph has nine vertices having degrees 2, 2, 2, 3, 3, 3, 4, 4, and 5. How many edges are there? How many faces are there?

10. Show that adding or deleting loops, parallel edges, or edges in series does not affect the planarity of a graph.

11H. Show that any graph having four or fewer vertices is planar.

12. Show that any graph having five or fewer vertices and a vertex of degree 2 is planar.

13H. Show that in any simple, connected, planar graph $e \le 3v - 6$.

14. Give an example of a simple, connected, nonplanar graph for which $e \le 3v - 6$.

15H. Use Exercise 13 to show that K_5 is not planar.

★16. Show that if a simple graph G has 11 or more vertices, then either G or its complement $\overline{G}$ is not planar.

A **coloring** of a graph G by the colors $C_1, C_2, \ldots, C_n$ assigns to each vertex a color C_i so that any vertex has a color different from any adjacent vertex. For example, the graph shown here is colored with three colors. The rest of the exercises deal with coloring planar graphs.

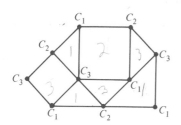

A **planar map** is a planar graph where the faces are interpreted as countries, the edges are interpreted as borders between countries, and the vertices represent the intersections of borders. The problem of coloring a planar map G, so that no countries with adjoining boundaries have the same color, can be reduced to the problem of coloring a graph by first constructing the **dual graph** G' of G in the following way. The vertices of the dual graph G' consist of one point in each face of G, including the unbounded face. An edge in G' connects two vertices if the corresponding faces in G are separated by a boundary. Coloring the map G is equivalent to coloring the vertices of the dual graph G'.

17H. Find the dual of the map

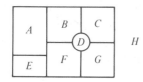

18. Show that the dual of a planar map is a planar graph.

19H. Show that any coloring of the map of Exercise 17, excluding the unbounded region, requires at least three colors.

20. Color the map of Exercise 17, excluding the unbounded region, using three colors.

21. Find the dual of the map

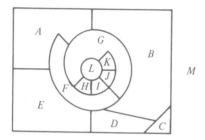

22. Show that any coloring of the map of Exercise 21, excluding the unbounded region, requires at least four colors.

23H. Color the map of Exercise 21, excluding the unbounded region, using four colors.

A **triangulation** of a simple, planar graph G is obtained from G by connecting as many vertices as possible while maintaining planarity and not introducing loops or parallel edges.

24. Find a triangulation of the graph

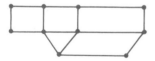

25H. Show that if a triangulation G' of a simple, planar graph G can be colored with n colors, so can G.

26. Show that in a triangulation of a simple, planar graph, $3f = 2e$.

Appel and Haken proved (see [Appel]) that every simple, planar graph can be colored with four colors. The problem had been posed in the mid-1800s and for years no one had succeeded in giving a proof. Those working on the four-color problem in recent years had one advantage their predecessors did not—the use of fast, electronic computers. The following exercises show how the proof begins.

Suppose there is a simple, planar graph that requires more than four colors to color. Then, among all such graphs, there is one with the fewest number of vertices. Let G be a triangulation of this graph. Then G also has a minimal number of vertices and by Exercise 25, G requires more than four colors to color.

27H. If the dual of a map has a vertex of degree 3, what must the original map look like?

28. Show that G cannot have a vertex of degree 3.

★29H. Show that G cannot have a vertex of degree 4.

★30H. Show that G has a vertex of degree 5.

The contribution of Appel and Haken was to show that only a finite number of cases involving the vertex of degree 5 needed to be considered and to analyze all of these cases and show that all could be colored using four colors. The reduction to a finite number of cases was facilitated by using the computer to help find the cases to be analyzed. The computer was then used again to analyze the resulting cases.

★31H. Show that any simple, planar graph can be colored using five colors.

3.7

Notes

Virtually any reference on discrete mathematics contains one or more chapters on graph theory. Books specifically on graph theory are [Berge; Busacker; Deo; Even, 1979; Harary; König; and Ore]. [Even, 1979] is recommended because of its algorithmic approach. [Deo] emphasizes applications and algorithms. Chapter 11, devoted to graph algorithms, contains a large number of references and some computer programs. [Bellman] contains a unified, algorithmic treatment of several graph problems, including the traveling salesperson problem and several puzzles.

Euler's original paper on the Königsberg bridges, edited by J. R. Newman, was reprinted as [Euler].

[Duda] is an introductory text on pattern recognition. For an expository article about pattern recognition in both machines and animals, see [Gose].

The solution presented to the Instant Insanity puzzle is due to [de Carte-blanche].

In [Gardner, 1959], Hamiltonian circuits are related to the Tower of Hanoi puzzle.

So-called **branch-and-bound methods** (see [Tucker]) often give solutions to the traveling salesperson problem more efficiently than will exhaustive search. [Parry] contains a computer program that uses a branch-and-bound method. For a good survey of results on the traveling salesperson problem, see [Bellmore].

Appel and Haken announced their solution to the four-color problem in [Appel].

COMPUTER EXERCISES

1. Consider
(a) A listing of edges of a graph given as pairs of positive integers

(b) The adjacency matrix
(c) The incidence matrix

Write a program that accepts as input any of (a), (b), or (c) and outputs the other two.

2.   Write a program that accepts as input the edges of a graph and determines whether or not the graph contains an Euler circuit.

3.   Write a program that randomly generates a 6 × 6 adjacency matrix. Have your program print the adjacency matrix, the number of edges, the number of loops, and the degree of each vertex.

4.   Write a program that solves an arbitrary Instant Insanity puzzle.

5.   Write a program that determines whether or not a graph is a bipartite graph. If it is a bipartite graph, the program should list the disjoint sets of vertices.

6.   Write a program that accepts as input the edges of a graph and then draws the graph using a computer graphics display.

7.   [*Project*] Prepare a report on a computer language developed especially to handle graphs. References may be found in Section 11-10 of [Deo].

8.   Write a program that finds an Euler circuit in a connected graph in which all vertices have even degree.

9.   Write a program that lists all simple paths between two given vertices.

10. Write a program that, given the edges of a graph and an ordered set of edges, determines whether the set of edges is a path, a simple path, a circuit, or a simple circuit.

11. Implement Dijkstra's shortest-path algorithm, Algorithm 3.4.1, as a program. The program should find a shortest path and its length.

12. Write a program that finds a longest path between two vertices in a circuit-free, connected graph and the length of this path.

13. Write a program that, given the incidence matrix of a simple graph G, computes the incidence matrix of the complement $\overline{G}$.

CHAPTER 4

Trees

Trees form one of the most useful subclasses of graphs. Computer science, in particular, makes extensive use of trees. In computer science, trees are useful in organizing and relating data as, for example, in a data base (see Example 4.1.5). Trees also arise in theoretical problems such as the optimal time for sorting (see Section 4.6).

In Section 4.1, after giving some basic definitions, we look at several applications of trees. Section 4.2 gives several alternative characterizations of trees and defines binary trees, an important subclass of trees. Spanning trees are discussed in Sections 4.3 and 4.4. Tree traversals, ways of systematically visiting all vertices in a tree, are the subject of Section 4.5. Section 4.6, devoted to sorting, considers specific sorting algorithms as well as the question of how fast we can sort. Section 4.7 is concerned with game trees—structures that facilitate the generation of game-playing algorithms for certain two-person games.

4.1

Examples

Intuitively, a **tree** is a graph that can be drawn to look like an inverted, natural tree (see Figure 4.1.1).

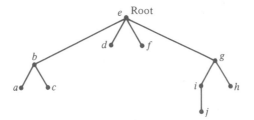

FIGURE 4.1.1

DEFINITION 4.1.1. A (*free*) *tree* is a simple graph having the property that there is a unique path between every pair of vertices.

A *rooted tree* is a tree in which a particular vertex is designated the root.

EXAMPLE 4.1.2. The graph of Figure 4.1.2 is a free tree and the graph of Figure 4.1.1 is a rooted tree. Figure 4.1.1 is obtained from Figure 4.1.2 by designating vertex *e* in Figure 4.1.2 as the root.

Notice that in Figure 4.1.2 there is a unique path between every pair of vertices. For example, the unique path from *c* to *h* is (*c, b, e, g, h*).

Figure 4.1.1 shows the way a rooted tree is usually drawn. First, we place the root at the top. Under the root and on the same level, we place the vertices that can be reached from the root on a path of length 1. Under each of these vertices and on the same level, we place the vertices that can be reached from the root on a path of length 2. We continue in this way until the entire tree is drawn. Since the path from the root to any given vertex is unique, each vertex is on a uniquely determined level. We call the level of the root level 0. The vertices under the root are said to be on level 1, and so on. Thus the **level of a vertex** *v* is the length of the path from the root to *v*. The **height** of a (rooted) tree is the maximum level that occurs.

EXAMPLE 4.1.3. The vertices *a, b, c, d, e, f, g, h, i,* and *j* of Figure 4.1.1 are on (respectively) levels 2, 1, 2, 1, 0, 1, 1, 2, 2, and 3. The height of the tree is 3.

EXAMPLE 4.1.4. A rooted tree is often used to specify hierarchical relationships. When a tree is used in this way, if vertex *a* is on a level one

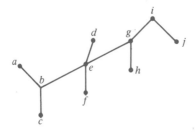

FIGURE 4.1.2

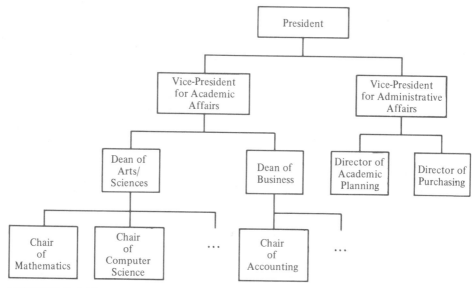

FIGURE 4.1.3

less than the level of vertex *b* and *a* and *b* are adjacent, then *a* is "just above" *b* and a logical relationship exists between *a* and *b*: *a* dominates *b* or *b* is subordinate to *a* in some way. A partial example of such a tree, which is the administrative organizational chart of a hypothetical university, is given in Figure 4.1.3.

EXAMPLE 4.1.5   Hierarchical Definition Trees.   Figure 4.1.4 is an example of a **hierarchical definition tree**. Such trees are used to show logical relationships among records in a database. The tree of Figure 4.1.4 might be used as a model for setting up a database to maintain records about books housed in several libraries.

EXAMPLE 4.1.6   Eight-Coins Puzzle.   Eight coins are identical in appearance. One coin is either heavier or lighter than the others, which all weigh the same. Develop a method of finding the odd coin and determining whether it is heavier or lighter than the others. You are permitted to use only a pan balance, which compares the weights of two sets of coins.

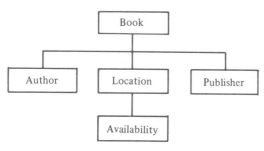

FIGURE 4.1.4

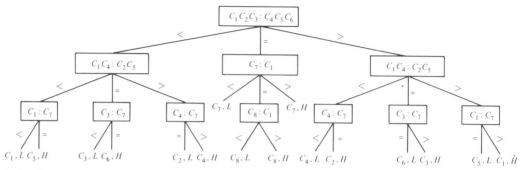

FIGURE 4.1.5

Any solution can be represented as a **decision tree** (see Figure 4.1.5). The coins are labeled C_1, C_2, . . . , C_8. We begin at the root and perform the indicated comparisons. For example, we start by comparing $C_1C_2C_3$ with $C_4C_5C_6$. After each comparison, we follow the appropriate edge. An edge labeled "<" (respectively, "=", ">") beneath a box labeled $A{:}B$ means that this edge is to be chosen if the weight of A is less than (respectively, equal to, greater than) the weight of B. For example, if the set $C_1C_2C_3$ is lighter than the set $C_4C_5C_6$, we would follow the edge labeled "<". Ultimately, we reach a vertex that identifies the odd coin and tells whether it is lighter or heavier than the others.

Let us look at a specific case. Suppose that C_8 is heavier than the others. Beginning at the root we find that

$$C_1C_2C_3 = C_4C_5C_6;$$

thus we follow the edge labeled "=" from the root. Next, we find that

$$C_7 = C_1;$$

thus we follow the edge labeled "=". Finally, we find that

$$C_8 > C_1$$

and the decision tree correctly tells us that C_8 is the heavy coin.

Notice that the procedure given in Figure 4.1.5 requires only three weighings. We can show that the odd coin cannot be found in two weighings. Since any of the eight coins is potentially the odd coin and the odd coin can be either light or heavy, there are 16 possible outcomes. A tree of height 2 where each decision vertex has the maximum three possible branches can represent at most nine outcomes (see Figure 4.1.6); therefore, the odd coin cannot be found in two weighings.

EXAMPLE 4.1.7 Huffman Codes. The most common way to represent characters internally in a computer is by using fixed-length bit strings. For example, ASCII (American Standard Code for Information Interchange)

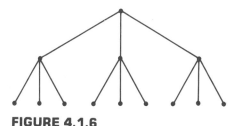

FIGURE 4.1.6

represents each character by a string of seven bits. Examples are given in Table 4.1.1.

Huffman codes, which represent characters by variable-length bit strings, provide alternatives to ASCII and other fixed-length codes. The idea is to use short bit strings to represent the most frequently used characters and to use longer bit strings to represent less frequently used characters. In this way it is generally possible to represent strings of characters, such as text, programs, and so on, in less space than if ASCII were used. Because of limited memory, some hand-held computers have used Huffman codes (see [Williams]).

A Huffman code is most easily defined by a rooted tree (see Figure 4.1.7). To decode a bit string, we begin at the root and move down the tree until a character is encountered. The bit, 0 or 1, tells us whether to move right or left. As an example, let us decode the string

$$01010111. \qquad (4.1.1)$$

We begin at the root. Since the first bit is 0, the first move is right. Next, we move left and then right. At this point, we encounter the first character, R. To decode the next character, we begin again at the root. The next bit is 1, so we move left and encounter the next character, A. The last bits 0111 decode as T. Therefore, the bit string (4.1.1) represents the word RAT.

Given a tree that defines a Huffman code, such as Figure 4.1.7, any bit string [e.g., (4.1.1)] can be uniquely decoded even though the characters are represented by variable-length bit strings. For the Huffman code defined by

TABLE 4.1.1

Character	ASCII Code
A	100 0001
B	100 0010
C	100 0011
1	011 0001
2	011 0010
!	010 0001
*	010 1010

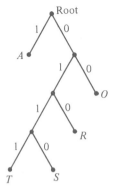

FIGURE 4.1.7

the graph of Figure 4.1.7, the character A is represented by a bit string of length 1, whereas S and T are represented by bit strings of length 4.

It is possible to construct a Huffman code from a table giving the frequency of occurrence of the characters to be represented so that the code constructed represents strings of characters in minimal space, provided that the strings to be represented have character frequencies identical to the character frequencies in the table (see [Standish]).

EXERCISES

1H. What does it mean for free trees T_1 and T_2 to be isomorphic?

2. Define: rooted trees T_1 and T_2 are isomorphic.

3H. Draw all distinct (nonisomorphic) free trees having three vertices.

4. Draw all distinct free trees having four vertices.

5H. Draw all distinct free trees having five vertices.

6. Draw all distinct free trees having six vertices.

7H. Draw all distinct rooted trees having three vertices.

8. Draw all distinct rooted trees having four vertices.

9H. Find the level of each vertex in the tree shown.

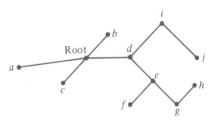

10. Find the height of the tree of Exercise 9.

11H. Draw the free tree of Figure 4.1.2 as a rooted tree with a as a root. What is the height of the resulting tree?

12. Draw the free tree of Figure 4.1.2 as a rooted tree with b as a root. What is the height of the resulting tree?

13. Give an example similar to Example 4.1.4 of a tree that is used to specify hierarchical relationships.

14. Give an example different from Example 4.1.5 of a hierarchical definition tree.

Exercises 15–17 refer to four coins, identical in appearance. One coin is either heavier or lighter than the others, which all weigh the same. You are permitted to use only a pan balance, which compares the weights of two sets of coins.

★15H. Show that three weighings are required to find the odd coin and determine whether it is heavier or lighter than the others.

16. Draw a tree like that in Figure 4.1.5 which shows how to find the odd coin in three weighings and determine whether it is heavier or lighter than the others.

17H. Draw a tree like that in Figure 4.1.5 which shows how to find the odd coin in two weighings without determining whether it is heavier or lighter than the others.

18. Eight coins are identical in appearance. One coin is possibly heavier or lighter than the others, which all weigh the same. You are permitted to use only a pan balance, which compares the weights of two sets of coins. Draw a tree like that in Figure 4.1.5 which shows how to determine if there is an odd coin and if there is one identifies it and determines whether it is heavier or lighter than the others. Only three weighings are permitted.

Exercises 19 and 20 refer to 12 coins, identical in appearance. One coin is either heavier or lighter than the others, which all weigh the same. You are permitted to use only a pan balance, which compares the weights of two sets of coins.

19H. Show that identifying the odd coin and determining whether it is heavier or lighter cannot be done in two weighings.

★20. Draw a tree like that in Figure 4.1.5 which shows how to find the odd coin in three weighings and determine whether it is heavier or lighter than the others.

Decode each bit string using the Huffman code given.

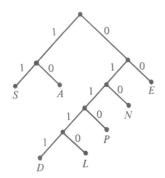

21H.   011000010

22.   01110100110

23H.   01111001001110

Encode each word using the Huffman code above.

24.   DEN   **25H.**   NEED   **26.** LEADEN

27H.   Design a Huffman code for the set of letters in the table. Attempt to use short bit codes for the most frequently used letters.

Letter	Percent of Time Occurring
I	7.5
U	20.0
B	2.5
S	27.5
C	5.0
H	10.0
M	2.5
P	25.0

28.   (a) Use the code developed in Exercise 27 to encode the following words (which have frequencies consistent with the table of Exercise 27): *BUS, CUPS, MUSH, PUSS, SIP, PUSH, CUSS, HIP, PUP, PUPS, HIPS*.
(b) Compute the number of bits needed to encode the words in part (a).

29H.   Can you lower the number of bits computed in Exercise 28(b) by redesigning your code of Exercise 27?

30.   What factors in addition to the amount of memory used should be considered when choosing a code, such as ASCII or a Huffman code, to represent characters in a computer?

31.   What techniques in addition to the use of Huffman codes might be used to save memory when storing text?

32.   Show that any tree with two or more vertices has a vertex of degree 1.

33H.   Show that a tree is a planar graph.

34.   Show that a tree is a bipartite graph.

35H.   Show that the vertices of a tree can be colored with two colors so that each edge is incident on vertices of different colors.

The **eccentricity** of a vertex v in a tree T is the maximum length of a path that begins at v.

36.   Find the eccentricity of each vertex in the tree of Figure 4.1.2.

A vertex v in a tree T is a **center** for T if the eccentricity of v is minimal.

37H.   Find the center(s) of the tree of Figure 4.1.2.

★38.   Show that a tree has either one or two centers.

★39H.   Show that if a tree has two centers they are adjacent.

40.   Define the radius r and the diameter d of a tree using the concepts of eccentricity and center. Is it always true, according to your definitions, that $2r = d$?

4.2
___

Properties of Trees

A tree is connected, since there is a path between any two vertices. If a tree T contained a circuit C, since T is a simple graph, C could not be a loop, so C would contain at least two vertices, a and b. Since a and b would lie in a circuit, there would be at least two paths from a to b, which is impossible. Therefore, a tree cannot contain a circuit. The converse is also true; every connected, circuit-free graph is a tree. The next theorem gives this characterization of trees as well as others.

THEOREM 4.2.1.   *Let T be a simple graph with n vertices. The following are equivalent.*
   (a) T is a tree.
   (b) T is connected and contains no circuits.
   (c) T is connected and has n − 1 edges.
   (d) T contains no circuits and has n − 1 edges.

   Proof.   We showed that (a) implies (b) before the statement of the theorem.
   Suppose that T is connected and contains no circuits. We will prove that T has $n - 1$ edges by induction on n.
   If $n = 1$, T consists of one vertex and zero edges, so the result is true if $n = 1$.
   Now suppose that the result holds for a connected, circuit-free graph with n vertices. Let T be a connected, circuit-free graph with $n + 1$ vertices. Choose a path P of maximum length. Since T is circuit-free, P is not a circuit, so P contains a vertex v of degree 1 (see Figure 4.2.1). Let T^* be T with v and the edge incident on v removed. Then T^* is connected and circuit-free and, because T^* contains n vertices, by induction T^* contains $n - 1$

FIGURE 4.2.1

edges. Therefore, T contains n edges. The inductive argument is complete. We have shown that (b) implies (c).

Now suppose that T is connected and has $n - 1$ edges. We must show that T contains no circuits.

Suppose that T contains at least one circuit. Since removing an edge from a circuit does not disconnect a graph, we may remove edges, but no vertices, from circuit(s) in T until the resulting graph T^* is circuit-free. Now, T^* is a simple, connected graph with n vertices that contains no circuits. We may use our just proven result, (b) implies (c), to conclude that T^* has $n - 1$ edges. But now T has more than $n - 1$ edges. This is a contradiction. Therefore, T contains no circuits. We have shown that (c) implies (d).

Finally, suppose that T contains no circuits and has $n - 1$ edges. We must show that T is a tree, that is, that T has a unique path between every pair of distinct vertices.

We will first show that T is connected. Suppose, by way of contradiction, that T is not connected. Let

$$T_1, T_2, \ldots, T_k$$

be the components of T. Since T is not connected, $k > 1$. Suppose that T_i has n_i vertices. Each T_i is connected and circuit-free, so we may use our previously proven result, (b) implies (c), to conclude that T_i has $n_i - 1$ edges. Now

$$
\begin{aligned}
n - 1 &= (n_1 - 1) + (n_2 - 1) + \cdots + (n_k - 1) \quad \text{(counting edges)} \\
&< (n_1 + n_2 + \cdots + n_k) - 1 \quad \text{(since } k > 1) \\
&= n - 1, \quad \text{(counting vertices)}
\end{aligned}
$$

which is impossible. Therefore, T is connected.

Suppose that there are distinct paths P and $(v_0, \ldots, v_n)$ from a to b in T. Then, P followed by $(v_n, \ldots, v_0)$ is a circuit, which is impossible. Thus there is a unique path between every pair of vertices in T. Therefore, T is a tree. We have shown that (d) implies (a) and hence all the conditions are equivalent. ∎

The following terminology is useful in discussing a rooted tree.

DEFINITION 4.2.2. Let T be a rooted tree with root v_0. Suppose that x, y, and z are vertices in T and that $(v_0, v_1, \ldots, v_n)$ is a path in T. Then
  (a) v_{n-1} is the *parent* of v_n.
  (b) $v_0, \ldots, v_{n-1}$ are *ancestors* of v_n.
  (c) v_n is a *child* of v_{n-1}.
  (d) If x is an ancestor of y, then y is a *descendant* of x.
  (e) If x and y are children of z, then x and y are *siblings*.
  (f) If x has no children, x is a *terminal vertex* (or a *leaf*).
  (g) If x is not a terminal vertex, x is an *internal* (or *branch*) *vertex*.

(h) The subgraph of T consisting of x and all its descendants, with x designated as a root, is the *subtree of T rooted at x*.

EXAMPLE 4.2.3. In the rooted tree of Figure 4.2.2,
(a) The parent of i is d.
(b) The ancestors of q are k, d, and a.
(c) The children of a are b, c, d, and e.
(d) The descendants of b are f, g, l, m, and n.
(e) p and q are siblings.
(f) The terminal vertices are e, h, j, l, m, n, o, p, and q.
(g) The internal vertices are a, b, c, d, f, g, i, and k.
(h) The subtree rooted at g is shown in Figure 4.2.3.

Sometimes we wish to take into account the order of the children in a tree. Special cases are given in the next definition.

DEFINITION 4.2.4. A *binary tree* is a rooted tree in which every vertex has either a left child, a right child, a left and a right child, or no children. A *full binary tree* is a binary tree in which every vertex has either a left and a right child or no children.

EXAMPLE 4.2.5. The graph of a single elimination tournament is a full binary tree (see Figure 4.2.4). The contestants' names are listed on the left. Winners progress to the right. Eventually, there is a single winner at the root. If the number of contestants is not a power of 2, some contestants receive byes. In Figure 4.2.4, contestant 7 has a first-round bye.

A fundamental result about full binary trees is our next theorem.

THEOREM 4.2.6. *If T is a full binary tree with i internal vertices, then T has $i + 1$ terminal vertices and $2i + 1$ total vertices.*

Proof. The vertices of T consist of the vertices that are children (of some parent) and the vertices that are not children (of any parent). There is one nonchild—the root. Since there are i internal vertices, each having two chil-

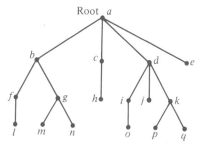

FIGURE 4.2.2

FIGURE 4.2.3

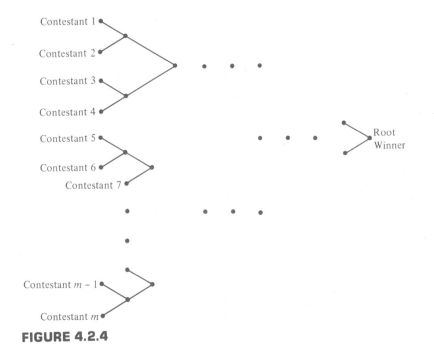

FIGURE 4.2.4

dren, there are $2i$ children. Thus the total number of vertices of T is $2i + 1$ and the number of terminal vertices is

$$(2i + 1) - i = i + 1.$$ ∎

EXAMPLE 4.2.7. If there are 60 contestants in a single elimination tournament, how many matches are played?

As in Example 4.2.5, such a tournament can be represented as a full binary tree. The number of contestants is the same as the number of terminal vertices and the number of matches i is the same as the number of internal vertices. Thus, by Theorem 4.2.6,

$$60 + i = 2i + 1$$

so that $i = 59$.

Our next result about binary trees relates the number of terminal vertices to the height.

THEOREM 4.2.8. *If a binary tree of height h has t terminal vertices, then*

$$\lg t \leq h. \tag{4.2.1}$$

Proof. We will prove the equivalent inequality

$$t \leq 2^h \tag{4.2.2}$$

by induction on h. Inequality (4.2.1) is obtained from (4.2.2) by taking the logarithm to the base 2 of both sides of (4.2.2).

If $h = 0$, the binary tree consists of a single vertex. In this case, $t = 1$ and thus (4.2.2) is true.

Assume that the result holds for a binary tree whose height is less than h. Let T be a binary tree of height $h > 0$ with t terminal vertices. Suppose first that the root of T has only one child. If we eliminate the root and the edge incident on the root, the resulting tree has height $h - 1$ and the same number of terminals as T. By induction, $t \leq 2^{h-1}$. Since $2^{h-1} < 2^h$, (4.2.2) is established for this case.

Now suppose that the root of T has children v_1 and v_2. Let T_i be the subtree rooted at v_i and suppose that T_i has height h_i and t_i terminal vertices, $i = 1$, 2. By induction,

$$t_i \leq 2^{h_i}, \qquad i = 1, 2. \tag{4.2.3}$$

The terminal vertices of T consist of the terminal vertices of T_1 and T_2. Hence

$$t = t_1 + t_2. \tag{4.2.4}$$

Combining (4.2.3) and (4.2.4), we obtain

$$t = t_1 + t_2 \leq 2^{h_1} + 2^{h_2} \leq 2^{h-1} + 2^{h-1} = 2^h.$$

The inductive step has been verified and the proof is complete. ∎

EXAMPLE 4.2.9. The binary tree in Figure 4.2.5 has height $h = 3$ and the number of terminals $t = 8$. For this tree, the inequality (4.2.1) becomes an equality.

Binary trees are useful for representing order relationships in a set of data. The structure used is called a **binary search tree**.

DEFINITION 4.2.10. A *binary search tree* is a binary tree T in which data are associated with the vertices. The data are arranged so that, for any vertex V in T, each data item in the left (respectively, right) subtree of V is less than (respectively, greater than) the data item in V. For ordinary words, we use alphabetical order.

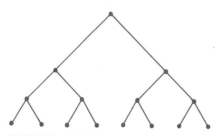

FIGURE 4.2.5

EXAMPLE 4.2.11. The words ONCE UPON A TIME THERE WAS AN OLD MAN may be placed in a binary search tree as shown in Figure 4.2.6.

Our next algorithm shows how to find a data item in a binary search tree.

ALGORITHM 4.2.12 Searching a Binary Search Tree. We are given a binary search tree T with root ROOT. If V is a vertex in T, LEFT(V) gives the left child. If V has no left child, LEFT(V) = λ. RIGHT(V) is defined similarly. VALUE(V) gives the data item stored in vertex V. The data are arranged so that, for any vertex V in T, each data item in the left (respectively, right) subtree of V is less than (respectively, greater than) the data item in V.

Given a data item W, this algorithm returns the vertex V containing W or returns λ if W is not in the tree.

    1. [Initialization.] $P := $ ROOT.

    2. [Found?] If $P = \lambda$, stop (search unsuccessful); otherwise, if VALUE(P) = W, stop (success—P is the vertex containing W).

    3. [Move.] If $W >$ VALUE(P), set $P := $ RIGHT(P) (move right) and go to step 2; otherwise, set $P := $ LEFT(P) (move left) and go to step 2.

Let T be a binary search tree with n vertices and let T^* be the full binary tree obtained from T by adding left and right children to existing vertices in T wherever possible. In Figure 4.2.7, we show the full binary tree that is produced from the binary search tree of Figure 4.2.6. The added vertices are drawn as boxes. An unsuccessful search in T corresponds to arriving at an added (box) vertex in T^*. If we define the time needed to run Algorithm 4.2.12 as the number of times statement 2 is executed, in the worst case, Algorithm 4.2.12 requires h steps, where h is the height of T^*. Since T^* is a binary tree, by Theorem 4.2.8, $\lg t \leq h$, where t is the number of terminal vertices. The full binary tree

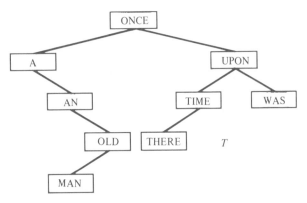

FIGURE 4.2.6

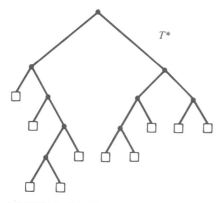

FIGURE 4.2.7

T^* has n internal vertices, so by Theorem 4.2.6, $t = n + 1$. Thus the worst case will require at least $\lg t = \lg(n + 1)$ steps. To minimize the worst-case run time, we should minimize the height of T. Exercise 28 shows that if this is done, the worst case requires exactly $\lceil \lg(n + 1) \rceil$ steps. The worst case never requires more than $n + 1$ steps. In this case we would execute the second statement n times to inspect every vertex in T and once to terminate the algorithm.

EXERCISES

1H.   Draw all distinct full binary trees having seven vertices.
2.   Draw all distinct full binary trees having nine vertices.
3H.   Draw all distinct binary trees having three vertices.
4.   Draw all distinct binary trees having four vertices.

In Exercises 5–14, draw a graph having the given properties or explain why no such graph exists.

5H.   Six edges; eight vertices
6.   Connected; six edges; eight vertices
7H.   Circuit-free; four edges; six vertices
8.   Tree; all vertices of degree 2
9H.   Tree; five vertices having degrees 1, 1, 2, 2, 4
10.   Tree; six vertices having degrees 1, 1, 1, 1, 3, 3
11H. Tree; four internal vertices; six terminal vertices
12.   Full binary tree; four internal vertices; five terminal vertices
13H. Full binary tree; height = 3; nine terminal vertices

14.   Full binary tree; height $= 4$; nine terminal vertices

Answer the questions in Exercises 15–23 for the two trees shown.

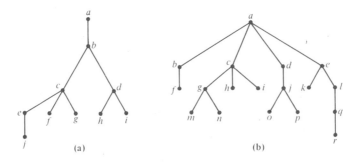

(a)   (b)

15H.   Find the parents of c and of h.

16.   Find the ancestors of c and of j.

17H.   Find the children of d and of e.

18.   Find the descendants of c and of e.

19H.   Find the siblings of f and of h.

20.   Find the terminal vertices.

21H.   Find the internal vertices.

22.   Draw the subtree rooted at j.

23H.   Draw the subtree rooted at e.

24.   Place the words FOURSCORE AND SEVEN YEARS AGO OUR FORE-
          FATHERS BROUGHT FORTH in a binary search tree of minimum height.

25H.   If Algorithm 4.2.12 is applied to the binary search tree constructed in Exercise
          24, what would the worst-case run time be?

26.   Do Exercises 24 and 25 using the words NEVER MET A MAN I DID NOT
          LIKE.

27H.   Write an algorithm that stores n distinct words in a binary search tree. The
          words are to be placed in the tree successively in the order given.

★28.   Write an algorithm that stores n distinct words in a binary search tree T of
          minimal height. Show that the derived tree T^*, as described in the text, has
          height $\lceil \lg (n + 1) \rceil$.

29H.   True or false? Let T be a binary tree. If for every vertex V in T the data item
          in V is greater than the data item in the left child of V and the data item in V
          is less than the data item in the right child of V, then T is a binary search tree.
          If true, prove it; otherwise, give a counterexample.

A **forest** is a simple graph with no circuits.

30. Show that a forest is a union of trees.

31H. If a forest F consists of m trees and has n vertices, how many edges does F have?

32. Prove that the following are equivalent about a simple graph T.
(a) T is a tree.
(b) T is connected but the removal of any edge (but no vertices) from T disconnects T.
(c) T is connected and, if an edge is added between any two vertices, exactly one circuit is created.

33H. Let G be a connected graph. Show that G is a tree if and only if every vertex of degree 2 or more is an articulation point. ("Articulation point" is defined before Exercise 47, Section 3.3.)

34. A **full *m*-ary tree** is a rooted tree such that every parent has m ordered children. If T is a full m-ary tree with i internal vertices, how many vertices does T have? How many terminal vertices does T have? Prove your results.

35H. Give an algorithm for constructing a full binary tree with $n > 1$ terminal vertices.

36. Give a recursive version of Algorithm 4.2.12.

37H. Find the maximum height of a full binary tree having t terminal vertices.

Define a binary tree T to be **balanced** if, for every vertex v in T, the heights of the left and right subtrees of v differ by at most 1. (Here the height of an empty tree is defined to be -1.)
State whether each tree in Exercises 38–41 is balanced or not.

38.

39H.

40.

41H.

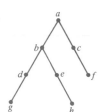

42. Can a balanced binary tree have four internal vertices?

43H. Draw a balanced binary tree with the fewest vertices having height 4.

Spanning Trees

In this section we consider the problem of finding a subgraph T of a graph G such that T is a tree containing all of the vertices of G. Such a tree is called a **spanning tree**. We will see that the methods of finding spanning trees may be applied to other problems as well.

> **DEFINITION 4.3.1.** A tree T is a *spanning tree* of a graph G if T is a subgraph of G that contains all of the vertices of G.

> **EXAMPLE 4.3.2.** A spanning tree of the graph G of Figure 4.3.1 is drawn with heavy lines.

> **EXAMPLE 4.3.3.** In general, a graph will have several spanning trees. Another spanning tree of the graph G of Figure 4.3.1 is shown in Figure 4.3.2.

Suppose that a graph G has a spanning tree T. Let a and b be vertices of G. Since a and b are also vertices in T and T is a tree, there is a path P from a to b. However, P also serves as a path from a to b in G; thus G is connected. The converse is also true.

> **THEOREM 4.3.4.** *A graph has a spanning tree if and only if G is connected.*

> **Proof.** We have already shown that if G has a spanning tree, then G is connected.
> Suppose that G is connected. If G is circuit-free, by Theorem 4.2.1, G is a tree.
> Suppose that G contains a circuit. We remove an edge (but no vertices) from this circuit. The graph produced is still connected. If it is circuit-free, we stop. If it contains a circuit, we remove an edge from this circuit. Continuing in this way, we eventually produce a circuit-free, connected subgraph

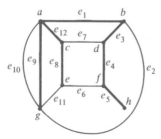

FIGURE 4.3.1

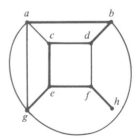

FIGURE 4.3.2

T. By Theorem 4.2.1, *T* is a tree. Since *T* contains all the vertices of *G*, *T* is a spanning tree of *G*. ∎

An algorithm for finding a spanning tree based on the proof of Theorem 4.3.4 would not be very efficient; it would involve the time-consuming process of finding circuits. We can do much better. We shall illustrate the first algorithm for finding a spanning tree by an example and then state the algorithm.

EXAMPLE 4.3.5. Find a spanning tree for the graph *G* of Figure 4.3.1.

We will use a method called **breadth-first search** (Algorithm 4.3.6). The idea of breadth-first search is to process all the vertices on a given level before moving to the next higher level.

First, select an ordering, say *abcdefgh*, of the vertices of *G*. Select the first vertex *a* and label it the root. Let *T* consist of the single vertex *a*. Add to *T* all edges (a, x), for $x = b$ to *h*, that do not produce a circuit when added to *T*. We would add to *T* edges (a,b), (a, c), and (a, g). (We could use either of the parallel edges incident on *a* and *g*.) Repeat this procedure with the vertices on level 1 by examining each in order:

b: Include (b, d).
c: Include (c, e).
g: None.

Repeat this procedure with the vertices on level 2:

d: Include (d, f).
e: None.

Repeat this procedure with the vertices on level 3:

f: Include (f, h).

Since no edges can be added to the single vertex *h* on level 4, the procedure ends. We have found the spanning tree shown in Figure 4.3.1.

We formalize the method of Example 4.3.5 as Algorithm 4.3.6.

ALGORITHM 4.3.6   Breadth-First Search for a Spanning Tree.
Let *G* be a connected graph with the vertices of *G* ordered

$$v_1, v_2, \ldots, v_n.$$

This algorithm finds a spanning tree *T* using the breadth-first search procedure. In this algorithm *S* denotes a sequence.

1. [Initialize.] Set $S = (v_1)$ and let *T* consist of the vertex v_1. Call v_1 the root.

2. [Add edges.] For each $x \in S$ in order, add edge (x, y) to *T* for each vertex $y \in G$ in order, provided that (x, y) does not produce a circuit if added to *T*. If no edges can be added, stop. (*T* is a spanning tree.)

3. [Update S.] Replace S by the children (in T) of S ordered consistently with the original ordering. Go to 2.

Exercise 17 is to give an argument to show that Algorithm 4.3.6 is correct.

Algorithm 4.3.6 can be used to test whether an arbitrary graph G with n vertices is connected (see Exercise 32). We simply apply Algorithm 4.3.6 to produce a tree T. Then G is connected if and only if T has n vertices.

An alternative to breadth-first search is **depth-first search**, which proceeds to successive levels in a tree at the earliest possible opportunity.

ALGORITHM 4.3.7   Depth-First Search for a Spanning Tree.   Let G be a connected graph with vertices ordered

$$v_1, v_2, \ldots, v_n.$$

This algorithm finds a spanning tree T using the depth-first search procedure.

1. [Initialization.] Set $w := v_1$ and let T be the graph consisting of the vertex v_1. Call v_1 the root.

2. [Add an edge.] Choose the edge (w, v_k), with minimum k, where adding (w, v_k) to T does not create a circuit. If no such edge exists, go to step 3; otherwise, add (w, v_k) to T; set $w := v_k$; go to step 2.

3. [Done?] If $w = v_1$, stop. (T is a spanning tree.)

4. [Backtrack.] Let x be the parent of w (in T). Set $w := x$. Go to step 2.

Exercise 18 is to give an argument to show that Algorithm 4.3.7 is correct.

EXAMPLE 4.3.8.   Use depth-first search (Algorithm 4.3.7) to find a spanning tree for the graph of Figure 4.3.2 with the vertex ordering $abcdefgh$.

We select the first vertex a and call it the root (see Figure 4.3.2). Next, we add the edge (a, x), with minimal x, to our tree. In our case we add the edge (a, b).

We repeat this process. We add the edges (b, d), (d, c), (c, e), (e, f), and (f, h). At this point, we cannot add an edge of the form (h, x), so we backtrack to the parent f of h and try to add an edge of the form (f, x). Again, we cannot add an edge of the form (f, x), so we backtrack to the parent e of f. This time we succeed in adding the edge (e, g). At this point, no more edges can be added, so we finally backtrack to the root and the procedure ends.

Because of step 4 in Algorithm 4.3.7, where we retreat along an edge toward the initially chosen root, depth-first search is also called **backtracking**. In the following example, we use backtracking to solve a puzzle.

EXAMPLE 4.3.9   Four-Queens Problem.   The four-queens problem is to place four tokens on a 4×4 grid so that no two tokens are on the same row, column, or diagonal. Construct a backtracking algorithm to solve the four-queens problem. (To use chess terminology, this is the problem of placing four queens on a 4 × 4 board so that no queen attacks another queen.)

The idea of the algorithm is to place tokens successively in the columns. When it is impossible to place a token in a column, we backtrack and adjust the token in the preceding column.

ALGORITHM 4.3.10 Solving the Four-Queens Problem Using Backtracking. This algorithm uses backtracking to search for an arrangement of four tokens on a 4 × 4 array B so that no two tokens are on the same row, column, or diagonal. In the B-array, a 1 signifies the presence of a token and a 0 signifies the absence of a token. I and J give the current row and column positions. When the algorithm terminates, SOL is YES if there

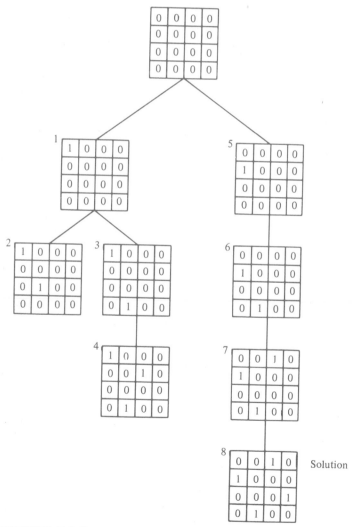

FIGURE 4.3.3

is a solution and SOL is NO if there is not a solution. In case there is a solution, the array B contains the solution.

   1. [Initialization.] Set all entries of the B-array to 0. Set SOL $:=$ YES; $I := 1; J := 1.$
   2. [Legal move?] If setting $B(I, J) := 1$ generates a legal move (no 1's are on the same row, column, or diagonal), set $B(I, J) := 1$ and go to step 6.
   3. [Search for legal move.] Set $I := I + 1$. If $I < 5$, go to step 2.
   4. [Backtrack.] Set $J := J - 1$ (move to a previous column). If $J = 0$ (no solution exists), set SOL $:=$ NO and stop; otherwise, set $I := 1.$
   5. [Revise previous token.] If $B(I, J) = 1$ (found previous token), set $B(I, J) := 0$ and go to step 3; otherwise, set $I := I + 1$ and go to step 5.
   6. [Solution?] If $J = 4$, stop (a solution is found).
   7. [Next column.] Set $J := J + 1$ and $I := 1$. Go to step 2.

The tree that Algorithm 4.3.10 generates is shown in Figure 4.3.3. The numbering indicates the order in which the vertices are generated. The solution is found at vertex 8.

Backtracking or depth-first search is especially attractive in a problem such as that in Example 4.3.9, where all that is desired is one solution. Since a solution, if one exists, is found at a terminal vertex, by moving to the terminal vertices as rapidly as possible, in general we can avoid generating some unnecessary vertices.

EXERCISES

1H.   Use breadth-first search (Algorithm 4.3.6) with the vertex ordering $hgfedcba$ to find a spanning tree for graph G of Figure 4.3.1.

2.   Use breadth-first search (Algorithm 4.3.6) with the vertex ordering $hfdbgeca$ to find a spanning tree for graph G of Figure 4.3.1.

3H.   Use breadth-first search (Algorithm 4.3.6) with the vertex ordering $chbgadfe$ to find a spanning tree for graph G of Figure 4.3.1.

4.   Use depth-first search (Algorithm 4.3.7) with the vertex ordering $hgfedcba$ to find a spanning tree for graph G of Figure 4.3.1.

5H.   Use depth-first search (Algorithm 4.3.7) with the vertex ordering $hfdbgeca$ to find a spanning tree for graph G of Figure 4.3.1.

6.   Use depth-first search (Algorithm 4.3.7) with the vertex ordering $dhcbefag$ to find a spanning tree for graph G of Figure 4.3.1.

In Exercises 7–9, find a spanning tree for each graph.

7H.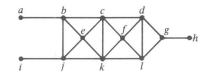

8.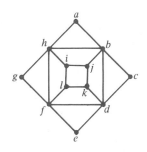

9H.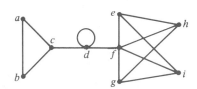

10. Show that there is no solution to the two-queens or the three-queens problem.

11H. Find a solution to the five-queens and six-queens problems.

★12. True or false? If G is a connected graph and T is a spanning tree for G, there is an ordering of the vertices of G such that Algorithm 4.3.6 produces T as a spanning tree. If true, prove it; otherwise, give a counterexample.

★13H. True or false? If G is a connected graph and T is a spanning tree for G, there is an ordering of the vertices of G such that Algorithm 4.3.7 produces T as a spanning tree. If true, prove it; otherwise, give a counterexample.

14. Find all nonisomorphic (as free trees and not as rooted trees) spanning trees for each graph in Exercises 7–9.

15H. Show, by an example, that Algorithm 4.3.6 can produce identical spanning trees for a connected graph G from two distinct vertex orderings of G.

16. Show, by an example, that Algorithm 4.3.7 can produce identical spanning trees for a connected graph G from two distinct vertex orderings of G.

17H. Prove that Algorithm 4.3.6 is correct.

18. Prove that Algorithm 4.3.7 is correct.

19H. Under what conditions is an edge in a connected graph G contained in every spanning tree of G?

★20. Let T and T' be two spanning trees of a connected graph G. Suppose that an edge x is in T but not in T'. Show that there is an edge y in T' but not in T such that $(T - \{x\}) \cup \{y\}$ and $(T' - \{y\}) \cup \{x\}$ are spanning trees of G.

The **8's puzzle** consists of eight movable squares and one empty position as shown.

1	2	3
4		6
7	5	8

An adjacent square can be moved to the empty position. For example, the pattern above can be transformed to

1	2	3
4	5	6
7	8	

21.   Write a backtracking algorithm that transforms an arbitrary position of the 8's puzzle to the position of the figure immediately above or decides that no such transformation exists.

Define a relation on the positions of the 8's puzzle as P_1RP_2 if P_1 can be transformed to P_2.

22.   Show that R is an equivalence relation.

★23.   How many equivalence classes are there?

★24.   Give a representative of each equivalence class.

★25.   Devise a simple test to determine whether P_1RP_2.

Define an $n \times n$ array of the numbers $1, 2, \ldots, n^2$ to be a **magic square** if the same sum is obtained by summing any row or column or either of the two main diagonals. A 3×3 magic square is shown.

8	1	6
3	5	7
4	9	2

26.   Show that the common sum of an $n \times n$ magic square is $n(n^2 + 1)/2$.

27.   Write a backtracking algorithm to find a 4×4 magic square.

28. Let T be a spanning tree for a graph G. Show that if an edge in G, but not in T, is added to T, a unique circuit is produced.

A circuit as described in Exercise 28 is called a **fundamental circuit**. The **fundamental circuit matrix** of a graph G has its rows indexed by the fundamental circuits of G relative to a spanning tree T for G and its columns indexed by the edges of G. The ijth entry is 1 if edge j is in the ith fundamental circuit and 0 otherwise. For example, the fundamental circuit matrix of the graph G of Figure 4.3.1 relative to the spanning tree shown in Figure 4.3.1 is

$$
\begin{array}{c}
\\
(abdca) \\
(efdbace) \\
(ageca) \\
(aga) \\
(abga)
\end{array}
\begin{array}{c}
e_7\ \ e_6\ \ e_{11}\ e_{10}\ \ e_2\ \ e_1\ \ e_3\ \ e_4\ \ e_5\ \ e_8\ \ e_9\ \ e_{12} \\
\left(\begin{array}{cccccccccccc}
1 & 0 & 0 & 0 & 0 & 1 & 1 & 0 & 0 & 0 & 0 & 1 \\
0 & 1 & 0 & 0 & 0 & 1 & 1 & 1 & 0 & 1 & 0 & 1 \\
0 & 0 & 1 & 0 & 0 & 0 & 0 & 0 & 0 & 1 & 1 & 1 \\
0 & 0 & 0 & 1 & 0 & 0 & 0 & 0 & 0 & 0 & 1 & 0 \\
0 & 0 & 0 & 0 & 1 & 1 & 0 & 0 & 0 & 0 & 1 & 0
\end{array}\right)
\end{array}
$$

Find the fundamental circuit matrix of each graph. The spanning tree to be used is drawn with heavy lines.

29H.

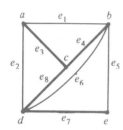

30.

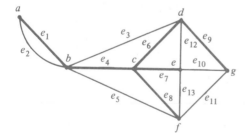

31H.

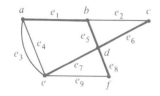

32. Write a breadth-first search algorithm to test whether a graph is connected.

33H. Write a depth-first search algorithm to test whether a graph is connected.

34. Write a breadth-first search algorithm that finds all solutions to the four-queens problem.

★**35.**   Write a backtracking algorithm that determines whether a graph can be colored with n colors and, if there is such a coloring, finds it.

★**36.**   Write a backtracking algorithm that determines whether a graph has a Hamiltonian circuit and, if there is a Hamiltonian circuit, finds it.

4.4

Minimal Spanning Trees

The weighted graph G of Figure 4.4.1 shows six cities and the costs of building roads between certain pairs of cities. We want to build the lowest-cost road system that will connect the six cities. The solution can be represented by a subgraph. This subgraph must be a spanning tree since it must contain all the vertices (so that each city is in the road system), it must be connected (so that any city can be reached from any other), and it must have a unique path between each pair of vertices (since a graph containing multiple paths between a vertex pair could not represent a minimum-cost system). Thus what is needed is a spanning tree the sum of whose weights is a minimum. Such a tree is called a **minimal spanning tree**.

> **DEFINITION 4.4.1.** Let G be a weighted graph. A *minimal spanning tree* of G is a spanning tree of G with minimum weight.

> **EXAMPLE 4.4.2.** The tree T' shown in Figure 4.4.2 is a spanning tree for graph G of Figure 4.4.1. The weight of T' is 20. This tree is not a minimal spanning tree since spanning tree T shown in Figure 4.4.3 has weight 12. We will see later that T is a minimal spanning tree for G.

The algorithm to find a minimal spanning tree that we will discuss is known as **Prim's Algorithm** (Algorithm 4.4.3). This algorithm builds a tree by iteratively, adding edges until a minimal spanning tree is obtained. At each iteration, we connect a minimum weight edge that does not complete a circuit to the current tree. Another algorithm to find a minimal spanning tree, known as **Kruskal's Algorithm**, is presented in the exercises (see Exercises 20–26).

> **ALGORITHM 4.4.3 Prim's Algorithm.** This algorithm finds a minimal spanning tree in a connected, weighted graph G with vertices $v_1, \ldots,$
> v_n.

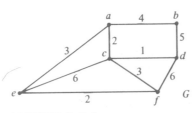

FIGURE 4.4.1

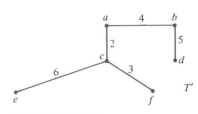

FIGURE 4.4.2

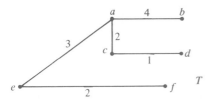

FIGURE 4.4.3

1. [Initialization.] Let T be the tree consisting of the vertex v_1 and no edges.

2. [Done?] If T has $n - 1$ edges, stop. (T is a minimal spanning tree.)

3. [Add edge.] Among all the edges not in T that are incident on a vertex in T and do not complete a circuit if added to T, select one having minimum weight and add it to T. If more than one edge has the same minimum weight, select (v_i, v_j) with the smallest i, say v_{i_0}. If two or more edges (v_{i_0}, v_j) have the same minimum weight, select the edge with the smallest j. Go to step 2.

EXAMPLE 4.4.4.   Apply Prim's Algorithm to the graph of Figure 4.4.1 with vertex ordering a, b, c, d, e, f to obtain a minimal spanning tree.

At step 1, we select the vertex a. At this point, T consists of the vertex a and no edges. Since T does not have five edges, we proceed to step 3. We select the edge (a, c) incident on a which has minimum weight and add it to T. Again, T does not have five edges, so we move to step 3.

Among the edges

$$\{(a, b), (a, e), (c, d), (c, e), (c, f)\}$$

not in T that are incident on a vertex in T and do not complete a circuit if added to T, we select (c, d), which has minimum weight, and add it to T. At this point T consists of the edges (a, c) and (c, d). Since T does not have five edges, we proceed to step 3.

Among the edges

$$\{(a, b), (a, e), (c, e), (c, f), (d, b), (d, f)\}$$

not in T that are incident on a vertex in T and do not complete a circuit if added to T, both (a, e) and (c, f) have minimum weight 3. In this case we select (a, e), since a precedes c in the vertex ordering given. At this point T consists of the edges $(a, c,)$, (c, d), and (a, e). Since T does not have five edges, we proceed to step 3.

Among the edges

$$\{(a, b), (c, f), (d, b), (d, f), (e, f)\}$$

not in T that are incident on a vertex in T and do not complete a circuit if added to T, we select (e, f), which has minimum weight, and add it to T. At this point T consists of the edges (a, c), (c, d), (a, e), and (e, f). Since T does not have five edges, we proceed to step 3.

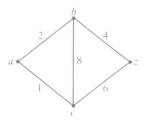

FIGURE 4.4.4

Among the edges

$$\{(a, b), (d, b)\}$$

not in T that are incident on a vertex in T and do not complete a circuit if added to T, we select (a, b), which has minimum weight, and add it to T. At this point T consists of the edges (a, c), (c, d), (a, e), (e, f), and (a, b). Since T has five edges, the algorithm terminates. The minimal spanning T is shown in Figure 4.4.3.

Prim's Algorithm furnishes an example of a **greedy algorithm**. A greedy algorithm is an algorithm that optimizes the choice at each iteration without regard to previous choices. The principle can be summarized as "doing the best locally." In Prim's Algorithm, since we want a minimal spanning tree, at each iteration we simply add an available edge with minimum weight.

Optimizing at each iteration does not necessarily give an optimal solution to the original problem. We will show shortly (Theorem 4.4.5) that Prim's Algorithm is correct—we do obtain a minimal spanning tree. As an example of a greedy algorithm that does not lead to an optimal solution, consider a "shortest-path algorithm" in which at each step we select an available edge having minimum weight incident on the most recently obtained vertex. If we apply this algorithm to the weighted graph of Figure 4.4.4 to find a shortest path from a to z, we would select the edge (a, c) and then the edge (c, z). Unfortunately, this is not the shortest path from a to z.

We next show that Prim's Algorithm is correct.

THEOREM 4.4.5. *Prim's Algorithm (Algorithm 4.4.3) is correct; that is, at the termination of Algorithm 4.4.3, T is a minimal spanning tree.*

Proof. In this proof we will specify a tree by listing its edges.

By construction, at the termination of Algorithm 4.4.3, T is a connected, circuit-free subgraph of G containing all the vertices of G; hence T is a spanning tree of G.

To show that T is a minimal spanning tree, we will use induction to show that at the kth iteration of Algorithm 4.4.3, T is contained in a minimal spanning tree. It will then follow that at termination, T is a minimal spanning tree. We let T_i denote the tree produced by Algorithm 4.4.3 at the ith iteration.

If $k = 1$, T_1 consists of a single vertex. In this case T_1 is contained in every minimal spanning tree. We have verified the Basis Step.

Next, assume that at the $(k - 1)$st iteration of Algorithm 4.4.3, T_{k-1} is contained in a minimal spanning tree T'. Let V be the set of vertices in T_{k-1}. Algorithm 4.4.3 selects an edge (i, j) of minimum weight where $i \in V$ and $j \notin V$ and adds it to T_{k-1} to produce T_k. If $(i, j) \in T'$, then T_k is contained in the minimal spanning tree T'. If $(i, j) \notin T'$, $T' \cup \{(i, j)\}$ contains a circuit C. Choose an edge $(x, y) \in C$, different from (i, j), with $x \in V$ and $y \notin V$. Then

$$w(x, y) \geq w(i, j). \tag{4.4.1}$$

Because of (4.4.1), the graph $T'' = [T' \cup \{(i, j)\}] - \{(x, y)\}$ has weight less than or equal to the weight of T'. Since T'' is a spanning tree, T'' is a minimal spanning tree. Since T_k is contained in T'', the Inductive Step has been verified. The proof is complete. ∎

Our version of Prim's Algorithm examines $O(n^3)$ edges in the worst case (see Exercise 7) to find a minimal spanning tree for a graph having n vertices. It is possible (see Exercise 9) to implement Prim's Algorithm so that only $O(n^2)$ edges are examined. Since K_n has $O(n^2)$ edges, the latter version is optimal.

EXERCISES

In Exercises 1–5, find the minimal spanning tree given by Algorithm 4.4.3 for each graph for the vertex ordering $a, b, c, \ldots$.

1H.

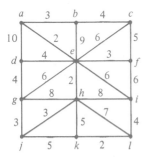

2.

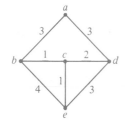

3H.

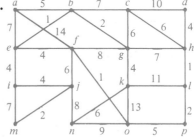

4.

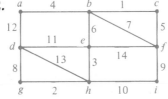

5H.

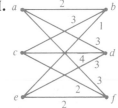

6. Let G be a connected, weighted graph with vertices $1, \ldots, n$. Let D be an

$n \times n$ matrix where $D(i, j)$ is the weight of edge (i, j) and $D(i, j)$ is ∞ if there is no edge (i, j). Restate Prim's Algorithm in terms of operations on the matrix D.

7H. Show that Algorithm 4.4.3 examines $O(n^3)$ edges in the worst case.

Exercises 8–10 refer to the version of Prim's Algorithm given below.

ALGORITHM 4.4.6 Alternate Version of Prim's Algorithm. This algorithm finds a minimal spanning tree in a connected, weighted graph G with vertices $v_1, \ldots, v_n$. At each step some vertices have temporary labels and some have permanent labels. The label of vertex v_i is denoted L_i. The algorithm returns the minimal spanning tree T.

1. Let T be the null tree. Set $L_1 := 0$. This label is permanent. For $j = 2, \ldots, n$, set $L_j := w(v_1, v_j)$. These labels are temporary.

2. Pick the smallest temporary label L_i. Make L_i permanent. Add the edge (v_i, v_j), whose weight is L_i and where v_j has a permanent label, to T.

3. If there are no temporary labels, stop; otherwise, for each temporary label L_k, set

$$L_k := \min \{L_k, w(v_i, v_k)\}.$$

Go to step 2.

8. Show how Algorithm 4.4.6 finds a minimal spanning tree for the graphs of Exercises 1–5.

9H. Show that Algorithm 4.4.6 examines $O(n^2)$ edges in the worst case.

10. Prove that Algorithm 4.4.6 is correct; that is, at the termination of Algorithm 4.4.6, T is a minimal spanning tree.

11H. Show that an algorithm that finds a minimal spanning tree in a connected, weighted graph G must examine every edge in G.

★12. Show that if all weights in a connected graph G are distinct, G has a unique minimal spanning tree.

In Exercises 13–16, decide if the statement is true or false. If the statement is true, prove it; otherwise, give a counterexample. In each exercise, G is a connected, weighted graph.

13H. If all the weights in G are distinct, distinct spanning trees of G have distinct weights.

14. If e is an edge in G whose weight is less than the weight of every other edge, e is in every minimal spanning tree of G.

15H. If Algorithm 4.4.3 is altered as follows, the resulting algorithm finds a minimal spanning tree.

3. Choose the smallest i with v_i in T.

4. Set $v := v_i$. Select the edge (v, v_j), where v_j is not in T, with smallest weight, add it to T, and go to step 2. If there is no such edge, choose the vertex v_j in T with smallest j greater than i. Set $i := j$ and go to step 4.

16. If T is a minimal spanning tree of G, there is a vertex ordering of G so that Algorithm 4.4.3 produces T.

17H. Let G be a connected, weighted graph. Show that if, as long as possible, we remove an edge from G having maximum weight whose removal does not disconnect G, the result is a minimal spanning tree for G.

★18. Write an algorithm that finds a maximal spanning tree in a connected, weighted graph.

19. Prove that your algorithm in Exercise 18 is correct.

Kruskal's Algorithm finds a minimal spanning tree in a connected, weighted graph G having n vertices as follows. The graph T initially consists of the vertices of G and no edges. At each iteration, we add an edge to T having minimum weight which does not complete a circuit in T. When T has $n - 1$ edges, we stop.

20. Formally state Kruskal's Algorithm.

21H. Show how Kruskal's Algorithm finds minimal spanning trees for the graphs of Exercises 1–5.

22. Show that Kruskal's Algorithm is correct; that is, at the termination of Kruskal's Algorithm, T is a minimal spanning tree.

★23. Let G be a connected, weighted graph with vertices $1, \ldots, n$. Let D be an $n \times n$ matrix where $D(i, j)$ is the weight of edge (i, j) and $D(i, j)$ is ∞ if there is no edge (i, j).

Restate Kruskal's Algorithm in terms of operations on the matrix D.

24. How many edges does Kruskal's Algorithm, as implemented in Exercise 20, examine?

25. Can you revise your implementation of Kruskal's Algorithm of Exercise 20 so that your result in Exercise 24 is improved?

26. Let V be a set of n vertices and let s be a "dissimilarity function" on $V \times V$ (see Example 3.1.6). Let G be the complete, weighted graph having vertices V and weights $w(v_i, v_j) = s(v_i, v_j)$. Modify Kruskal's Algorithm so that it groups data into classes. This modification is known as the **method of nearest neighbors** (see [Duda]).

Exercises 27 and 28 refer to the following situation. Suppose that we have stamps of various denominations and that we want to choose the minimum number of stamps to make a given amount of postage. Consider a greedy algorithm that selects stamps by choosing as many of the largest denomination as possible, then as many of the second largest denomination as possible, and so on.

27H. Show that if the available denominations are 1, 8, and 10 cents, the algorithm described above does not always produce the fewest number of stamps to make a given amount of postage.

★**28.** Show that if the available denominations are 1, 5, and 25 cents, the algorithm described above produces the fewest number of stamps to make any given amount of postage.

4.5

Tree Traversal

Breadth-first search and depth-first search provide ways to "walk" a tree, that is, to traverse a tree in a systematic way so that each vertex is visited exactly once. In this section we consider three additional tree traversal methods. We define these traversals recursively.

ALGORITHM 4.5.1 Preorder Traversal. This algorithm processes the vertices of a binary tree using preorder traversal. The algorithm is recursive. It receives a binary tree T as input. It is assumed that PT is the root of T.
    1. [Trivial?] If PT is empty, then return.
    2. [Process root.] Process PT.
    3. [Process left subtree.] Call Algorithm 4.5.1 with input the subtree whose root is the left child of PT.
    4. [Process right subtree.] Call Algorithm 4.5.1 with input the subtree whose root is the right child of PT and return.

Let us examine Algorithm 4.5.1 for some simple cases. If the binary tree is empty, nothing is processed since, in this case, the algorithm simply returns at step 1.

Suppose that the tree T consists of a single vertex ROOT. We set $PT := $ ROOT and call Algorithm 4.5.1. Nothing happens at step 1, since PT is not empty. At step 2, we process the root. At step 3, we replace T by the subtree whose root is the left child of PT. However, this is an empty tree and we just saw that when Algorithm 4.5.1 is called with an empty tree, nothing is processed. Similarly, at step 4, when we call Algorithm 4.5.1, nothing is processed. We then return. In this case, we processed the root and returned.

Now suppose that we have tree T of Figure 4.5.1. We set $PT := $ ROOT and call Algorithm 4.5.1. Nothing happens at step 1, since PT is not empty. At step 2 we process the root. At step 3 we replace T by the subtree whose root is the left child of PT (see Figure 4.5.2). We then call Algorithm 4.5.1. We just saw

FIGURE 4.5.1

FIGURE 4.5.2

that in case the tree consists of a single vertex, Algorithm 4.5.1 processes that vertex. Similarly, at step 4, when we call Algorithm 4.5.1, we process vertex *C*. Thus the vertices are processed in the order *ABC*.

EXAMPLE 4.5.2. In what order are the vertices of the tree of Figure 4.5.3 processed if preorder traversal is used?

Following steps 2, 3, and 4—root/left/right—of Algorithm 4.5.1, the traversal proceeds as shown in Figure 4.5.4. Thus the order of processing is *ABCDEFGHIJ*.

Inorder traversal and postorder traversal are obtained by changing the position of step 2 (root) in Algorithm 4.5.1. "Pre," "in," and "post" refer to the position of the root in the traversal; that is, "preorder" means root first, "inorder" means root second, and "postorder" means root last.

ALGORITHM 4.5.3 Inorder Traversal. This algorithm processes the vertices of a binary tree using inorder traversal. The algorithm is recursive. It receives a binary tree *T* as input. It is assumed that *PT* is the root of *T*.
1. [Trivial?] If *PT* is empty, then return.
2. [Process left subtree.] Call Algorithm 4.5.3 with input the subtree whose root is the left child of *PT*.
3. [Process root.] Process *PT*.
4. [Process right subtree.] Call Algorithm 4.5.3 with input the subtree whose root is the right child of *PT* and return.

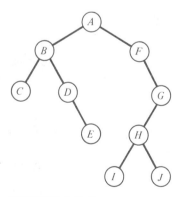

FIGURE 4.5.3

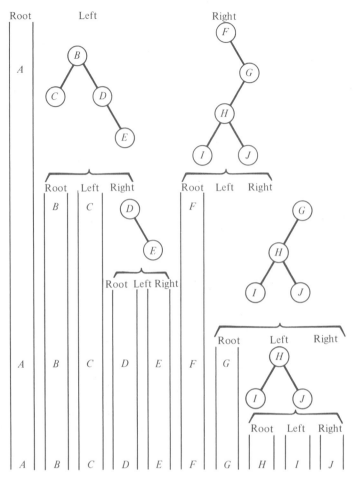

FIGURE 4.5.4

EXAMPLE 4.5.4. In what order are the vertices of the binary tree of Figure 4.5.3 processed if inorder traversal is used?

Following steps 2, 3, and 4—left/root/right—of Algorithm 4.5.3, we obtain the inorder listing *CBDEAFIHJG*.

ALGORITHM 4.5.5 Postorder Traversal. This algorithm processes the vertices of a binary tree using postorder traversal. The algorithm is recursive. It receives a binary tree *T* as input. It is assumed that *PT* is the root of *T*.

  1. [Trivial?] If *PT* is empty, then return.

  2. [Process left subtree.] Call Algorithm 4.5.5 with input the subtree whose root is the left child of *PT*.

  3. [Process right subtree.] Call Algorithm 4.5.5 with input the subtree whose root is the right child of *PT*.

  4. [Process root.] Process *PT* and return.

EXAMPLE 4.5.6. In what order are the vertices of the binary tree of Figure 4.5.3 processed if postorder traversal is used?

Following steps 2, 3, and 4—left/right/root—of Algorithm 4.5.5, we obtain the postorder listing *CEDBIJHGFA*.

Notice that preorder traversal may be obtained by following the route shown in Figure 4.5.5, and that reverse postorder traversal may be obtained by following the route shown in Figure 4.5.6.

If data are stored in a binary search tree, as described in Section 4.2, inorder traversal will process the data in order, since the sequence left/root/right agrees with the ordering of the data in the tree.

In the remainder of this section we consider binary tree representations of arithmetic expressions. Such representations facilitate the computer evaluation of expressions.

We will restrict our operators to $+$, $-$, $*$, and $/$. An example of an expression involving these operators is

$$(A+B)*C-D/E. \tag{4.5.1}$$

This standard way of representing expressions is called the **infix form of an expression**. The variables A, B, C, D, and E are referred to as **operands**. The **operators** $+$, $-$, $*$, and $/$ operate on pairs of operands or expressions. In the infix form of an expression, an operator appears between its operands.

An expression such as (4.5.1) can be represented as a binary tree. The terminal vertices correspond to the operands, and the internal vertices correspond to the operators. The expression (4.5.1) would be represented as shown in Figure 4.5.7. In the binary tree representation of an expression, an operator operates on its left and right subtrees. For example, in the subtree whose root is $/$ in Figure 4.5.7, the divide operator operates on the operands D and E; that is, D is to be divided by E. In the subtree whose root is $*$ in Figure 4.5.7, the

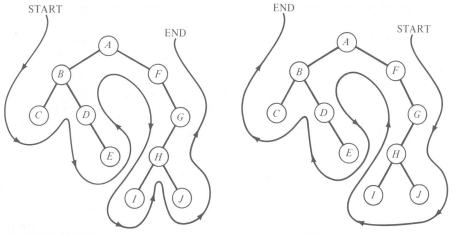

FIGURE 4.5.5   **FIGURE 4.5.6**

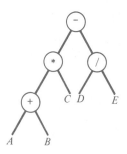

FIGURE 4.5.7

multiplication operator operates on the subtree headed by $+$, which itself represents an expression, and C.

In a binary tree we distinguish the left and right subtrees of a vertex. The left and right subtrees of a vertex correspond to the left and right operands or expressions. This left/right distinction is important in expressions. For example, $4-6$ and $6-4$ are different.

If we traverse the binary tree of Figure 4.5.7 using inorder, and insert a pair of parentheses for each operation, we obtain

$$(((A+B)*C)-(D/E)).$$

This form of an expression is called the **fully parenthesized form of the expression**. In this form we do not need to specify which operations (such as multiplication) are to be performed before others (such as addition), since the parentheses unambiguously dictate the order of operations.

If we traverse the tree of Figure 4.5.7 using postorder, we obtain

$$AB+C*DE/-.$$

This form of the expression is called the **postfix form of the expression** (or **reverse Polish notation**). In postfix, the operator follows its operands. For example, the first three symbols $AB+$ indicate that A and B are to be added. Advantages of the postfix form over the infix form are that in postfix no parentheses are needed and no conventions are necessary regarding the order of operations. The expression will be unambiguously evaluated. For these reasons, and others, many compilers translate infix expressions to postfix form. Also, some hand calculators require expressions to be entered in postfix form.

A third form of an expression can be obtained by applying preorder traversal to a binary tree representation of an expression. In this case, the result is called the **prefix form of the expression** (or **Polish notation**). As in postfix, no parentheses are needed and no conventions are necessary regarding the order of operations. The prefix form of (4.5.1), obtained by applying preorder traversal to the tree of Figure 4.5.7, is

$$-*+ABC/DE.$$

EXERCISES

In Exercises 1–5, list the order in which the vertices are processed using preorder, inorder, and postorder traversal.

1H.

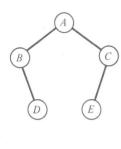

2.

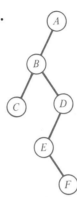

3H.

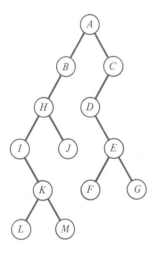

4.

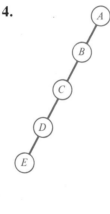

5H.

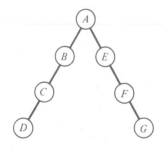

In Exercises 6–10, represent the expression as a binary tree and write the prefix and postfix forms of the expression.

6. $(A+B)*(C-D)$

7H. $((A-C)*D)/(A+(B-D))$

8. $(A*B + C*D) - (A/B - (D + E))$

9H. $(((A + B)*C + D)*E) - ((A + B)*C + D)$

10. $(A*B - C/D + E) + (A - B - C - D*D)/(A + B + C)$

In Exercises 11–15, represent the postfix expression as a binary tree and write the prefix form, the usual infix form, and the fully parenthesized infix form of the expression.

11H. $AB + C -$

12. $ABC + -$

13H. $ABCD + */E -$

14. $ABC**CDE + / -$

15H. $AB + CD*EF/ - -A*$

In Exercises 16–21, find the value of the postfix expression if $A = 1$, $B = 2$, $C = 3$, and $D = 4$.

16. $ABC + -$

17H. $AB + C -$

18. $AB + CD*AA/ - -B*$

19H. $ABC**ABC + + -$

20. $ABAB* + *D*$

21H. $ADBCD* - + *$

22. Show, by example, that distinct binary trees with vertices A, B, and C can give rise to the same preorder listing ABC.

23H. Show that there is a unique binary tree with six vertices whose preorder vertex listing is $ABCEFD$ and whose inorder vertex listing is $ACFEBD$.

★24. Write an algorithm that reconstructs the binary tree given its preorder and inorder vertex orderings.

25H. Give examples of distinct binary trees, B_1 and B_2, each with two vertices, with the preorder vertex listing of B_1 equal to the preorder listing of B_2 and the postorder vertex listing of B_1 equal to the postorder listing of B_2.

26. Let P_1 and P_2 be permutations of $ABCDEF$. Is there a binary tree with vertices A, B, C, D, E, and F whose preorder listing is P_1 and whose inorder listing is P_2? Explain.

27H. Prove that Algorithm 4.5.1 terminates.

28. Write a recursive algorithm that prints the contents of the terminal vertices of a binary tree from left to right.

29H. Write a recursive algorithm that interchanges all left and right children of a binary tree.

30. Write a recursive algorithm that initializes each vertex of a binary tree to the number of its descendants.

In the remaining exercises, every expression involves only the operands A, B, . . . , Z and the operators $+$, $-$, $*$, $/$.

★31H. Give a test that determines whether or not a string of symbols is a valid postfix expression.

★32. Write an algorithm that constructs the binary tree representation of an infix expression.

33H. Write an algorithm that, given the binary tree representation of an expression, outputs the fully parenthesized infix form of the expression.

★34. Write an algorithm that constructs the binary tree representation of a postfix expression.

35H. Write an algorithm that converts a postfix expression to the infix form.

36. Write an algorithm that converts an infix expression to the postfix form.

★37. Write a recursive algorithm that evaluates a postfix expression given the values of the operands.

4.6

Sorting

The importance of sorting in data processing is evident. Imagine trying to find a telephone number in the Chicago directory if the names were not sorted! As a result, a great deal of attention has been given to sorting. Many sorting algorithms have been devised, each having attractive features in certain situations. In this section we discuss two sorting algorithms. Others are given in the Exercises (see Algorithm 4.6.9 and Exercises 16–22 and 32). We will also show that sorting n items requires at least $O(n \lg n)$ comparisons in the worst case.

The sorting problem is easily described. Given n items

$$x_1, \ldots, x_n,$$

arrange them in increasing (or decreasing) order. We will discuss sorting algorithms that repeatedly compare elements and, based on the result of the comparison, permute the order until the elements are sorted. A simple example is the bubble sort.

ALGORITHM 4.6.1 Bubble Sort. Given an array

$$S(1), \ldots, S(N),$$

this algorithm sorts the array in increasing order.
  1. [Trivial?] If $N = 1$, stop.
  2. [Initialize I.] $I := 1$. (The items $S(K)$, with $K < I$, are sorted.)

3. [Initialize J.] $J := N - 1$. (The pair $S(J)$ and $S(J + 1)$ are to be compared.)

4. [Compare.] If $S(J + 1) < S(J)$, then (exchange) TEMP $:= S(J + 1)$; $S(J + 1) := S(J)$; $S(J) :=$ TEMP.

5. [End J-loop?] If $J > I$, then $J := J - 1$; go to 4.

6. [End I-loop?] If $I = N - 1$, then, stop; otherwise, $I := I + 1$; go to 3.

EXAMPLE 4.6.2. Figure 4.6.1 shows how the bubble sort sorts the sequence

$$10 \quad 3 \quad 6 \quad 1.$$

As the algorithm progresses, the smaller numbers at the bottom are "bubbled up" to the correct positions. Note, for example, how 1 is bubbled up.

In the bubble sort, the permutation that results from the comparison at step 4 is particularly simple. If the pair tested is not in order, the items are exchanged; otherwise, the items are left in their original positions.

We will determine the number of comparisons required by the bubble sort. When $I = 1$, the number of comparisons made, which is the same as the number of iterations of the J-loop (steps 3, 4, 5), is $N - 1$. When $I = 2$, the number of comparisons made is $N - 2$, and so on. Thus the total number of comparisons made in the bubble sort is

$$(N - 1) + (N - 2) + \cdots + 2 + 1 = (N - 1)N/2 = O(N^2).$$

Any algorithm that repeatedly compares elements and, based on the result of the comparison, permutes the order until the elements are sorted, can be described by a binary tree. Each vertex corresponds to a comparison and each of the two edges under a vertex shows the possible permutations that can result from the comparison. For example, the bubble sort, Algorithm 4.6.1 for $n = 3$, can be described by the binary tree of Figure 4.6.2. Since there are $n!$ permutations of n items, a tree that describes an algorithm which sorts n items must have at least $n!$ terminal vertices. This fact, combined with Theorem 4.2.8, may be used to show that any comparison-based sorting algorithm requires at least $O(n \lg n)$ comparisons in the worst case.

THEOREM 4.6.3. *If $f(n)$ is the number of comparisons needed to sort n items in the worst case by an algorithm that repeatedly compares elements*

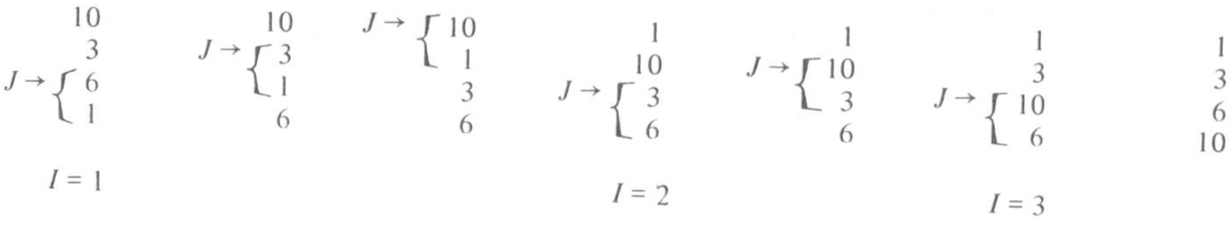

FIGURE 4.6.1

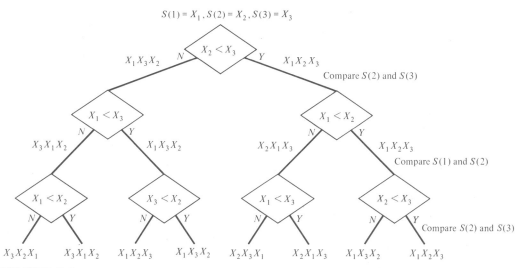

$S(1) = X_1, S(2) = X_2, S(3) = X_3$

FIGURE 4.6.2

and, based on the result of the comparison, permutes the order until the elements are sorted, then

$$\tfrac{1}{2}n[(\lg n) - 1] \le f(n).$$

Proof. Let T be the binary tree, as described before the statement of the theorem, which represents the algorithm for an input of size n. Then T has $n!$ terminal vertices. By Theorem 4.2.8, the height h of T satisfies

$$\lg n! \le h. \tag{4.6.1}$$

Since the vertices of T correspond to comparisons, in the worst case, the algorithm requires h comparisons. Therefore,

$$h = f(n). \tag{4.6.2}$$

Now,

$$n! = n(n - 1)(n - 2) \cdots 2 \cdot 1$$

$$\ge n(n - 1) \cdots \left\lceil \frac{n}{2} \right\rceil$$

$$\ge \frac{n}{2} \frac{n}{2} \cdots \frac{n}{2} = \left(\frac{n}{2}\right)^{n/2}.$$

Therefore,

$$\frac{n}{2} \lg \frac{n}{2} = \lg \left(\frac{n}{2}\right)^{n/2} \le \lg n! \tag{4.6.3}$$

Finally,

$$\tfrac{1}{2}n[(\lg n) - 1] = \tfrac{1}{2}n[\lg n - \lg 2]$$

$$= \frac{n}{2} \lg \left(\frac{n}{2}\right). \qquad (4.6.4)$$

Now (4.6.1)–(4.6.4) give the desired result. ∎

Theorem 4.6.3 implies that we cannot improve on $O(n \lg n)$ comparisons, in the worst case of a sorting algorithm. (See Exercise 23 for a precise statement of this fact.) For example, there is no sorting algorithm that uses $O(n)$ comparisons in the worst case. If there were and $f(n)$ is the number of comparisons needed in the worst case, we would have

$$\tfrac{1}{2}n[(\lg n) - 1] \le f(n) \le Cn$$

for all but finitely many n. This implies that

$$\tfrac{1}{2}[(\lg n) - 1] \le C$$

for all but finitely many n, which is impossible.

Since the bubble sort (Algorithm 4.6.1) requires $O(n^2)$ comparisons to sort n items, it is not optimal. On the other hand, Algorithm 4.6.6, based on merging two sorted lists (Algorithm 4.6.4), achieves the optimal $O(n \lg n)$ comparisons in the worst case.

ALGORITHM 4.6.4 Merging Two Arrays. Given two arrays

$$A(1), \ldots, A(M), \qquad B(1), \ldots, B(N),$$

sorted in increasing order, this algorithm creates an array

$$C(1), \ldots, C(M + N)$$

consisting of the elements of A and B arranged in increasing order.

1. [Initialization.] $I := 1; J := 1; K := 1.$ (I is the position in the A-array; J is the position in the B-array; K is the position in the C-array.)

2. [Done with A or B?] If $I > M$ (all elements in A are in C) or $J > N$ (all elements in B are in C), go to step 5.

3. [Put smallest in C.] If $A(I) < B(J)$, then $C(K) := A(I); I := I + 1;$ otherwise, $C(K) := B(J); J := J + 1.$

4. [Increment K.] $K := K + 1;$ go to step 2.

5. [Done with A?] If $I > M$, go to step 7.

6. [Copy end of A array.] (At this point, all elements of B have been put into C.) $C(K) := A(I); I := I + 1; K := K + 1;$ go to step 5.

7. [Done with B?] If $J > N$, stop.

8. [Copy end of B array.] (At this point, all elements of A have been put into C.) $C(K) := B(J); J := J + 1; K := K + 1,$ go to step 7.

EXAMPLE 4.6.5. Figure 4.6.3 shows how Algorithm 4.6.4 would merge the sequences

$$1 \quad 3 \quad 4 \qquad 2 \quad 4 \quad 5 \quad 6.$$

In Algorithm 4.6.4, the comparison occurs at step 3. Step 3 is in a loop consisting of steps 2, 3, and 4. This loop will execute as long as $I \leq M$ and $J \leq N$. Since initially I and J are both 1, the loop will execute at most $M + N - 1$ times. Thus Algorithm 4.6.4 requires at most $M + N - 1$ comparisons.

We will use Algorithm 4.6.4 to construct a divide-and-conquer sorting algorithm. Suppose that the array to be sorted,

$$S(1), \ldots, S(N),$$

is divided into two arrays

$$S(1), \ldots, S\left(\left\lfloor \frac{N}{2} \right\rfloor\right), \qquad S\left(\left\lfloor \frac{N}{2} \right\rfloor + 1\right), \ldots, S(N)$$

and that each of these arrays is sorted. We could then use Algorithm 4.6.4 to merge these two arrays into a single, sorted array. This is the idea behind the merge sort. We give a recursive version.

ALGORITHM 4.6.6 Merge Sort. This recursive algorithm sorts an array into increasing order by using Algorithm 4.6.4 which merges two sorted arrays. The input is the array $S(1), \ldots, S(N)$.

1. [Trivial?] If $N = 1$, return.

2. [Divide array and sort.] Set $M := \lfloor N/2 \rfloor$. Call Algorithm 4.6.6 with input $S(1), \ldots, S(M)$. Call Algorithm 4.6.6 with input $S(M + 1), \ldots, S(N)$.

3. [Merge.] Call Algorithm 4.6.4 with

$$A: S(1), \ldots, S(M) \qquad \text{and} \qquad B: S(M + 1), \ldots, S(N).$$

Copy C, which is returned by Algorithm 4.6.4, into S and return.

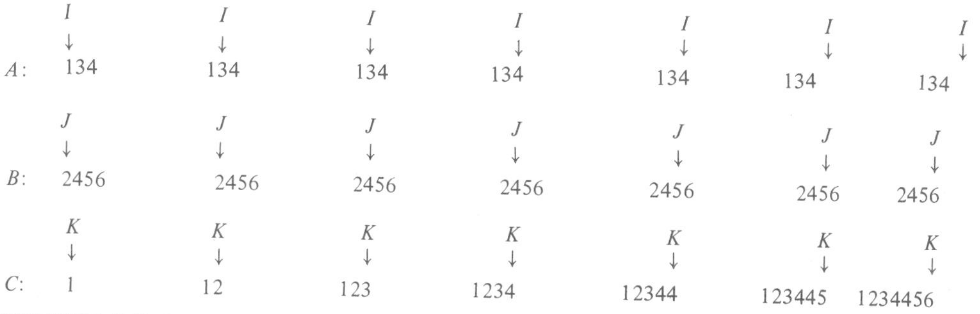

	I	I	I	I	I	I	I
	↓	↓	↓	↓	↓	↓	↓
A:	134	134	134	134	134	134	134
	J	J	J	J	J	J	J
	↓	↓	↓	↓	↓	↓	↓
B:	2456	2456	2456	2456	2456	2456	2456
	K	K	K	K	K	K	K
	↓	↓	↓	↓	↓	↓	↓
C:	1	12	123	1234	12344	123445	1234456

FIGURE 4.6.3

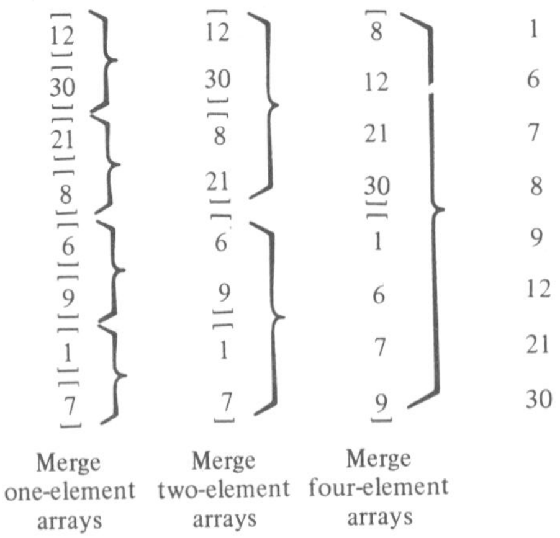

Merge one-element arrays　　Merge two-element arrays　　Merge four-element arrays

FIGURE 4.6.4

Let us examine Algorithm 4.6.6 for some simple cases. If $N = 1$, Algorithm 4.6.6 simply returns at step 1. Next, suppose that $N = 2$. At step 2, the array is divided into two arrays containing the single elements $S(1)$ and $S(2)$. Since Algorithm 4.6.6 returns on input consisting of a single element, at step 2, when Algorithm 4.6.6 is called, nothing changes. At step 3, we merge $S(1)$ and $S(2)$. Now the array $S(1)$, $S(2)$ is sorted.

EXAMPLE 4.6.7.　Figure 4.6.4 shows how Algorithm 4.6.6 sorts the sequence

$$12 \quad 30 \quad 21 \quad 8 \quad 6 \quad 9 \quad 1 \quad 7.$$

Our next theorem shows that Algorithm 4.6.6 is optimal.

THEOREM 4.6.8.　*Merge sort (Algorithm 4.6.6) requires at most $4 N \lg N$ comparisons to sort N items in the worst case.*

Proof.　Let a_N be the number of comparisons required by Algorithm 4.6.6 to sort N items in the worst case. Then, $a_1 = 0$. If $N > 1$, at step 2 at most

$$a_{\lfloor N/2 \rfloor} + a_{\lfloor (N+1)/2 \rfloor}$$

comparisons are required. At step 3, at most

$$\left\lfloor \frac{N}{2} \right\rfloor + \left\lfloor \frac{N+1}{2} \right\rfloor - 1 = N - 1$$

comparisons are required. Therefore,

$$a_N \leq a_{\lfloor N/2 \rfloor} + a_{\lfloor (N+1)/2 \rfloor} + N. \tag{4.6.5}$$

We convert (4.6.5) to a recurrence relation by replacing "≤" by "=."
More precisely, we *define* the sequence $b_1, b_2, \ldots$ by the equations

$$b_1 = 0$$

$$b_N = b_{\lfloor N/2 \rfloor} + b_{\lfloor (N+1)/2 \rfloor} + N.$$

If $N = 2^k$ is a power of 2, the second equation above becomes

$$b_{2^k} = 2b_{2^{k-1}} + 2^k.$$

We may solve this last equation by using iteration (see Section 2.6):

$$b_{2^k} = 2b_{2^{k-1}} + 2^k = 2[2b_{2^{k-2}} + 2^{k-1}] + 2^k$$

$$= 2^2 b_{2^{k-2}} + 2 \cdot 2^k = 2^2[2b_{2^{k-3}} + 2^{k-2}] + 2 \cdot 2^k$$

$$= 2^3 b_{2^{k-3}} + 3 \cdot 2^k = \cdots = 2^k b_{2^0} + k2^k = k2^k. \quad (4.6.6)$$

We can use this result to show that $b_N \leq 4 N \lg N$, for $N = 1, 2, 3, \ldots$.
Since this inequality holds for $N = 1, 2$, we may assume that $N > 2$. Using
induction, it can be shown (see Exercise 8) that

$$b_N \leq b_{N+1'} \quad (4.6.7)$$

for $N = 1, 2, 3, \ldots$. Let N be a positive integer greater than 2. Choose k
so that $2^k < N \leq 2^{k+1}$. Using (4.6.6) and (4.6.7), we obtain

$$b_N \leq b_{2^{k+1}} = (k + 1)2^{k+1} \leq (k + k)2^{k+1} = 4(2^k k) \leq 4 N \lg N.$$

Another induction argument (see Exercise 9) shows that $a_N \leq b_N$, for
$N = 1, 2, 3, \ldots$. Therefore, for any N,

$$a_N \leq b_N \leq 4 N \lg N. \quad \blacksquare$$

Even though merge sort, Algorithm 4.6.6, is optimal, it may not be the
algorithm of choice for a particular sorting problem. Factors such as the average-
case performance of the algorithm, the number of items to be sorted, available
memory, the data structures to be used, whether the items to be sorted are in
memory or reside on peripheral storage devices such as disks or tapes, whether
the items to be sorted are already "nearly" sorted, and the hardware to be used
must be taken into account.

EXERCISES

1H. Show how merge sort (Algorithm 4.6.6) sorts the sequence 1   9   7   3.

2. Show how merge sort (Algorithm 4.6.6) sorts the sequence 2   3   7   2   8
9   7   5   4.

3. Represent bubble sort (Algorithm 4.6.1) for size $n = 4$ by a binary tree.

★4. Use induction to prove that

$$n! > \left(\frac{n}{2}\right)^{n/2}$$

5H. Suppose that we have two arrays both of size n and both sorted in increasing order.

(a) Under what conditions does the maximum number of comparisons occur in Algorithm 4.6.4?

(b) Under what conditions does the minimum number of comparisons occur in Algorithm 4.6.4?

6. What is the minimum number of comparisons required by Algorithm 4.6.6 to sort an array of size 6?

7H. What is the maximum number of comparisons required by Algorithm 4.6.6 to sort an array of size 6?

8. Let b_N be as in the proof of Theorem 4.6.8. Show that $b_N \leq b_{N+1}$ for $N = 1$, $2, 3, \ldots$.

9H. Let a_N and b_N be as in the proof of Theorem 4.6.8. Show that $a_N \leq b_N$ for $N = 1, 2, 3, \ldots$.

10. Let a_N denote the number of comparisons required by merge sort in the worst case. Show that $a_N \leq 3N \lg N$ for $N = 1, 2, 3, \ldots$.

11H. Suppose that algorithm A requires $\lceil n \lg n \rceil$ comparisons to sort n items and algorithm B requires $\lceil n^2/4 \rceil$ comparisons to sort n items. For which n is algorithm B superior to algorithm A?

Exercises 12–15 concern the **Quicksort** algorithm (Algorithm 4.6.9).

ALGORITHM 4.6.9 Quicksort. This algorithm recursively sorts the array

$$S(1), \ldots, S(N)$$

of N items into increasing order.

1. [Trivial?] If $N \leq 1$, return. ($N = 0$ means that the array is empty.)

2. [Initialization.] Set $I := 1$; $J := N$. (I moves right and J moves left along the array as elements are interchanged. When $I = J$, the element at the Ith position will be greater than all elements to the left and less than all elements to the right.)

3. [Move J?] If $S(J) > S(1)$, then $J := J - 1$; go to 3.

4. [Move I?] If $I = J$, go to step 6; otherwise, $I := I + 1$.

5. [Interchange?] If $S(I) < S(1)$, go to 4; otherwise, (interchange) TEMP $:= S(I)$; $S(I) := S(J)$; $S(J) := $ TEMP; go to step 3.

6. [I met J.] (Interchange.) TEMP $:= S(1)$; $S(1) := S(I)$; $S(I) := $ TEMP.

7. [Sort subproblems.] Call Algorithm 4.6.9 with input the array $S(1)$, $\ldots, S(I - 1)$. Call Algorithm 4.6.9 with input the array $S(I + 1), \ldots$, $S(N)$. Return.

12. Show how Quicksort would sort the array 4 3 1 6 7 2 5 8.

13H. Draw the binary tree representation of Quicksort for the case $n = 3$.

14. Show that Quicksort sorts n items.

15H. Which sequences produce the maximum number of comparisons in Quicksort? How many comparisons are needed?

> Exercises 16–22 refer to **tournament sort** described below.
>
> TOURNAMENT SORT.   We are given a sequence
>
> $$S(1), \ldots , S(2^k)$$
>
> to sort in increasing order.
>
> We will build a binary tree with terminal vertices labeled $S(1), \ldots , S(2^k)$. An example is shown.

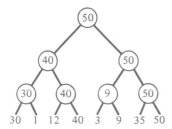

Working left to right, create a parent for each pair and label it with the maximum of the children. Continue in this way until reaching the root. At this point, the largest value, M, has been found.

To find the second largest value, first pick a value V less than all the items in the sequence. Replace the terminal vertex W containing M with V. Relabel the vertices by following the path from W to the root, as shown.

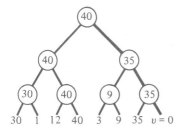

At this point, the second largest value is found. Continue until the sequence is ordered.

16. Why is the name "tournament" appropriate?

17H. Draw the two trees that would be created after the last tree above when tournament sort is applied.

18. How many comparisons does tournament sort require to find the largest element?

19H. Show that any algorithm that finds the largest value among n items requires at least $n - 1$ comparisons.

20. How many comparisons does tournament sort require to find the second largest element?

21. Write tournament sort as an algorithm.

22. Show that if n is a power of 2, tournament sort requires $O(n \lg n)$ comparisons.

23H. Show that if $g(n)$ is a positive function on the positive integers, which is significantly smaller than $n \lg n$, there is no comparison-based sorting algorithm that is $O(g(n))$ in the worst case. Specifically, assume that for any $\epsilon > 0$, no matter how small, there exists a positive integer N such that, if $n > N$, then

$$\frac{g(n)}{n \lg n} < \epsilon.$$

Show that there is no comparison-based sorting algorithm that is $O(g(n))$ in the worst case.

Exercises 24–30 refer to the following situation. We let P_n denote a particular problem of size n. If P_n is divided into subproblems of sizes i and j, there is an algorithm that combines the solutions of these two subproblems into a solution to P_n in time at most $2 + \lg (ij)$. Assume that a problem of size 1 is already solved.

24. Write a recursive algorithm to solve P_n similar to Algorithm 4.6.6.

25H. Let a_n be the worst-case time to solve P_n by the algorithm of Exercise 24. Show that

$$a_n \le a_{\lfloor n/2 \rfloor} + a_{\lfloor (n+1)/2 \rfloor} + 2 \lg n.$$

26. Let b_n be the recurrence relation obtained from Exercise 25 by replacing "$\le$" by "$=$." Assume that $b_1 = a_1 = 0$. Show that if n is a power of 2,

$$b_n = 4n - 2 \lg n - 4.$$

27H. Show that $a_n \le b_n$ for $n = 1, 2, 3, \ldots$.

28. Show that $b_n \le b_{n+1}$ for $n = 1, 2, 3, \ldots$.

29H. Show that $a_n \le 8n$ for $n = 1, 2, 3, \ldots$.

30H. Let G be a weighted graph with e edges. Show that Kruskal's algorithm (given before Exercise 20, Section 4.4) requires at least $O(e \lg e)$ comparisons in the worst case to find a minimal spanning tree for G.

31. [*Project*] Investigate other sorting algorithms. Consider specifically complexity, empirical studies, and special features of the algorithms (see [Knuth, 1973 Vol. 3]).

Game Trees

Trees are useful in the analysis of games such as tic-tac-toe, chess, and checkers, in which players alternate moves. In this section we show how trees can be used to develop game-playing strategies. This kind of approach is used in the development of many computer programs that allow human beings to play against computers or even computer against computer.

As an example of the general approach, consider the game of nim. Initially, there are n piles, each containing a number of identical tokens. Players alternate moves. A move consists of removing one or more tokens from any one pile. The player who removes the last token loses. As a specific case, consider an initial distribution consisting of two piles: one containing three tokens and one containing two tokens. All possible move sequences can be listed in a **game tree** (see Figure 4.7.1). The first player is represented by a box and the second player is represented by a circle. Each vertex shows a particular position in the game. In our game, the initial position is shown as $\binom{3}{2}$. A path represents a sequence of moves. If a position is shown in a square, it is the first player's move; if a position is shown in a circle, it is the second player's move. A terminal vertex represents the end of the game. In nim, if the terminal vertex is a circle, the first player removed the last token and lost the game. If the terminal vertex is a box, the second player lost.

The analysis begins with the terminal vertices. We label each terminal vertex with the value of the position to the first player. If the terminal vertex is a circle, since the first player lost, this position is worthless to the first player and we assign it the value 0 (see Figure 4.7.2). If the terminal vertex is a box, since the first player won, this position is valuable to the first player and we label it with a value greater than 0, say 1 (see Figure 4.7.2). At this point, all terminal vertices have been assigned values.

Now consider the problem of assigning values to the internal vertices. Suppose, for example, that we have an internal box, all of whose children have been assigned a value. For example, if we have the situation shown in Figure 4.7.3, the first player (box) should move to the position represented by vertex B, since this position is the most valuable. In other words, box moves to a position represented by a child with the maximum value. We assign this maximum value to the box vertex.

Consider the situation from the second (circle) player's point of view. Suppose that we have the situation shown in Figure 4.7.4. Circle should move to the position represented by vertex C, since this position is least valuable to box and therefore most valuable to circle. In other words, circle moves to a position represented by a child with the minimum value. We assign this minimum value to the circle vertex. The process by which circle seeks the minimum of its

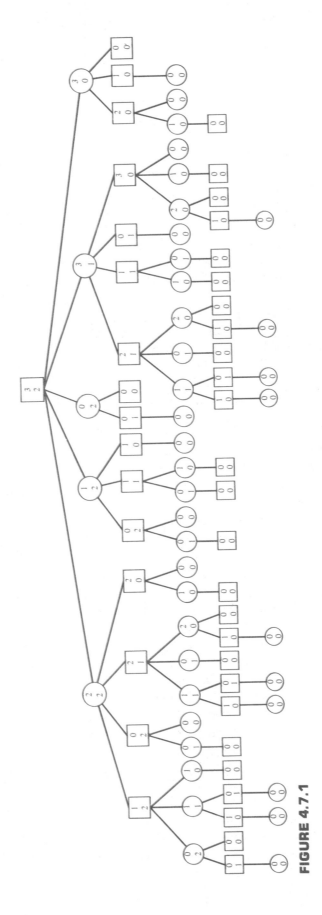

FIGURE 4.7.1

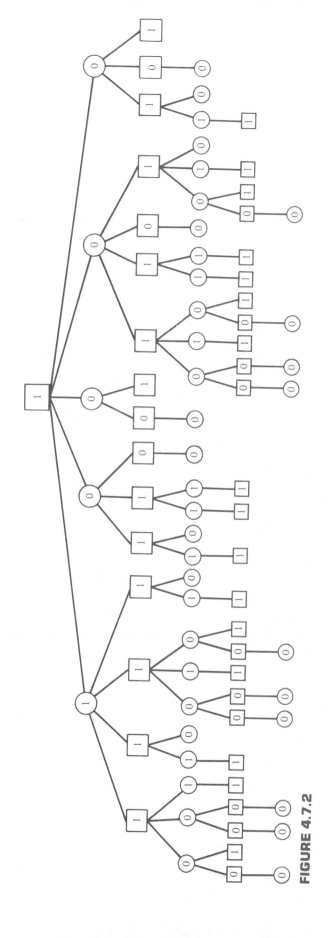

FIGURE 4.7.2

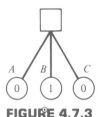

FIGURE 4.7.3

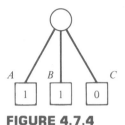

FIGURE 4.7.4

children and box seeks the maximum of its children is called the **minimax procedure**.

Working upward from the terminal vertices and using the minimax procedure, we can assign values to all of the vertices in the game tree (see Figure 4.7.2). These numbers represent the value of the game, at any position, to the first player. Notice that the root in Figure 4.7.2, which represents the original position, has a value of 1. This means that the first player can always win the game by using an optimal strategy. This optimal strategy is contained in the game tree: The first player always moves to a position that maximizes the value of the children. No matter what the second player does, the first player can always move to a vertex having value 1. Ultimately, a terminal vertex having value 1 is reached where the first player wins the game.

Interesting games, such as chess, have game trees so large that it is not feasible to use a computer to generate the entire tree. Conversely, any game having a tree small enough for a computer to generate would not be sufficiently challenging to sustain anyone's interest for very long. Nevertheless, the concept of a game tree is still useful for analyzing nontrivial games.

When using a game tree, we should use a depth-first search. If the game tree is so large that it is not feasible to reach a terminal vertex, we limit the level to which depth-first search is carried out. The search is said to be an **n-level search** if we limit the search to n levels below the given vertex. Since the vertices at the lowest level may not be terminal vertices, some method must be found to assign them a value. Here is where the specifics of the game must be dealt with. An **evaluation function** E is constructed that assigns each possible game position P the value $E(P)$ of the position to the first player. After the vertices at the lowest level are assigned values by using the function E, the minimax procedure can be applied to generate the values of the other vertices. We illustrate these concepts with an example.

EXAMPLE 4.7.1.   Find the value of the root in tic-tac-toe using a two-level, depth-first minimax search. Use the evaluation function E, which assigns a position the value

$$NX - NO$$

where NX (respectively, NO) is the number of rows, columns, or diagonals containing an X (respectively, O) that X (respectively, O) might complete. For example, position P of Figure 4.7.5 has $NX = 2$, since X might complete

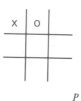

FIGURE 4.7.5

the column or the diagonal, and $NO = 1$, since O can only complete a column. Therefore,

$$E(P) = 2 - 1 = 1.$$

In Figure 4.7.6, we have drawn the game tree for tic-tac-toe to level 2. We have omitted symmetric positions. We first assign the vertices at level 2 the values given by E (see Figure 4.7.7). Next, we compute circle's values by minimizing over the children. Finally, we compute the value of the root by maximizing over the children. Using this analysis, the first move by the first player would be to the center square.

Evaluation of a game tree, or even a part of a game tree, can be a time-consuming task, so any technique that reduces the effort is welcomed. The most general technique is called **alpha-beta pruning**. In general, alpha-beta pruning allows us to bypass many vertices in a game tree, yet still find the value of a vertex. The value obtained is the same as if we had evaluated all the vertices.

As an example, consider the game tree in Figure 4.7.8. Suppose that we want to evaluate vertex A using a two-level, depth-first search. We begin by evaluating the vertices, E, F, and G. The values shown are obtained from an evaluation function. Vertex B is 2, the minimum of its children. At this point, we know that the value x of A must be at least 2, since the value of A is the maximum of its children; that is,

$$x \geq 2. \tag{4.7.1}$$

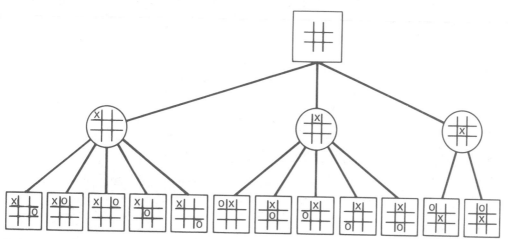

FIGURE 4.7.6

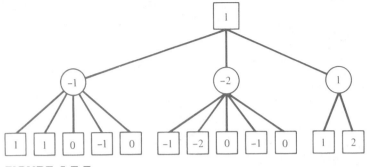

FIGURE 4.7.7

This lower bound for A is called an **alpha value** of A. The next vertices to be evaluated are H, I, and J. When I evaluates to 1, we know that the value y of C cannot exceed 1, since the value of C is the minimum of its children; that is

$$y \le 1. \tag{4.7.2}$$

It follows from (4.7.1) and (4.7.2) that whatever the value of y is, it will not affect the value of x; thus we need not concern ourselves further with the subtree headed by vertex C. We say that an **alpha cutoff** occurs. We next evaluate the children of D and then D itself. Finally, we find that the value of A is 3.

To summarize, an alpha cutoff occurs at a box vertex v when a grandchild w of v has a value less than or equal to the alpha value of v. The subtree whose root is the parent of w may be deleted (pruned). This deletion will not affect the value of v. An alpha value for a vertex v is only a lower bound for the value of v. The alpha value of a vertex is dependent on the current state of the search and changes as the search progresses.

Similarly, a **beta cutoff** occurs at a circle vertex v when a grandchild w of v has a value greater than or equal to the beta value of v. The subtree whose root is the parent of w may be pruned. This deletion will not affect the value of v. A **beta value** for a vertex v is only an upper bound for the value of v. The beta value of a vertex is dependent on the current state of the search and changes as the search progresses.

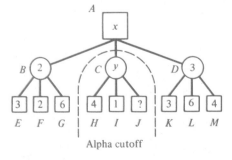

FIGURE 4.7.8

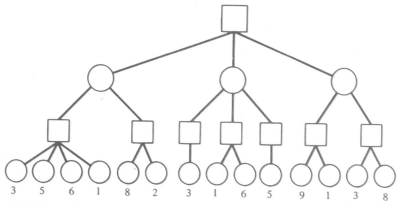

FIGURE 4.7.9

EXAMPLE 4.7.2. Evaluate the root of the tree of Figure 4.7.9 using depth-first search with alpha-beta pruning. For each vertex whose value is computed, write the value in the vertex. Place a check by the root of each subtree that is pruned. The value of each terminal vertex is written under the vertex.

We begin by evaluating vertices A, B, C, and D (see Figure 4.7.10). Next, we find that the value of E is 6. This results in a beta value of 6 for F. Next, we evaluate vertex G. Since its value is 8 and 8 exceeds the beta value of F, we obtain a beta cutoff and prune the subtree with root H. The value of F is 6. This results in an alpha value of 6 for I. Next, we evaluate vertices J and K. Since the value 3 of K is less than the alpha value 6 of I, an alpha cutoff occurs and the subtree with root L may be pruned. Next, we evaluate M, N, O, P, Q, R, and S. No further pruning is possible. Finally, we determine that the root I has value 8.

It has recently been shown (see [Pearl]) that for game trees in which every parent has n children and in which the terminal values are randomly ordered,

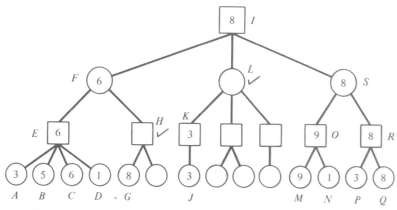

FIGURE 4.7.10

for a given amount of time, the alpha-beta procedure permits a search depth greater by $\frac{4}{3}$ than the pure minimax procedure, which evaluates every vertex. Pearl also shows that for such game trees, the alpha-beta procedure is optimal.

Other techniques have been combined with alpha-beta pruning to facilitate the search of a game tree. One idea is to order the children of the vertices to be evaluated so that the most promising moves are examined first (see Exercises 21–24). Another idea is to allow a variable-depth search in which the search backtracks when it reaches an unpromising position as measured by some function.

Some game-playing programs have been quite successful. A computer program beat the top human player in the world in backgammon (see [Berliner]). The best computer chess programs play at a level considerably above average human players. Success has been more elusive in card games. Although computers have been programmed to bid bridge hands at a competitive level, no computer program yet plays bridge hands (declarer or defender) at a very respectable level.

EXERCISES

1H. Draw the complete game tree for a version of nim in which the initial position consists of one pile of six tokens and a turn consists of taking one, two, or three tokens. Assign values to all vertices so that the resulting tree is analogous to Figure 4.7.2. Assume that the last player to take a token loses. Will the first or second player, playing an optimal strategy, always win? Describe an optimal strategy for the winning player.

2. Draw the complete game tree for nim in which the initial position consists of two piles of three tokens each. Omit symmetric positions. Assume that the last player to take a token loses. Assign values to all vertices so that the resulting tree is analogous to Figure 4.7.2. Will the first or second player, playing an optimal strategy, always win? Describe an optimal strategy for the winning player.

3H. Draw the complete game tree for nim in which the initial position consists of two piles, one containing three tokens and the other containing two tokens. Assume that the last player to take a token wins. Assign values to all vertices so that the resulting tree is analogous to Figure 4.7.2. Will the first or second player, playing an optimal strategy, always win? Describe an optimal strategy for the winning player.

4. Draw the complete game tree for nim in which the initial position consists of two piles of three tokens each. Omit symmetric positions. Assume that the last player to take a token wins. Assign values to all vertices so that the resulting tree is analogous to Figure 4.7.2. Will the first or second player, playing an optimal strategy, always win? Describe an optimal strategy for the winning player.

5H. Draw the complete game tree for the version of nim described in Exercise 1. Assume that the last person to take a token wins. Assign values to all vertices so that the resulting tree is analogous to Figure 4.7.2. Will the first or second player, playing an optimal strategy, always win? Describe an optimal strategy for the winning player.

Exercises 6 and 7 refer to nim and nim'. Nim is the game using n piles of tokens as described in this section in which the last player to move loses. Nim' is the game using n piles of tokens as described in this section except that the last player to move wins. We fix n piles each with a fixed number of tokens. We assume that at least one pile has at least two tokens.

★6. Show that the first player can always win nim if and only if the first player can always win nim'.

★7H. Given a winning strategy for a particular player for nim, describe a winning strategy for this player for nim'.

Evaluate each vertex in each game tree. The values of the terminal vertices are given.

8.

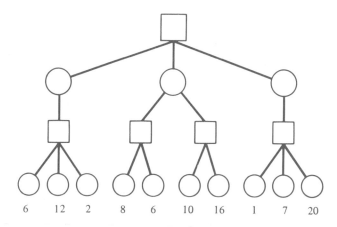

9H.

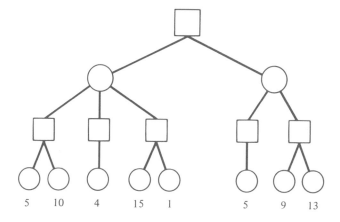

10.

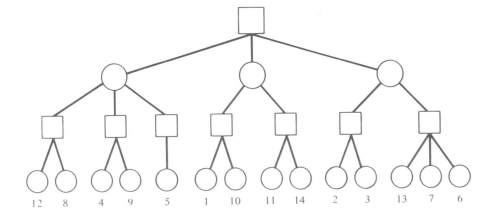

11H.

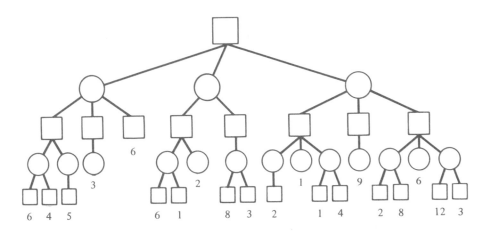

12.

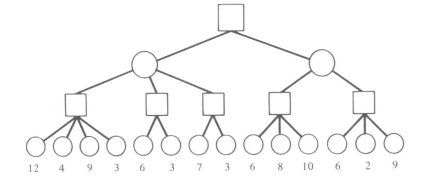

13H. Evaluate the root of each of the trees of Exercises 8–12 using a depth-first search with alpha-beta pruning. For each vertex whose value is computed, write the value in the vertex. Place a check by the root of each subtree that is pruned. The value of each terminal vertex is written under the vertex.

In Exercises 14–16, determine the value of the tic-tac-toe position using the evaluation function of Example 4.7.1.

14.

		O
	X	

15H.

		O
	X	O
X		

16.

O	X	O
	X	
	O	X

17H. Assume that the first player moves to the center square in tic-tac-toe. Draw a two-level game tree, with the root having an X in the center square. Omit symmetric positions. Evaluate all the vertices using the evaluation function of Example 4.7.1. Where will O move?

★18. Would a two-level search program based on the evaluation function E of Example 4.7.1 play a perfect game of tic-tac-toe? If not, can you alter E so that a two-level search program will play a perfect game of tic-tac-toe?

19H. Write an algorithm that evaluates vertices of a game tree to level n using depth-first search. Assume the existence of an evaluation function E.

★20. Write an algorithm that evaluates the root of a game tree using an n-level, depth-first search with alpha-beta pruning. Assume the existence of an evaluation function E.

The following approach often leads to more pruning than pure alpha-beta minimax. First, perform a two-level search. At this point, all the children of the root will have values. Next, order the children of the root with the most promising moves first. Now, use an n-level depth-first search with alpha-beta pruning.

Carry out this procedure for $n = 4$ for each game tree of Exercises 21–23. Place a check by the root of each subtree that is pruned. The value of each vertex, as given by the evaluation function, is given under the vertex.

21H.

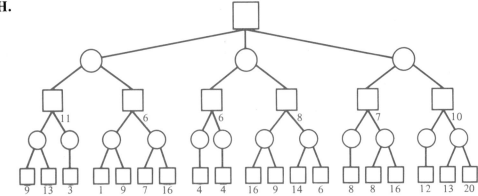

22.

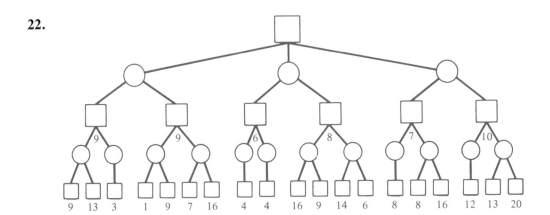

23.

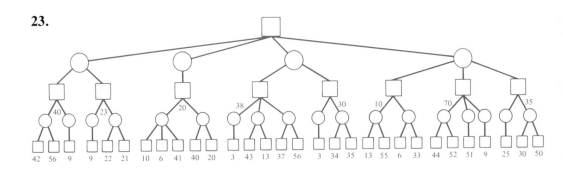

24. Write an algorithm to carry out the procedure described before Exercise 21.

Mu Torere is a two-person game played by the Maoris (see [Bell]). The board is an eight-pointed star (see below) with a circular area in the center known as the *putahi*. The first player has four black tokens and the second player has four white tokens. The initial position is shown. A player who cannot make a move loses. Players alternate moves. At most one token can occupy a point of the star or the putahi. A move consists of:

(a) Moving to an adjacent point

(b) Moving from the putahi to a point

(c) Moving from a point to the putahi provided that one or both of the adjacent points contain the opponent's pieces

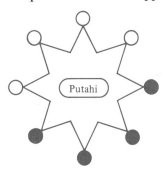

★**25.** Develop an evaluation function for Mu Torere.

★**26.** Combine the evaluation function in Exercise 25 with a two-level search of the game tree to obtain a game-playing algorithm for Mu Torere. Evaluate the game-playing ability of this algorithm.

★**27.** Can the first player always win in Mu Torere?

★**28.** Can the first player always tie in Mu Torere?

★**29.** [*Project*] Develop an evaluation function for Kalah. (See [Ainslie] for the rules.)

★**30.** Develop a game-playing algorithm for Kalah based on the evaluation function of Exercise 29. Evaluate the game-playing ability of this algorithm.

4.8

Notes

The following are recommended references on trees: [Berge; Berztiss; Busacker; Deo; Harary; Horowitz, 1976; Knuth, 1973 Vol. 1; Lipschutz, 1976; Liu, 1985; Ore; Sahni; Standish; Tremblay; and Tucker].

See [Tsichritzis] for the use of trees in hierarchical data bases.

Decision trees are covered in [Standish].

The use of Huffman codes in hand-held computers is detailed in [Williams]. See [Standish] for additional information on Huffman codes.

[Golomb] describes backtracking and contains several examples and applications.

A nice discussion of minimal spanning tree algorithms can be found in [Baase].

The best reference on sorting is [Knuth, 1973 Vol. 3]. [Nievergelt] contains a good summary of the major sorting algorithms. In [Grillo], the major sorting algorithms are written in BASIC and some time comparisons are made.

Good references on game trees are [Horowitz, 1976; Nievergelt; Nilsson; and Slagle]. In [Frey], the minimax procedure is applied to a simple game. Various methods to speed up the search of the game tree are discussed and compared. Programs are given in BASIC. [Berlekamp] contains a general theory of games as well as analyses of many specific games.

COMPUTER EXERCISES

1. Write a program that tests if a graph is a tree.

2. Write a program that, given the adjacency matrix of a free tree and a vertex, draws the rooted tree using a computer graphics display.

3. Write a computer program to encode and decode strings of characters according to the Huffman code given in the table.

Character	Huffman Code
(blank)	0
E	1100
T	1001
A	11111
O	11110
N	11100
R	11011
I	11010
S	10110
H	10101
D	111011
L	101111
F	101001
C	101000
M	100011
U	100010
G	100001
Y	100000
P	1110101
W	1011101
B	1011100
V	11101001
K	1110100011
X	1110100001
J	1110100000
Q	11101000101
Z	11101000100

4. Use your program of Exercise 3 to encode some sample text. Compare the number of bits used to encode your text with the number of bits necessary to encode your text in ASCII.

5. Write a program that, given a frequency table for characters, constructs an optimum Huffman code. (See [Standish] for the algorithm.)

6. Write a program that, given a tree T, computes the eccentricity of each vertex in T and finds the center(s) of T.

7. Write a program that, given a rooted tree and a vertex v,
   (a) Finds the parent of v
   (b) Finds the ancestors of v
   (c) Finds the children of v
   (d) Finds the descendants of v
   (e) Finds the siblings of v
   (f) Determines whether v is a terminal vertex

8. Write a program that accepts strings and puts them into a binary search tree.

9. Implement Algorithms 4.2.12, 4.3.6, 4.3.7, and 4.4.3 as programs.

10. Write a program that determines whether a graph is connected.

11. Write a program that finds the components of a graph.

12. Write a program to solve the n-queens problem.

13. Write a program that transforms a given position of the 8's puzzle to another given position or decides that no such transformation exists. (See Exercise 21, Section 4.3.)

14. The 15's puzzle is similar to the 8's puzzle (see Exercise 21, Section 4.3) except that there are 15 movable squares numbered 1 to 15 with one empty position on a 4 × 4 grid. Write a program that transforms a given position of the 15's puzzle to another given position or decides that no such transformation exists.

15. Write a backtracking program to find an $n \times n$ magic square.

16. Write a program that, given a graph G and a spanning tree for G, computes the fundamental circuit matrix of G.

17. Write a backtracking program that determines whether a graph can be colored with n colors and, if it can be colored with n colors, produces a coloring.

18. Write a backtracking program that determines whether a graph has a Hamiltonian circuit and, if there is a Hamiltonian circuit, finds it.

19. Implement Kruskal's Algorithm (given before Exercise 20, Section 4.4) as a program.

20. Implement preorder, postorder, and inorder tree traversals as programs.

21. Write a program that reconstructs a binary tree given its preorder and inorder vertex orderings.

22. Implement Exercises 28–30 and 32–37, Section 4.5, as programs.

23. Implement the bubble sort (Algorithm 4.6.1) as a program.

24. Write a program to merge two arrays (Algorithm 4.6.4).

25. Implement the merge sort (Algorithm 4.6.6) as a program.

26. Implement Quicksort (Algorithm 4.6.9) as a program.

27. Implement Tournament Sort as a program (see Exercises 16–22, Section 4.6).

28. Generate some random data and sort it using bubble sort, merge sort, Quicksort, and Tournament Sort (Exercises 23 and 25–27). Compare the times required by the sorting methods.

29. Write a program to generate the complete game tree for nim in which the initial position consists of two piles of four tokens each. Assume that the last player to take a token loses.

30. Implement the minimax procedure as a program.

31. Implement the minimax procedure with alpha-beta pruning as a program.

32. Implement the method of playing tic-tac-toe in Example 4.7.1 as a program.

33. [*Project*] Develop a computer program to play a game that has relatively simple rules. Suggested games are Othello, The Mill, Battleship, and Kalah. (See [Ainslie] and [Freeman] for rules and strategies.)

CHAPTER 5

Network Models and Petri Nets

In this chapter we discuss two topics, network models and Petri nets, that make use of directed graphs. The major portion of the chapter is devoted to the problem of maximizing the flow through a network. The network might be a transportation network through which commodities flow or a pipeline network through which oil flows, or any number of other possibilities. In each case the problem is to find a maximal flow. Many other problems, which on the surface seem not to be flow problems, can, in fact, be modeled as network flow problems.

Maximizing the flow in a network is a problem that belongs both to graph theory and to operations research. The traveling salesperson problem furnishes another example of a problem in graph theory and operations research. **Operations research** studies the very broad category of problems of optimizing the performance of a system. Typical problems studied in operations research are network problems, allocation of resources problems, and personnel assignment problems.

Petri nets model systems in which processing can occur concurrently. The model provides a framework for dealing with questions such as whether the system will deadlock and whether the capacities of components of the system will be exceeded.

Network Models

Consider the directed graph in Figure 5.1.1, which represents an oil pipeline network. Oil is unloaded at the dock a and pumped through the network to the refinery z. The vertices b, c, d, and e represent intermediate pumping stations. The directed edges represent subpipelines of the system and show the direction the oil can flow. The labels on the edges show the capacities of the subpipelines. The problem is to find a way to maximize the flow from the dock to the refinery and to compute the value of this maximum flow. Figure 5.1.1 provides an example of a **transport network**.

> **DEFINITION 5.1.1.**   A *transport network* (or more simply *network*) is a simple, weighted, directed graph G satisfying:
>
>   (a) There is exactly one vertex in G, called the *source*, having no incoming edges.
>   (b) There is exactly one vertex in G, called the *sink*, having no outgoing edges.
>   (c) The weight C_{ij} of the directed edge (i, j), called the *capacity* of (i, j), is a nonnegative number.
>   (d) The undirected graph obtained from G by ignoring the directions of the edges is connected.

> **EXAMPLE 5.1.2.**   The graph of Figure 5.1.1 is a transport network. The source is vertex a and the sink is vertex z. The capacity of edge (a, b), C_{ab}, is 3 and the capacity of edge (b, c), C_{bc}, is 2.

Throughout this chapter, if G is a network, we will denote the source by a and the sink by z.
  A **flow** in a network assigns a flow in each directed edge which does not exceed the capacity of that edge. Moreover, it is assumed that the flow into a vertex v, which is neither the source nor the sink, is equal to the flow out of v. The next definition makes these ideas precise.

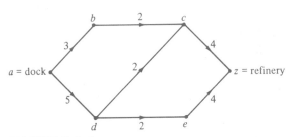

FIGURE 5.1.1

DEFINITION 5.1.3. Let G be a transport network. Let C_{ij} denote the capacity of the directed edge (i, j). A *flow F* in G assigns each directed edge (i, j) a nonnegative number F_{ij} such that:

(a) $F_{ij} \leq C_{ij}$.

(b) For each vertex j, which is neither the source nor the sink,

$$\sum_i F_{ij} = \sum_i F_{ji}. \qquad (5.1.1)$$

[In a sum such as (5.1.1), unless specified otherwise, the sum is assumed to be taken over all vertices i. Also, if (i, j) is not an edge, we set $F_{ij} = 0$.]

We call F_{ij} the *flow in edge* (i, j). For any vertex j, we call

$$\sum_i F_{ij}$$

the *flow into j* and we call

$$\sum_i F_{ji}$$

the *flow out of j*.

The property expressed by equation (5.1.1) is called **conservation of flow**. In the oil-pumping example of Figure 5.1.1, conservation of flow means that oil is neither used nor supplied at pumping stations b, c, d, and e.

EXAMPLE 5.1.4. The assignments,

$$F_{ab} = 2, \quad F_{bc} = 2, \quad F_{cz} = 3, \quad F_{ad} = 3,$$

$$F_{dc} = 1, \quad F_{de} = 2, \quad F_{ez} = 2,$$

define a flow for the network of Figure 5.1.1. For example, the flow into vertex d,

$$F_{ad} = 3,$$

is the same as the flow out of vertex d,

$$F_{dc} + F_{de} = 1 + 2 = 3.$$

In Figure 5.1.2 we have redrawn the network of Figure 5.1.1 to show the

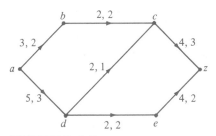

FIGURE 5.1.2

flow of Example 5.1.4. An edge e is labeled x, y if the capacity of e is x and the flow in e is y. This notation will be used throughout this chapter.

Notice that in Example 5.1.4, the flow out of the source a,

$$F_{ab} + F_{ad},$$

is the same as the flow into the sink z,

$$F_{cz} + F_{ez};$$

both values are 5. The next theorem shows that it is always true that the flow out of the source equals the flow into the sink.

THEOREM 5.1.5. *Given a flow F in a network, the flow out of the source a equals the flow into the sink z; that is,*

$$\sum_i F_{ai} = \sum_i F_{iz}.$$

Proof. Let V be the set of vertices. We have

$$\sum_{j \in V} \left(\sum_{i \in V} F_{ij} \right) = \sum_{j \in V} \left(\sum_{i \in V} F_{ji} \right),$$

since each double sum is

$$\sum_{e \in E} F_e$$

where E is the set of edges. Now

$$0 = \sum_{j \in V} \left(\sum_{i \in V} F_{ij} - \sum_{i \in V} F_{ji} \right)$$

$$= \left(\sum_{i \in V} F_{iz} - \sum_{i \in V} F_{zi} \right) + \left(\sum_{i \in V} F_{ia} - \sum_{i \in V} F_{ai} \right)$$

$$+ \sum_{\substack{j \in V \\ j \neq a, z}} \left(\sum_{i \in V} F_{ij} - \sum_{i \in V} F_{ji} \right)$$

$$= \sum_{i \in V} F_{iz} - \sum_{i \in V} F_{ai}$$

since $F_{zi} = 0 = F_{ia}$, for all $i \in V$, and (Definition 5.1.3b)

$$\sum_{i \in V} F_{ij} - \sum_{i \in V} F_{ji} = 0 \qquad \text{if } j \in V - \{a, z\}.$$
∎

In light of Theorem 5.1.5, we can make the following definition.

DEFINITION 5.1.6. Let F be a flow in a network G. The value

$$\sum_i F_{ai} = \sum_i F_{iz}$$

is called the *value of the flow F.*

EXAMPLE 5.1.7. The value of the flow in the network of Figure 5.1.2 is 5.

The problem for a transport network G may be stated: Find a maximal flow in G; that is, among all possible flows in G, find a flow F so that the value of F is a maximum. In the next section we will give an algorithm that efficiently solves this problem. We will conclude this section by giving additional examples.

EXAMPLE 5.1.8. Figure 5.1.3 represents a pumping network in which water for two cities, A and B, is delivered from three wells, w_1, w_2, and w_3. The capacities of the intermediate systems are shown on the edges. Vertices b, c, and d represent intermediate pumping stations. Model this system as a transport network.

The graph of Figure 5.1.3 is almost a transport network. It is not a transport network since there are multiple sources and sinks. We can produce an equivalent transport network by tying together the sources into a **supersource** and tying together the sinks into a **supersink** (see Figure 5.1.4). In Figure 5.1.4, ∞ represents an unlimited capacity.

EXAMPLE 5.1.9. It is possible to go from city A to city C directly or by going through city B. During the period 6:00 to 7:00 P.M., the average trip times are

$$A \text{ to } B\text{---}15 \text{ minutes}$$

$$B \text{ to } C\text{---}30 \text{ minutes}$$

$$A \text{ to } C\text{---}30 \text{ minutes.}$$

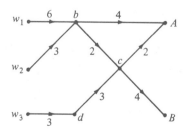

FIGURE 5.1.3

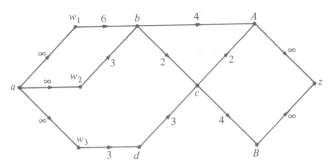

FIGURE 5.1.4

The maximum capacities of the routes are

$$A \text{ to } B\text{---}3000 \text{ vehicles}$$

$$B \text{ to } C\text{---}2000 \text{ vehicles}$$

$$A \text{ to } C\text{---}4000 \text{ vehicles.}$$

Represent the flow of traffic from A to C during the period 6:00 to 7:00 P.M. as a network.

A vertex will represent a city at a particular time (see Figure 5.1.5). An edge connects X, t_1 to Y, t_2 if we can leave city X at t_1 P.M. and arrive at city Y at t_2 P.M. The capacity of an edge is the capacity of the route. Edges of infinite capacity connect A, t_1 to A, t_2 and B, t_1 to B, t_2 to indicate that any number of cars can wait at city A or city B. Finally, we introduce a supersource and supersink.

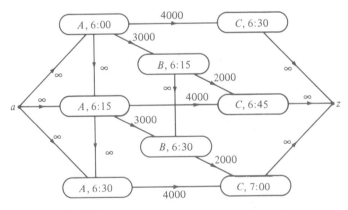

FIGURE 5.1.5

EXERCISES

In Exercises 1–3, fill in the missing edge flows so that the result is a flow in the given network. Determine the values of the flows.

1H.

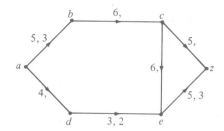

2.

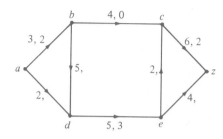

3H.

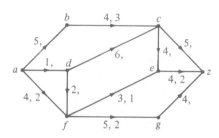

4. The accompanying graph represents a pumping network in which oil for three refineries, A, B, and C, is delivered from three wells, w_1, w_2, and w_3. The capacities of the intermediate systems are shown on the edges. Vertices b, c, d, e, and f represent intermediate pumping stations. Model this system as a network.

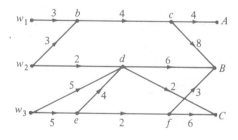

5H. Model the system of Exercise 4 as a network assuming that well w_1 can pump at most 2 units, well w_2 at most 4 units, and well w_3 at most 7 units.

6. Model the system of Exercise 5 as a network assuming, in addition to the limitations on the wells, that city A requires 4 units, city B requires 3 units, and city C requires 4 units.

7H. Model the system of Exercise 6 as a network assuming, in addition to the limitations on the wells and the requirements by the cities, that the intermediate pumping station d can pump at most 6 units.

8.   There are two routes from city A to city D. One route passes through city B and the other route passes through city C. During the period 7:00 to 8:00 A.M., the average trip times are

$$A \text{ to } B \text{---}30 \text{ minutes}$$

$$A \text{ to } C \text{---}15 \text{ minutes}$$

$$B \text{ to } D \text{---}15 \text{ minutes}$$

$$C \text{ to } D \text{---}15 \text{ minutes}.$$

The maximum capacities of the routes are

$$A \text{ to } B \text{---}1000 \text{ vehicles}$$

$$A \text{ to } C \text{---}3000 \text{ vehicles}$$

$$B \text{ to } D \text{---}4000 \text{ vehicles}$$

$$C \text{ to } D \text{---}2000 \text{ vehicles}.$$

Represent the flow of traffic from A to D during the period 7:00 to 8:00 A.M. as a network.

9H.   In the system shown, we want to maximize the flow from a to z. The capacities are shown on the edges. The flow between two vertices, neither of which is a or z, can be in either direction. Model this system as a network.

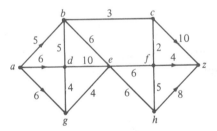

10.   Give an example of a network with exactly two maximal flows, where each F_{ij} is a nonnegative integer.

5.2

A Maximal Flow Algorithm

If G is a transport network, a **maximal flow** in G is a flow with maximum value. In general, there will be several flows having the same maximum value. In this section we give an algorithm for finding a maximal flow. The basic idea is simple—start with some initial flow and iteratively increase the value of the flow until no more improvement is possible. The resulting flow will then be a maximal flow.

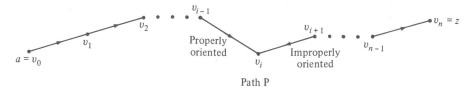

FIGURE 5.2.1

We can take the initial flow to be the flow in which the flow in each edge is zero. To increase the value of a given flow, we must find a path from the source to the sink and increase the flow along this path.

It is helpful at this point to introduce some terminology. Throughout this section, G denotes a network with source a, sink z, and capacity C. Momentarily, consider the edges of G to be undirected and let

$$P = (v_0, v_1, \ldots, v_n), \quad v_0 = a, \quad v_n = z,$$

be a path from a to z in this undirected graph. (All paths in this section are with reference to the underlying undirected graph.) If an edge e in P is directed from v_{i-1} to v_i, we say that e is **properly oriented (with respect to P)**; otherwise, we say that e is **improperly oriented (with respect to P)** (see Figure 5.2.1).

If we can find a path P from the source to the sink in which every edge in P is properly oriented and the flow in each edge is less than the capacity of the edge, it is possible to increase the value of the flow.

EXAMPLE 5.2.1. Consider the path from a to z in Figure 5.2.2. All the edges in P are properly oriented. The value of the flow in this network can be increased by 1, as shown in Figure 5.2.3.

It is also possible to increase the flow in certain paths from the source to the sink in which we have properly and improperly oriented edges. Let P be a path from a to z and let x be a vertex in P that is neither a nor z (see Figure 5.2.4). There are four possibilities for the orientations of the edges e_1 and e_2 incident on x. In case (a), both edges are properly oriented. In this case, if we increase the flow in each edge by Δ, the flow into x will still equal the flow out of x. In case (b), if we increase the flow in e_2 by Δ, we must *decrease* the flow in e_1 by Δ so that the flow into x will still equal the flow out of x. Case (c) is similar to case (b), except that we increase the flow in e_1 by Δ and decrease the flow in e_2 by Δ. In case (d), we decrease the flow in both edges by Δ. In every case, the resulting edge assignments give a flow. Of course, to carry out these alter-

FIGURE 5.2.2

FIGURE 5.2.3

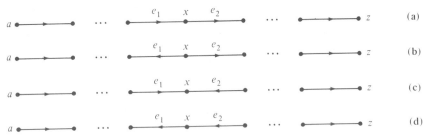

FIGURE 5.2.4

ations, we must have flow less than capacity in a properly oriented edge and a nonzero flow in an improperly oriented edge.

EXAMPLE 5.2.2.   Consider the path from a to z in Figure 5.2.5. Edges (a, b), (c, d), and (d, z) are properly oriented and edge (c, b) is improperly oriented. We decrease the flow by 1 in the improperly oriented edge (c, b) and increase the flow by 1 in the properly oriented edges (a, b), (c, d), and (d, z) (see Figure 5.2.6). The value of the new flow is 1 greater than the original flow.

We summarize the method of Examples 5.2.1 and 5.2.2 as a theorem.

THEOREM 5.2.3.   *Let P be a path from a to z in a network G satisfying:*
  (a) *For each properly oriented edge* (i, j) *in P,*

$$F_{ij} < C_{ij}.$$

  (b) *For each improperly oriented edge* (i, j) *in P,*

$$0 < F_{ij}.$$

Let

$$\Delta = \min X$$

where X consists of the numbers $C_{ij} - F_{ij}$, *for properly oriented edges* (i, j) *in P, and* F_{ij}, *for improperly oriented edges* (i, j) *in P. Define*

$$F^*_{ij} = \begin{cases} F_{ij} & \text{if } (i, j) \text{ is not in } P \\ F_{ij} + \Delta & \text{if } (i, j) \text{ is properly oriented in } P \\ F_{ij} - \Delta & \text{if } (i, j) \text{ is not properly oriented in } P. \end{cases}$$

Then F* is a flow whose value is Δ *greater than the value of F.*

FIGURE 5.2.5

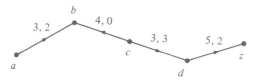

FIGURE 5.2.6

Proof. The argument that F^* is a flow is given just before Example 5.2.2. Since the edge (a, v) in P is increased by Δ, the value of F^* is Δ greater than the value of F. ∎

In the next section we will show that if there are no paths satisfying the conditions of Theorem 5.2.3, the flow is maximal. Thus it is possible to construct an algorithm based on Theorem 5.2.3. The outline is

1. Start with a flow (e.g., the flow in which the flow in each edge is 0).
2. Search for a path satisfying the conditions of Theorem 5.2.3. If no such path exists, stop; the flow is maximal.
3. Increase the flow through the path by Δ, where Δ is defined as in Theorem 5.2.3, and go to step 2.

In the formal algorithm, we search for a path satisfying the conditions of Theorem 5.2.3 while simultaneously keeping track of the quantities

$$C_{ij} - F_{ij}, F_{ij}.$$

ALGORITHM 5.2.4 Finding a Maximal Flow in a Network. We assume that G is a network with source a, sink z, and capacity C. The capacity of each edge is a nonnegative integer. The vertices of G are ordered $a = v_0$, . . . , $v_n = z$.

1. [Initialize flow.] Set $F_{ij} := 0$ for each edge (i, j).
2. [Label source.] Label vertex a $(, \infty)$.
3. [Sink labeled?] If the sink z is labeled, go to step 6.
4. [Next labeled vertex.] Choose the not yet examined, labeled vertex v_i with smallest index i. If none, stop (the flow is maximal); otherwise, set $v := v_i$.
5. [Label adjacent vertices.] Let (α, Δ) be the label of v. Examine each edge of the form (v, w), (w, v) [in the order (v, v_0), (v_0, v), (v, v_1), (v_1, v), . . .], where w is unlabeled. For an edge of the form (v, w), if $F_{vw} < C_{vw}$, label vertex w

$$(v, \min \{\Delta, C_{vw} - F_{vw}\});$$

if $F_{vw} = C_{vw}$, do not label w. For an edge of the form (w, v), if $F_{wv} > 0$, label vertex w

$$(v, \min \{\Delta, F_{wv}\});$$

if $F_{wv} = 0$, do not label w. Go to step 3.

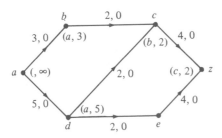

FIGURE 5.2.7

6. [Revise flow.] Let (γ, Δ) be the label of z. Let $w_0 = z$, $w_1 = \gamma$. If the label of w_i is (γ', Δ'), set $w_{i+1} = \gamma'$. Continue until $w_k = a$. Now

$$P: a = w_k, w_{k-1}, \ldots, w_1, w_0 = z$$

is a path from a to z. Change the flow of the edges in P as follows: If the edge e in P is properly oriented, increase the flow in e by Δ; otherwise, decrease the flow in e by Δ. Remove all labels from vertices and go to step 2.

A proof that Algorithm 5.2.4 terminates is left as an exercise (Exercise 21). If the capacities are allowed to be nonnegative rational numbers, the algorithm also terminates; however, if arbitrary nonnegative real capacities are allowed and we permit the edges in step 5 to be examined in any order, the algorithm may not terminate (see Exercise 22).

Algorithm 5.2.4 is often referred to as the **labeling procedure**. We will illustrate the algorithm with two examples.

EXAMPLE 5.2.5. Use Algorithm 5.2.4 to find a maximal flow for the network of Figure 5.1.1.

Let us order the vertices a, b, c, d, e, z. Initially, the flow is 0 in each edge (see Figure 5.2.7). Vertex a is labeled $(, \infty)$. At step 4 we find that a is the smallest labeled vertex not yet examined. At step 5 we examine edges (a, b) and (a, d). For the edge (a, b), we have

$$F_{ab} = 0 < C_{ab} = 3.$$

We label vertex b as $(a, 3)$, since

$$3 = \min \{\infty, 3 - 0\}.$$

Similarly, we label vertex d as $(a, 5)$.

Returning to step 3, since the sink is not labeled, we proceed to step 4. The smallest labeled vertex not yet examined is b. At step 5 we examine edge (b, c) and label vertex c as $(b, 2)$, since

$$2 = \min \{3, 2 - 0\}.$$

(The value 3 is obtained from the label of vertex b.)

Returning to step 3, since the sink is not labeled, we proceed to step 4.

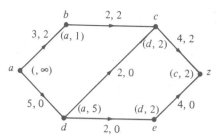

FIGURE 5.2.8

The smallest labeled vertex not yet examined is c. At step 5, we examine edge (c, z) and label vertex z as $(c, 2)$.

Returning to step 3, we find that the sink is labeled, so we proceed to step 6. At step 6,

$$P: a, b, c, z.$$

Since each edge of P is properly oriented, we increase the flow in each edge in P by $\Delta = 2$ to obtain Figure 5.2.8. (The value of Δ is obtained from the label of vertex z.) We then remove all the labels from all the vertices.

Returning to step 2, we label vertex a as $(, \infty)$ (see Figure 5.2.8). At step 4, we find that the smallest labeled vertex not yet examined is a. At step 5 we examine edges (a, b) and (a, d) and label vertex b as $(a, 1)$ and vertex d as $(a, 5)$.

Returning to step 3, since the sink is not labeled, we proceed to step 4. The smallest labeled vertex not yet examined is b. At step 5, we examine edge (b, c). Since $F_{bc} = C_{bc}$, we do not label vertex c at this point.

Returning to step 3, since the sink is not labeled, we proceed to step 4. The smallest labeled vertex not yet examined is d. At step 5 we examine edge (d, c) and label vertex c as $(d, 2)$.

Continuing this labeling process until the sink is labeled (see Figure 5.2.8), we arrive at step 6, with the path

$$P: a, d, c, z.$$

Since each edge of P is properly oriented, we increase the flow of each edge in P by $\Delta = 2$ to obtain Figure 5.2.9.

You should check that the next iteration of the algorithm produces the

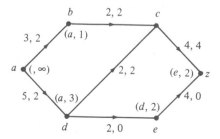

FIGURE 5.2.9

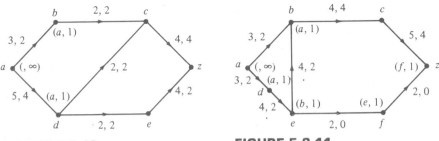

FIGURE 5.2.10   **FIGURE 5.2.11**

labeling shown in Figure 5.2.9. Augmenting the flow by $\Delta = 2$ produces Figure 5.2.10.

Returning to step 2, we label vertex a as $(, \infty)$. At step 4 we find that a is the smallest labeled vertex not yet examined. At step 5 we examine edges (a, b) and (a, d) and label both vertices b and d as $(a, 1)$ (see Figure 5.2.10).

Returning to step 3, since the sink is not labeled, we proceed to step 4. The smallest labeled vertex not yet examined is b. At step 5 we examine edge (b, c). Since $F_{bc} = C_{bc}$, we do not label vertex c.

Returning to step 3, since the sink is not labeled, we proceed to step 4. The smallest labeled vertex not yet examined is d. At step 5 we examine edges (d, c) and (d, e). Again, no additional vertices can be labeled.

Returning to step 3, since the sink is not labeled, we proceed to step 4. At this point all labeled vertices have been examined; thus the algorithm terminates. The flow of Figure 5.2.10 is maximal.

Our last example shows how to modify Algorithm 5.2.4 to generate a maximal flow from a given flow.

EXAMPLE 5.2.6.   Beginning with the flow of Figure 5.2.11, use Algorithm 5.2.4 to find a maximal flow for the network of Figure 5.2.11.

We replace the zero flow of step 1 with the flow given in Figure 5.2.11.

At step 4 the smallest labeled vertex not yet examined is a. At step 5 we examine edges (a, b) and (a, d) and label each of the vertices b and d as $(a, 1)$.

Again at step 4, the smallest labeled vertex not yet examined is b. At step 5 we examine edges (b, c) and (e, b). It is not possible to label vertex c, since $F_{bc} = C_{bc}$. Vertex e is labeled $(b, 1)$ since

$$1 = \min \{1, F_{eb} = 2\}.$$

Continuing, we ultimately label the sink (see Figure 5.2.11) and we obtain the path

$$P: a, b, e, f, z.$$

Edges (a, b), (e, f), and (f, z) are properly oriented so the flow in each is

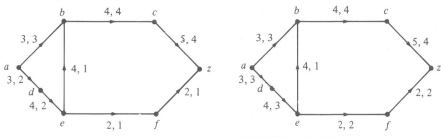

FIGURE 5.2.12   **FIGURE 5.2.13**

increased by 1. Since edge (e, b) is improperly oriented, its flow is decreased by 1. We obtain the flow of Figure 5.2.12.

Another iteration of the algorithm produces the maximal flow shown in Figure 5.2.13.

EXERCISES

In Exercises 1–3, a path from the source a to the sink z in a network is given. Find the maximum possible increase in the flow obtainable by altering the flows in the edges in the path.

1H.

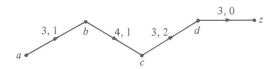

2.

3H.

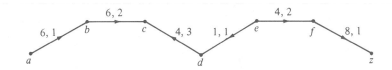

In Exercises 4–12, find a maximal flow in each network.

4.   Figure 5.1.4
5H.   Figure 5.1.5
6.   Exercise 4, Section 5.1
7H.   Exercise 5, Section 5.1
8.   Exercise 6, Section 5.1

9H.   Exercise 7, Section 5.1

10.   Exercise 8, Section 5.1

11H. Exercise 9, Section 5.1

12.

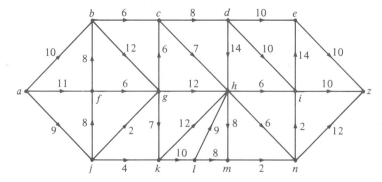

In Exercises 13–18, find a maximal flow in each network starting with the flow given.

13H. Figure 5.1.2

14.   Exercise 1, Section 5.1

15H. Exercise 2, Section 5.1

16.   Exercise 3, Section 5.1

17H. Figure 5.1.4 with flows

$$F_{a,w_1} = 2, \quad F_{w_1,b} = 2, \quad F_{bA} = 0, \quad F_{cA} = 0, \quad F_{Az} = 0,$$

$$F_{a,w_2} = 0, \quad F_{w_2,b} = 0, \quad F_{bc} = 2, \quad F_{cB} = 4, \quad F_{Bz} = 4,$$

$$F_{a,w_3} = 2, \quad F_{w_3,d} = 2, \quad F_{dc} = 2.$$

18.   Figure 5.1.4 with flows

$$F_{a,w_1} = 1, \quad F_{w_1,b} = 1, \quad F_{bA} = 4, \quad F_{cA} = 2, \quad F_{Az} = 6,$$

$$F_{a,w_2} = 3, \quad F_{w_2,b} = 3, \quad F_{bc} = 0, \quad F_{cB} = 1, \quad F_{Bz} = 1,$$

$$F_{a,w_3} = 3, \quad F_{w_3,d} = 3, \quad F_{dc} = 3.$$

★19. Is there a network G having the property that when Algorithm 5.2.4 is used to find a maximal flow in G starting with the zero flow at some point in step 5, there is an edge of the form (w, v) with $F_{wv} > 0$?

★20. Is there a network G having the property that when Algorithm 5.2.4 is used to find a maximal flow in G starting with the zero flow at some point in step 6, the path P has an improperly oriented edge?

21H. Show that Algorithm 5.2.4 terminates.

★**22.**   Give an example to show that if Algorithm 5.2.4 is modified so that nonnegative real capacities are allowed and the edges in step 5 can be examined in any order then Algorithm 5.2.4 may not terminate.

5.3

The Max Flow, Min Cut Theorem

In this section we show that at the termination of Algorithm 5.2.4, the flow in the network is maximal. Along the way we will define and discuss cuts in networks.

Let G be a network and consider the flow F at the termination of Algorithm 5.2.4. Some vertices are labeled and some are unlabeled. Let P $(\overline{P})$ denote the set of labeled (unlabeled) vertices. (Recall that $\overline{P}$ denotes the complement of P.) Then the source a is in P and the sink z is in $\overline{P}$. The set S of edges (v, w), with $v \in P$ and $w \in \overline{P}$, is called a **cut** and the sum of the capacities of the edges in S is called the **capacity of the cut**. We will see that this cut has a minimum capacity and, since a minimal cut corresponds to a maximal flow (Theorem 5.3.9), the flow F is maximal. We begin with the formal definition of cut.

Throughout this section, G is a network with source a and sink z. The capacity of edge (i, j) is C_{ij}.

> **DEFINITION 5.3.1.**   A *cut* $(P, \overline{P})$ *in* G consists of a set P of vertices and the complement $\overline{P}$ of P, with $a \in P$ and $z \in \overline{P}$.

> **EXAMPLE 5.3.2.**   Consider the network G of Figure 5.3.1. If we let $P = \{a, b, d\}$, then $\overline{P} = \{c, e, f, z\}$ and $(P, \overline{P})$ is a cut in G. As shown, we sometimes indicate a cut by drawing a dashed line to partition the vertices.

> **EXAMPLE 5.3.3.**   Figure 5.2.10 shows the labeling at the termination of Algorithm 5.2.4 for a particular network. If we let P $(\overline{P})$ denote the set of labeled (unlabeled) vertices, we obtain the cut shown in Figure 5.3.2.

We next define the capacity of a cut.

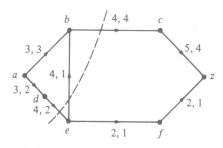

FIGURE 5.3.1

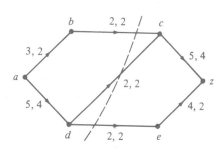

FIGURE 5.3.2

DEFINITION 5.3.4.   The *capacity of the cut* $(P, \overline{P})$ is the number

$$C(P, \overline{P}) = \sum_{i \in P} \sum_{j \in \overline{P}} C_{ij}.$$

EXAMPLE 5.3.5.   The capacity of the cut of Figure 5.3.1 is

$$C_{bc} + C_{de} = 8.$$

EXAMPLE 5.3.6.   The capacity of the cut of Figure 5.3.2 is

$$C_{bc} + C_{dc} + C_{de} = 6.$$

The next theorem shows that the capacity of any cut is always greater than or equal to the value of any flow.

THEOREM 5.3.7.   *Let F be a flow in G and let* $(P, \overline{P})$ *be a cut in G. Then the capacity of* $(P, \overline{P})$ *is greater than or equal to the value of F; that is,*

$$\sum_{i \in P} \sum_{j \in \overline{P}} C_{ij} \geq \sum_i F_{ai}. \tag{5.3.1}$$

(The notation Σ_i means the sum over all vertices i.)

   Proof.   Note that

$$\sum_{j \in P} \sum_{i \in P} F_{ji} = \sum_{j \in P} \sum_{i \in P} F_{ij},$$

since either side of the equation is merely the sum of F_{ij} over all $i, j \in P$.
   Now

$$\sum_i F_{ai} = \sum_{j \in P} \sum_i F_{ji} - \sum_{j \in P} \sum_i F_{ij}$$

$$= \sum_{j \in P} \sum_{i \in P} F_{ji} + \sum_{j \in P} \sum_{i \in \overline{P}} F_{ji} - \sum_{j \in P} \sum_{i \in P} F_{ij}$$

$$- \sum_{j \in P} \sum_{i \in \overline{P}} F_{ij}$$

$$= \sum_{j \in P} \sum_{i \in \overline{P}} F_{ji} - \sum_{j \in P} \sum_{i \in \overline{P}} F_{ij}$$

$$\leq \sum_{j \in P} \sum_{i \in \overline{P}} F_{ji} \leq \sum_{j \in P} \sum_{i \in \overline{P}} C_{ji}. \qquad \blacksquare$$

EXAMPLE 5.3.8.   In Figure 5.3.1, the value 5 of the flow is less than the capacity 8 of the cut.

A **minimal cut** is a cut having minimum capacity.

THEOREM 5.3.9 **Max Flow, Min Cut Theorem.** *Let F be a flow in G and let (P, P) be a cut in G. If equality holds in (5.3.1), then the flow is maximal and the cut is minimal. Moreover, equality holds in (5.3.1) if and only if*

(a) $F_{ij} = C_{ij}$ *for $i \in P, j \in \overline{P}$;*

and

(b) $F_{ij} = 0$ *for $i \in \overline{P}, j \in P$.*

Proof. The first statement follows immediately.

The proof of Theorem 5.3.7 shows that equality holds precisely when

$$\sum_{j \in P} \sum_{i \in \overline{P}} F_{ij} = 0$$

and

$$\sum_{j \in P} \sum_{i \in \overline{P}} F_{ji} = \sum_{j \in P} \sum_{i \in \overline{P}} C_{ji};$$

thus the last statement is also true. ∎

EXAMPLE 5.3.10. In Figure 5.3.2, the value of the flow and the capacity of the cut are both 6; therefore, the flow is maximal and the cut is minimal.

We can use Theorem 5.3.9 to show that Algorithm 5.2.4 produces a maximal flow.

THEOREM 5.3.11. *At termination, Algorithm 5.2.4 produces a maximal flow.*

Proof. Let P ($\overline{P}$) be the set of labeled (unlabeled) vertices of G at the termination of Algorithm 5.2.4. Consider an edge (i, j) where $i \in P, j \in \overline{P}$. Since i is labeled, we must have

$$F_{ij} = C_{ij};$$

otherwise, we would have labeled j at step 5. Now consider an edge (j, i), where $j \in \overline{P}, i \in P$. Since i is labeled, we must have

$$F_{ji} = 0;$$

otherwise, we would have labeled j at step 5. By Theorem 5.3.9, the flow at the termination of Algorithm 5.2.4 is maximal. ∎

EXERCISES

In Exercises 1–3, find the capacity of the cut $(P, \overline{P})$. Also, determine whether the cut is minimal.

1H. $P = \{a, d\}$ for Exercise 1, Section 5.1
2. $P = \{a, d, e\}$ for Exercise 2, Section 5.1
3H. $P = \{a, b, c, d\}$ for Exercise 3, Section 5.1

In Exercises 4–16, find a minimal cut in each network.

4.   Figure 5.1.1
5H.   Figure 5.1.4
6.   Figure 5.1.5
7H.   Exercise 1, Section 5.1
8.   Exercise 2, Section 5.1
9H.   Exercise 3, Section 5.1
10.   Exercise 4, Section 5.1
11H.   Exercise 5, Section 5.1
12.   Exercise 6, Section 5.1
13H.   Exercise 7, Section 5.1
14.   Exercise 8, Section 5.1
15H.   Exercise 9, Section 5.1
16.   Exercise 12, Section 5.2

Exercises 17–22 refer to a network G that, in addition to having nonnegative integer capacities C_{ij}, has nonnegative integer minimal edge flow requirements m_{ij}. That is, a flow F must satisfy

$$m_{ij} \leq F_{ij} \leq C_{ij},$$

for all edges (i, j).

17H.   Give an example of a network G, in which $m_{ij} \leq C_{ij}$ for all edges (i, j), for which no flow exists.

Define

$$C(\overline{P}, P) = \sum_{i \in \overline{P}} \sum_{j \in P} C_{ij}$$

$$m(P, \overline{P}) = \sum_{i \in P} \sum_{j \in \overline{P}} m_{ij}$$

$$m(\overline{P}, P) = \sum_{i \in \overline{P}} \sum_{j \in P} m_{ij}.$$

18.   Show that the value V of any flow satisfies

$$m(P, \overline{P}) - C(\overline{P}, P) \leq V \leq C(P, \overline{P}) - m(\overline{P}, P)$$

for any cut $(P, \overline{P})$.

19H.   Show that if a flow exists in G, a maximal flow exists in G with value

$$\min \{C(P, \overline{P}) - m(\overline{P}, P) \mid (P, \overline{P}) \text{ is a cut in } G\}.$$

20.   Assume that G has a flow F. Develop an algorithm for finding a maximal flow in G.

21H.   Show that if a flow exists in G, a minimal flow exists in G with value

$$\max \{m(P, \overline{P}) - C(\overline{P}, P) \mid (P, \overline{P}) \text{ is a cut in } G\}.$$

22.   Assume that G has a flow F. Develop an algorithm for finding a minimal flow in G.

23H.   True or false? If F is a flow in a network G and $(P, \overline{P})$ is a cut in G and the capacity of $(P, \overline{P})$ exceeds the value of the flow, F, then the cut $(P, \overline{P})$ is not minimal and the flow F is not maximal. If true, prove it; otherwise, give a counterexample.

5.4

Matching

In this section we consider the problem of matching elements in one set to elements in another set. We will see that this problem can be reduced to finding a maximal flow in a network. We begin with an example.

EXAMPLE 5.4.1.   Suppose that four persons A, B, C, and D apply for five jobs J_1, J_2, J_3, J_4, and J_5. Suppose that applicant A is qualified for jobs J_2 and J_5; applicant B is qualified for jobs J_2 and J_5; applicant C is qualified for jobs J_1, J_3, J_4, and J_5; and applicant D is qualified for jobs J_2 and J_5. Is it possible to find jobs for each applicant?

The situation can be modeled by the graph of Figure 5.4.1. The vertices represent the applicants and the jobs. An edge connects an applicant to a job for which the applicant is qualified. We can show that it is not possible to match a job to each applicant by considering applicants A, B, and D who are

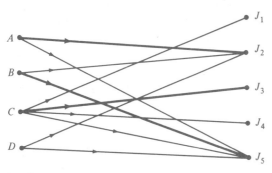

FIGURE 5.4.1

qualified for jobs J_2 and J_5. If A and B are assigned a job, none remains for D. Therefore, no assignments exist for A, B, C, and D.

In Example 5.4.1 a matching consists of finding jobs for qualified persons. A maximal matching finds jobs for the maximum number of persons. A maximal matching for the graph of Figure 5.4.1 is shown with heavy lines. A complete matching finds jobs for everyone. We showed that the graph of Figure 5.4.1 has no complete matching. The formal definitions follow.

DEFINITION 5.4.2. Let G be a directed, bipartite graph with disjoint vertex sets V and W in which the edges are directed from vertices in V to vertices in W. (Any vertex in G is either in V or in W.) A *matching* for G is a set of edges E with no vertices in common. A *maximal matching* for G is a matching E in which E contains the maximum number of edges. A *complete matching* for G is a matching E having the property that if $v \in V$, then $(v, w) \in E$ for some $w \in W$.

EXAMPLE 5.4.3. The matching for the graph of Figure 5.4.2, shown with heavy lines, is a maximal matching and a complete matching.

In the next example we illustrate how the matching problem can be modeled as a network problem.

EXAMPLE 5.4.4. Model the matching problem of Example 5.4.1 as a network.
    We first assign each edge in the graph of Figure 5.4.1 capacity 1 (see Figure 5.4.3). Next, we add a supersource a and edges of capacity 1 from a to each of A, B, C, and D. Finally, we introduce a supersink z and edges of capacity 1 from each of J_1, J_2, J_3, J_4, and J_5 to z. We call a network like that of Figure 5.4.3 a **matching network**.

The next theorem relates matching networks and flows.

THEOREM 5.4.5. *Let G be a directed, bipartite graph with disjoint vertex sets V and W in which the edges are directed from vertices in V to vertices in W. (Any vertex in G is either in V or in W.)*
    (a) A flow in the matching network gives a matching in G. The vertex

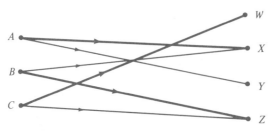

FIGURE 5.4.2

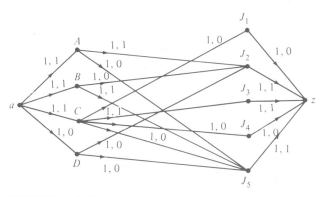

FIGURE 5.4.3

$v \in V$ *is matched with the vertex* $w \in W$ *if and only if the flow in edge* (v, w) *is* 1.

(b) *A maximal flow corresponds to a maximal matching.*

(c) *A flow whose value is* $|V|$ *corresponds to a complete matching.*

Proof. Let a (z) represent the source (sink) in the matching network and suppose that a flow is given.

Suppose that the edge (v, w), $v \in V$, $w \in W$, has flow 1. The only edge into vertex v is (a, v). This edge must have flow 1; thus the flow into vertex v is 1. Since the flow out of v is also 1, the only edge of the form (v, x) having flow 1 is (v, w). Similarly, the only edge of the form (x, w) having flow 1 is (v, w). Therefore, if E is the set of edges of the form (v, w) having flow 1, the members of E have no vertices in common; thus E is a matching for G.

Parts (b) and (c) follow from the fact that the number of vertices in V matched is equal to the value of the corresponding flow. ■

Since a maximal flow gives a maximal matching, Algorithm 5.2.4 applied to a matching network produces a maximal matching. In practice, the implementation of Algorithm 5.2.4 can be simplified by using the adjacency matrix of the graph (see Exercise 13).

EXAMPLE 5.4.6. The matching of Figure 5.4.1 is represented as a flow in Figure 5.4.3. Since the flow is maximal, the matching is maximal.

Next, we turn to the existence of a complete matching in a directed, bipartite graph G with vertex sets V and W. If $S \subseteq V$, we let

$$R(S) = \{w \in W \mid v \in S \text{ and } (v, w) \text{ is an edge in } G\}.$$

Suppose that G has a complete matching. If $S \subseteq V$ we must have

$$|S| \leq |R(S)|.$$

It turns out that if $|S| \leq |R(S)|$ for all subsets S of V, then G has a complete matching. This result was first given by the English mathematician Philip Hall and is known as **Hall's Marriage Theorem**, since if V is a set of men and W is a set of women and edges exist from $v \in V$ to $w \in W$ if v and w are compatible, the theorem gives a condition under which each man can marry a compatible woman.

THEOREM 5.4.7 Hall's Marriage Theorem. *Let G be a directed, bipartite graph with disjoint vertex sets V and W in which the edges are directed from vertices in V to vertices in W. (Any vertex in G is either in V or in W.) There exists a complete matching in G if and only if*

$$|S| \leq |R(S)| \qquad \text{for all } S \subseteq V. \tag{5.4.1}$$

Proof. We have already pointed out that, if there is a complete matching in G, condition (5.4.1) holds.

Suppose that condition (5.4.1) holds. Let $n = |V|$ and let $(P, \overline{P})$ be a minimal cut in the matching network. If we can show that the capacity of this cut is n, a maximal flow would have value n. The matching corresponding to this maximal flow would be a complete matching.

The argument is by contradiction. Assume that the capacity of the minimal cut $(P, \overline{P})$ is less than n. The capacity of this cut is the number of edges in the set

$$E = \{(x, y) \mid x \in P, y \in \overline{P}\}$$

(see Figure 5.4.4). A member of E is one of the three types

Type I: $(a, v), v \in V$.
Type II: $(v, w), v \in V, w \in W$.
Type III: $(w, z), w \in W$.

We will estimate the number of edges of each type.

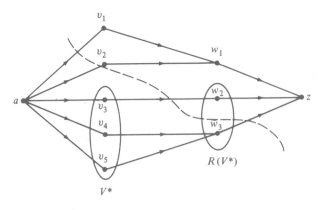

$$n = 5, W_1 = \{w_3\}, W_2 = \{w_2\}$$

FIGURE 5.4.4

If $V \subseteq \overline{P}$, the capacity of the cut is n (see Figure 5.4.5); thus

$$V^* = V \cap P$$

is nonempty. It follows that there are $n - |V^*|$ edges in E of type I.
We partition $R(V^*)$ into the sets

$$W_1 = R(V^*) \cap P \quad \text{and} \quad W_2 = R(V^*) \cap \overline{P}.$$

Then there are at least $|W_1|$ edges in E of type III. Thus there are less than

$$n - (n - |V^*|) - |W_1| = |V^*| - |W_1|$$

edges of type II in E. Since each member of W_2 contributes at most one type II edge,

$$|W_2| < |V^*| - |W_1|.$$

Thus

$$|R(V^*)| = |W_1| + |W_2| < |V^*|,$$

which contradicts (5.4.1). Therefore, a complete matching exists. ∎

EXAMPLE 5.4.8. For the graph in Figure 5.4.1, if $S = \{A, B, D\}$, we have $R(S) = \{J_2, J_5\}$ and

$$|S| = 3 > 2 = |R(S)|.$$

By Theorem 5.4.7, there is not a complete matching for the graph of Figure 5.4.1.

EXAMPLE 5.4.9. There are n computers and n disk drives. Each computer is compatible with m disk drives and each disk drive is compatible with m computers. Is it possible to match each computer with a compatible disk drive?

Let V be the set of computers and W be the set of disk drives. An edge exists from $v \in V$ to $w \in W$ if v and w are compatible. Notice that every vertex has degree m. Let $S = \{v_1, \ldots, v_k\}$ be a subset of V. Then there

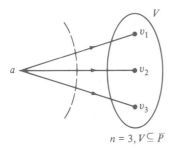

$n = 3, V \subseteq \overline{P}$

FIGURE 5.4.5

are km edges from the set S. If $R(S) = \{w_1, \ldots, w_j\}$, then $R(S)$ receives at most jm edges from S. Therefore,

$$km \leq jm.$$

Now

$$|S| = k \leq j = |R(S)|.$$

By Theorem 5.4.7 there is a complete matching. Thus it is possible to match each computer with a compatible disk drive.

EXERCISES

1H. Show that the flow in Figure 5.4.3 is a maximal by exhibiting a minimal cut whose capacity is 3.

2. Find the flow that corresponds to the matching of Figure 5.4.2. Show that this flow is maximal by exhibiting a minimal cut whose capacity is 3.

3H. Applicant A is qualified for jobs J_1 and J_4; B is qualified for jobs J_2, J_3, and J_6; C is qualified for jobs J_1, J_3, J_5, and J_6; D is qualified for jobs J_1, J_3, and J_4; and E is qualified for jobs J_1, J_3, and J_6.
(a) Model this situation as a matching network.
(b) Use Algorithm 5.2.4 to find a maximal matching.
(c) Is there a complete matching?

4. Applicant A is qualified for jobs J_1, J_2, J_4, and J_5; B is qualified for jobs J_1, J_4, and J_5; C is qualified for jobs J_1, J_4, and J_5; D is qualified for jobs J_1 and J_5; E is qualified for jobs J_2, J_3, and J_5; and F is qualified for jobs J_4 and J_5. Answer parts (a)–(c) of Exercise 3 for this situation.

5. Applicant A is qualified for jobs J_1, J_2, and J_4; B is qualified for jobs J_3, J_4, J_5, and J_6; C is qualified for jobs J_1 and J_5; D is qualified for jobs J_1, J_3, J_4, and J_8; E is qualified for jobs J_1, J_2, J_4, J_6, and J_8; F is qualified for jobs J_4 and J_6; and G is qualified for jobs J_3, J_5, and J_7. Answer parts (a)–(c) of Exercise 3 for this situation.

6H. Five students, V, W, X, Y, and Z, are members of four committees, C_1, C_2, C_3, and C_4. The members of C_1 are V, X, and Y; the members of C_2 are X and Z; the members of C_3 are V, Y, and Z; and the members of C_4 are V, W, X, and Z. Each committee is to send a representative to the administration. No student can represent two committees.
(a) Model this situation as a matching network.
(b) What is the interpretation of a maximal matching?
(c) What is the interpretation of a complete matching?
(d) Use Algorithm 5.2.4 to find a maximal matching.
(e) Is there a complete matching?

7H. Let G be a directed, bipartite graph with disjoint vertex sets V and W in which the edges are directed from vertices in V to vertices in W. (Any vertex in G is either in V or in W.) Let

$$M = \max\{\delta(w) \mid w \in W\}.$$

Show that if

$$M \leq \delta(v) \tag{5.4.2}$$

for every $v \in V$, then G has a complete matching. [$\delta(x)$ denotes the degree of x.]

8. Give an example of a bipartite graph G that has a complete matching but does not satisfy condition (5.4.2) of Exercise 7.

9H. Show that by suitably ordering the vertices, the adjacency matrix of a bipartite graph can be written

$$\begin{pmatrix} 0 & A \\ A^T & 0 \end{pmatrix},$$

where 0 is a matrix consisting only of 0's and A^T is the transpose of the matrix A.

In Exercises 10–12, G is a bipartite graph, A is the matrix of Exercise 9, and F is a flow in the associated matching network. Label each entry in A that represents an edge with flow 1.

10. What kind of labeling corresponds to a matching?

11H. What kind of labeling corresponds to a complete matching?

12. What kind of labeling corresponds to a maximal matching?

13. Restate Algorithm 5.2.4, applied to a matching network, in terms of operations on the matrix A of Exercise 9.

Let G be a directed, bipartite graph with disjoint vertex sets V and W in which the edges are directed from vertices in V to vertices in W. (Any vertex in G is either in V or in W.) We define the **deficiency of G** as

$$\delta(G) = \max\{|S| - |R(S)| \mid S \subseteq V\}.$$

14. Show that G has a complete matching if and only if $\delta(G) = 0$.

★15H. Show that the maximum number of vertices in V that can be matched with vertices in W is $|V| - \delta(G)$.

16. True or false? Any matching is contained in a maximal matching. If true, prove it; if false, give a counterexample.

Petri Nets

Consider the computer program shown in Figure 5.5.1. Normally, the instructions would be processed sequentially—first, $A = 1$, then $B = 2$, and so on. However, notice that there is no logical reason that prevents the first three instructions—$A = 1$; $B = 2$; $C = 3$—from being processed in any order or concurrently. With the continuing decline of the cost of computer hardware, and processors in particular, there is increasing interest in concurrent processing to achieve greater speed and efficiency. The use of **Petri nets**, graph models of concurrent processing, is one method of modeling and studying concurrent processing.

$A = 1$
$B = 2$
$C = 3$
$A = A + 1$
$C = B + C$
$B = A + C$

FIGURE 5.5.1

DEFINITION 5.5.1.   A *Petri net* is a directed graph $G = (V, E)$, where $V = P \cup T$ and $P \cap T = \emptyset$. Any edge e in E is incident on one member of P and one member of T. The set P is called the set of *places* and the set T is called the set of *transitions*.

Less formally, a Petri net is a directed, bipartite graph where the two classes of vertices are called places and transitions. In general, parallel edges are allowed in Petri nets; however, for simplicity, we will not permit parallel edges.

EXAMPLE 5.5.2.   An example of a Petri net is given in Figure 5.5.2. Places are typically drawn as circles and transitions as bars or rectangular boxes.

DEFINITION 5.5.3.   A *marking* of a Petri net assigns each place a nonnegative integer. A Petri net with a marking is called a *marked Petri net* (or sometimes just a Petri net).

If a marking assigns the nonnegative integer n to place p, we say that there are n **tokens** on p. The tokens are represented as black dots.

EXAMPLE 5.5.4.   An example of a marked Petri net is given in Figure 5.5.3.

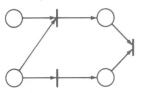

FIGURE 5.5.2

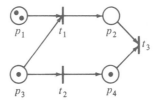

FIGURE 5.5.3

In modeling, the places represent **conditions**, the transitions represent **events**, and the presence of at least one token in a place (condition) indicates that that condition is met.

EXAMPLE 5.5.5.   In Figure 5.5.4 we have modeled the computer program of Figure 5.5.1. Here the events (transitions) are the instructions, and the places represent the conditions under which an instruction can be executed.

DEFINITION 5.5.6.   In a Petri net, if an edge is directed from place p to transition t, we say p is an *input place* for transition t. An *output place* is defined similarly. If every input place for a transition t has at least one token, we say that t is *enabled*. A *firing* of an enabled transition removes one token from each input place and adds one token to each output place.

EXAMPLE 5.5.7.   In the Petri net of Figure 5.5.3, places p_1 and p_3 are input places for transition t_1. Transitions t_1 and t_2 are enabled, but transition t_3 is not enabled. If we fire transition t_1, we obtain the marked Petri net of Figure 5.5.5. Transition t_3 is now enabled. If we then fire transition t_3, we obtain the net shown. At this point no transition is enabled and thus none may be fired.

DEFINITION 5.5.8.   If a sequence of firings transforms a marking M to a marking M', we say that M' is *reachable* from M.

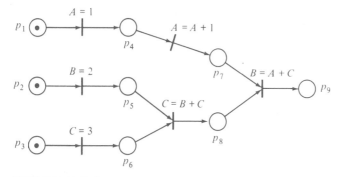

FIGURE 5.5.4

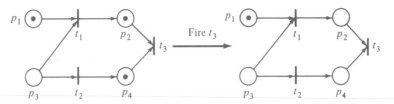

FIGURE 5.5.5

EXAMPLE 5.5.9. In Figure 5.5.6, M'' is reachable from M by first firing transition t_1 and then firing t_2.

In modeling, the firing of a transition simulates the occurrence of that event. Of course, an event can take place only if all of the conditions for its execution have been met; that is, the transition can be fired only if it is enabled. By putting tokens in places p_1, p_2, and p_3 in Figure 5.5.4, we show that the conditions for executing the instructions $A = 1$, $B = 2$, and $C = 3$ are met. The program is ready to be executed. Since the transitions $A = 1$, $B = 2$, and $C = 3$ are enabled, they can be fired in any order or concurrently. Transition $C = B + C$ is enabled only if places p_5 and p_6 have tokens. But these places

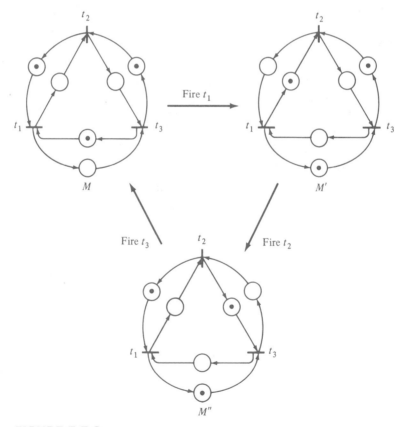

FIGURE 5.5.6

will have tokens only if transitions $B = 2$ and $C = 3$ have been fired. In other words, the conditions under which the event $C = B + C$ can occur is that $B = 2$ and $C = 3$ must have been executed. In this way we model the legal execution sequences of Figure 5.5.1 and the implicit concurrency within this program.

Among the most important properties studied in Petri net theory are **liveness** and **safeness**. Liveness is related to the absence of deadlocks and safeness is related to bounded memory capacity.

EXAMPLE 5.5.10. Two persons are sharing a computer system that has a disk drive D and a printer P. Each person needs both D and P. A possible Petri net model of this situation is shown in Figure 5.5.7. The marking indicates that both D and P are available.

Now suppose that person 1 requests D and then P (while person 2 requests neither). The occurrences of these events are simulated by first firing transition "request D" and then firing transition "request P" for person 1. The resulting Petri net is shown in Figure 5.5.8. When person 1 finishes the processing and releases D and P, simulated by firing the transitions "process" and then "release D and P," we return to the Petri net of Figure 5.5.7. If person 2 requests D and then P (while person 1 requests neither), we obtain a similar firing sequence.

Again, assume that we have the situation of Figure 5.5.7. Now suppose that person 1 requests D and then person 2 requests P. After the appropriate transitions are fired to simulate the occurrences of these events, we obtain

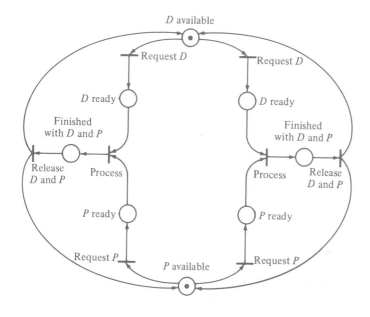

Person 1   Person 2

FIGURE 5.5.7

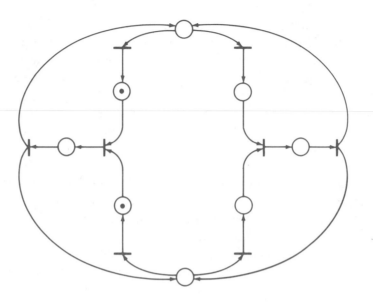

Person 1   Person 2

FIGURE 5.5.8

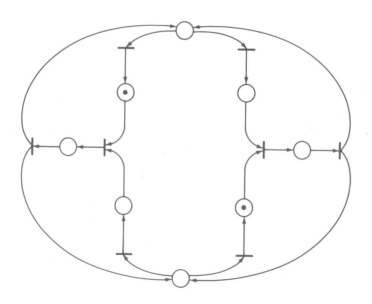

Person 1   Person 2

FIGURE 5.5.9

the Petri net of Figure 5.5.9. Notice that at this point, no transition can fire. Person 1 is waiting for person 2 to release P and person 2 is waiting for person 1 to release D. Activity within the system stops. We say that a deadlock occurs. Formally, we say that a marked Petri net is **deadlocked** if no transition can fire. Prevention of deadlocks within concurrent processing environments is a major practical concern.

Example 5.5.10 motivates the following definition.

DEFINITION 5.5.11. A marking M for a Petri net is *live* if, beginning from M, no matter what sequence of firings has occurred, it is possible to fire any given transition by proceeding through some additional firing sequence.

If a marking M is live for a Petri net P, then no matter what sequence of transitions is fired, P will never deadlock. Indeed, we can fire any transition by proceeding through some additional firing sequence.

EXAMPLE 5.5.12. The marking M of the net of Figure 5.5.6 is live. To see this, notice that the only transition for marking M that can be fired is t_1, which produces marking M'. The only transition for marking M' that can be fired is t_2, which produces marking M''. The only transition for marking M'' that can be fired is t_3 which returns us to marking M. Thus any firing sequence, starting with marking M, produces one of the markings M, M', or M'' and from there we can fire any transition t_1, t_2, or t_3 by proceeding as in Figure 5.5.6. Therefore, the marking M for the net of Figure 5.5.6 is live.

EXAMPLE 5.5.13. The marking shown in Figure 5.5.4 is not live since after transition $A = 1$ is fired, it can never fire again.

If a place is regarded as having limited capacity, **boundedness** assures us that no place will overflow.

DEFINITION 5.5.14. A marking M for a Petri net is *bounded* if there is some positive integer n having the property that in any firing sequence, no place ever receives more than n tokens. If a marking M is bounded and in any firing sequence, no place ever receives more than one token, we call M a *safe* marking.

If each place represents a register capable of holding one computer word and if an initial marking is safe, we are guaranteed that the memory capacity of the registers will not be exceeded.

EXAMPLE 5.5.15. The markings of Figure 5.5.6 are safe. The marking M of Figure 5.5.10 is not safe, since as shown, if transition t_1 is fired, place p_2 then has two tokens. By listing all the markings reachable from M, it can be verified that M is bounded and live (see Exercise 7).

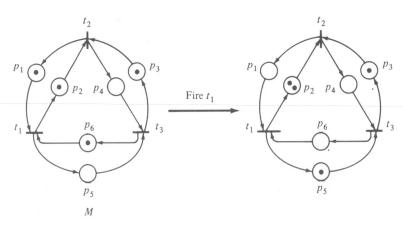

FIGURE 5.5.10

EXERCISES

In Exercises 1–3, model each program by a Petri net. Provide a marking that represents the situation prior to execution of the program.

1H.
$A = 1$
$B = 2$
$C = A + B$
$C = C + 1$

2.
$A = 2$
$B = A + A$
$C = 3$
$D = A + A$
$C = A + B + C$

3H.
$A = 1$
$S = 0$
10   $S = S + A$
$A = A + 1$
GOTO 10

4. Describe three situations involving concurrency that might be modeled as Petri nets.

5H. Give an example of a marked Petri net in which two transitions are enabled, but firing either one disables the other.

6. Consider the following algorithm for washing a lion.

    1. Get lion.
    2. Get soap.
    3. Get tub.
    4. Put water in tub.

5. Put lion in tub.
6. Wash lion with soap.
7. Rinse lion.
8. Remove lion from tub.
9. Dry lion.

Model this algorithm as a Petri net. Provide a marking that represents the situation prior to execution.

7H.   Show that the marking M of Figure 5.5.10 is live and bounded.

Answer the following questions for each marked Petri net in Exercises 8–12.
(a) Which transitions are enabled?
(b) Show the marking that results from firing t_1.
(c) Is M live?
(d) Is M safe?
(e) Is M bounded?
(f) Show or describe all markings reachable from M.
(g) Exhibit a marking (other than the marking that puts zero tokens in each place) not reachable from M.

8.

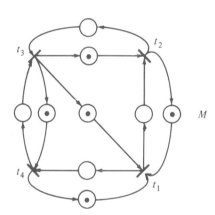

9H.

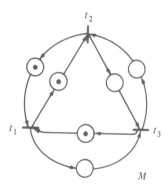

10.

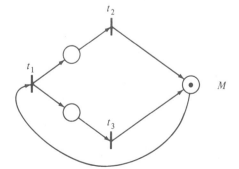

11H.

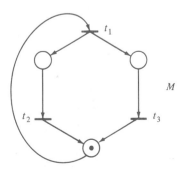

★**12.**

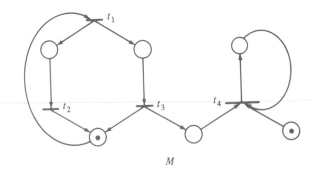

M

13H. Give an example of a Petri net with a marking that is safe, but not live.

14. Give an example of a Petri net with a marking that is bounded, but not safe.

15H. The **Dining Philosophers' Problem** (see [Dijkstra]) concerns five philosophers seated at a round table. Each philosopher either eats or meditates. The table is set alternately with one plate and one chopstick. Eating requires two chopsticks so that if each philosopher picks up the chopstick to the right of the plate, none can eat—the system will deadlock. Model this situation as a Petri net. Your model should be live so that the system will not deadlock and so that, at any point, any philosopher can potentially either eat or meditate.

16. Develop an alternative Petri net model for the situation of Example 5.5.10 that prevents deadlock.

　　　If each place in a marked Petri net P has one incoming and one outgoing edge, then P can be redrawn as a directed graph where vertices correspond to transitions and edges to places. The tokens are placed on the edges. Such a graph is called a **marked graph**. Here we show a marked Petri net and its representation as a marked graph.

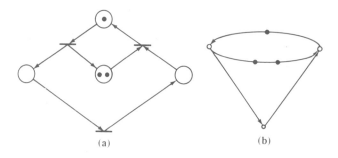

(a)　　　　　　　　　　(b)

17H. Which Petri nets in Exercises 8–12 can be redrawn as marked graphs?

18. Redraw the marked Petri net as a marked graph.

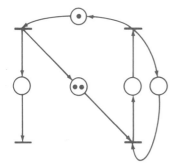

19H.   Redraw the marked Petri net as a marked graph.

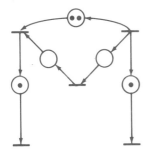

The **token count** of a simple directed circuit in a marked graph is the number of tokens on all the edges in the circuit.

20.   Show that the token count of a simple directed circuit does not change during any firing sequence.

★21H.   Show that a marking M for a marked graph G is live if and only if M places at least one token in each simple directed circuit in G.

★22.   Show that a live marking is safe for a marked graph G if and only if every edge in G belongs to a simple directed circuit with token count one.

23H.   Give an example of a marked graph with an unsafe marking in which every edge belongs to a simple directed circuit with token one.

24.   Let G be a marked graph. Show that each edge in G is contained in a simple directed circuit if and only if every marking for G is bounded.

★25.   Let G be a directed graph where, if we ignore the direction of the edges in G, G is connected as an undirected graph. Show that G has a live and safe marking if and only if given any two vertices v and w in G there is a directed path from v to w.

Notes

General references that contain sections on network models are [Berge; Berztiss; Busacker; Deo; Liu, 1968, 1985; and Tucker]. The classic work on networks is [Ford]; many of the results on networks, especially the early results, are due to Ford and Fulkerson, the authors of this book. [Hu, 1969 and 1982] are also recommended as references on network models.

Petri nets originated in C. Petri's doctoral dissertation [Petri] in 1962. Since then there has been much research on the properties of Petri nets and a great deal of interest in using Petri nets to model real systems. J. L. Peterson wrote the first text on Petri nets ([Peterson]) in 1981. Peterson's book contains an extensive, annotated bibliography. A recent expository paper with a generous number of references is [Johnsonbaugh].

The problem of finding a maximal flow in a network G, with source a, sink z, and capacities C_{ij} may be rephrased as follows:

$$\text{maximize} \sum_j F_{aj} \tag{5.6.1}$$

subject to

$$0 \leq F_{ij} \leq C_{ij} \qquad \text{for all } i, j;$$

$$\sum_i F_{ij} = \sum_i F_{ji} \qquad \text{for all } j. \tag{5.6.2}$$

Such a problem is an example of a **linear programming problem**. In a linear programming problem, we want to maximize (or minimize) a linear expression, such as (5.6.1), subject to linear inequality and equality constraints, such as (5.6.2). Although the **simplex algorithm** is normally an efficient way to solve a general linear programming problem, network transport problems are usually more efficiently solved using Algorithm 5.2.4. See [Hillier] for an exposition of the simplex algorithm.

Suppose that for each edge (i, j) in a network G, c_{ij} represents the cost of the flow of 1 unit through edge (i, j). Suppose that we want a maximal flow, with minimal cost

$$\sum_i \sum_j c_{ij} F_{ij}.$$

This problem, called the **transportation problem**, is again a linear programming problem and, like the maximal flow problem, a specific algorithm can be used to obtain a solution that is, in general, more efficient than the simplex algorithm (see [Hillier]).

COMPUTER EXERCISES

1. Write a program that accepts as input a network with a given flow and outputs all possible paths from the source to the sink on which the flow can be increased.

2. Implement Algorithm 5.2.4 as a program. Have the program output the minimal cut as well as the maximal flow.

3. Implement the algorithm of Exercise 20, Section 5.3, as a program.

4. Implement the algorithm of Exercise 22, Section 5.3, as a program.

5. Implement the algorithm of Exercise 13, Section 5.4, as a program. Allow the program to output optionally each intermediate labeled matrix A.

6. Write a program that computes $\delta(G)$, the deficiency of a network G.

7. Write a program that lists all firing sequences of length n in a Petri net beginning with a given marking.

8. Write a program that determines whether a marked graph is live or safe.

9. Write a program that accepts as input an initial city A, a destination city Z, and all possible routes from A to Z, including the travel times and maximum capacities of all routes over some particular time period, and outputs the maximal flow from A to Z (see Example 5.1.9).

CHAPTER 6

Boolean Algebras and Combinatorial Circuits

Several definitions honor the nineteenth-century mathematician George Boole— Boolean algebra, Boolean function, Boolean expression, and Boolean ring—to name a few. Boole is one of the persons in a long, historical chain who were concerned with formalizing and mechanizing the process of logical thinking. In fact, in 1854 Boole wrote a book entitled *The Laws of Thought*. Boole's contribution was the development of a theory of logic using symbols instead of words. For a discussion of Boole's work, see [Hailperin].

Almost a century after Boole's work, it was observed, especially by C. E. Shannon in 1938 (see [Shannon]), that Boolean algebra could be used to analyze electrical circuits. Thus Boolean algebra became an indispensable tool for the analysis and design of electronic computers in the succeeding decades. We explore the relationship of Boolean algebra to circuits throughout this chapter.

6.1

Combinatorial Circuits

In a digital computer, there are only two possibilities, written 0 and 1, for the smallest, indivisible object. All programs and data are ultimately reducible to combinations of bits. A variety of devices have been used throughout the years in digital computers to store bits. Electronic circuits allow these storage devices to communicate with each other. A bit in one part of a circuit is transmitted to another part of the circuit as a voltage. Thus two voltage levels are needed— for example, a high voltage can communicate 1 and a low voltage can communicate 0.

In this section we discuss **combinatorial circuits**. The output of a combinatorial circuit is uniquely defined for every combination of inputs. A combinatorial circuit has no memory; previous inputs and the state of the system do not affect the output of a combinatorial circuit. Circuits for which the output is a function, not only of the inputs, but also of the state of the system, are called **sequential circuits** and are considered Chapter 7.

Combinatorial circuits can be constructed using solid-state devices, called **gates**, which are capable of switching voltage levels (bits). We will begin by discussing AND, OR, and NOT gates.

DEFINITION 6.1.1.   An *AND gate* receives inputs x_1 and x_2, where x_1 and x_2 are bits, and produces output denoted $x_1 \wedge x_2$, where

$$x_1 \wedge x_2 = \begin{cases} 1 & \text{if } x_1 = 1 \text{ and } x_2 = 1 \\ 0 & \text{otherwise.} \end{cases}$$

An AND gate is drawn as shown in Figure 6.1.1.

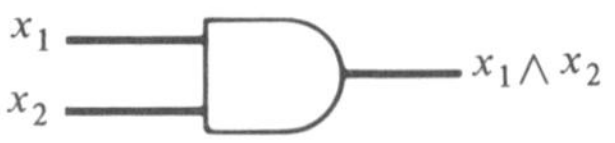

FIGURE 6.1.1

DEFINITION 6.1.2.   An *OR gate* receives inputs x_1 and x_2, where x_1 and x_2 are bits, and produces output denoted $x_1 \vee x_2$, where

$$x_1 \vee x_2 = \begin{cases} 1 & \text{if } x_1 = 1 \text{ or } x_2 = 1 \\ 0 & \text{otherwise.} \end{cases}$$

An OR gate is drawn as shown in Figure 6.1.2.

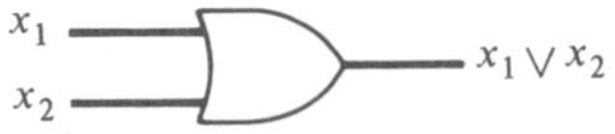

FIGURE 6.1.2

DEFINITION 6.1.3.   A *NOT gate* (or *inverter*) receives input x, where x is a bit, and produces output denoted $\bar{x}$, where

$$\bar{x} = \begin{cases} 1 & \text{if } x = 0 \\ 0 & \text{if } x = 1. \end{cases}$$

A NOT gate is drawn as shown in Figure 6.1.3.

FIGURE 6.1.3

The **logic table** of a combinatorial circuit lists all possible inputs together with the resulting outputs.

EXAMPLE 6.1.4. Following are the logic tables for the basic AND, OR, and NOT circuits (Figures 6.1.1–6.1.3).

x_1	x_2	$x_1 \wedge x_2$	x_1	x_2	$x_1 \vee x_2$	x	$\bar{x}$
1	1	1	1	1	1	1	0
1	0	0	1	0	1	0	1
0	1	0	0	1	1		
0	0	0	0	0	0		

We note that performing the operation AND (OR) is the same as taking the minimum (maximum) of the two bits x_1 and x_2.

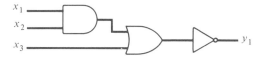

FIGURE 6.1.4

EXAMPLE 6.1.5. The circuit of Figure 6.1.4 is an example of a combinatorial circuit since the output y_1 is uniquely defined for each combination of inputs x_1, x_2, and x_3. The logic table for this combinatorial circuit follows.

x_1	x_2	x_3	y_1
1	1	1	0
1	1	0	0
1	0	1	0
1	0	0	1
0	1	1	0
0	1	0	1
0	0	1	0
0	0	0	1

Notice that all possible combinations of values for the inputs x_1, x_2, and x_3 are listed. For a given set of inputs, we can compute the value of the output y_1 by tracing the flow through the circuit. For example, the fourth line of the table gives the value of the output y_1 for the input values

$$x_1 = 1, \qquad x_2 = 0, \qquad x_3 = 0.$$

If $x_1 = 1$ and $x_2 = 0$, the output from the AND gate is 0 (see Figure 6.1.5). Since $x_3 = 0$, the inputs to the OR gate are both 0. Therefore, the output of the OR gate is 0. Since the input to the NOT gate is 0, it produces output $y_1 = 1$.

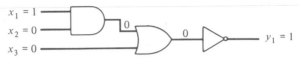

FIGURE 6.1.5

EXAMPLE 6.1.6. The circuit of Figure 6.1.6 is not a combinatorial circuit because the output y is not uniquely defined for each combination of inputs x_1 and x_2. For example, suppose that $x_1 = 1$ and $x_2 = 0$. If the output of the AND gate is 0, then $y = 0$. On the other hand, if the output of the AND gate is 1, then $y = 1$. Such a circuit might be used to store one bit.

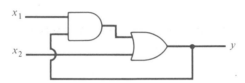

FIGURE 6.1.6

EXAMPLE 6.1.7. Individual combinatorial circuits may be interconnected. The combinatorial circuits C_1, C_2, and C_3 of Figure 6.1.7 may be combined, as shown, to obtain the combinatorial circuit C.

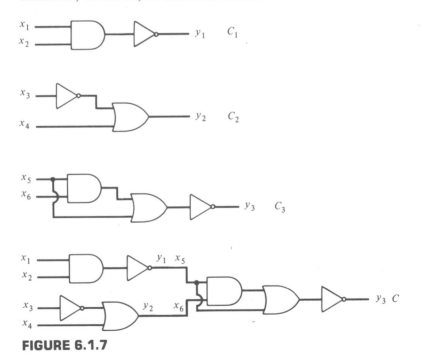

FIGURE 6.1.7

EXAMPLE 6.1.8. A combinatorial circuit with one output, like that in Figure 6.1.4, can be represented by an expression using the symbols $\wedge$, $\vee$, and $\overline{}$. We follow the flow of the circuit symbolically. First, x_1 and x_2 are ANDed (see Figure 6.1.8), which produces output $x_1 \wedge x_2$. This output is then ORed with x_3 to produce output $(x_1 \wedge x_2) \vee x_3$. This output is then NOTed. Thus the output y_1 may be written

$$y_1 = \overline{(x_1 \wedge x_2) \vee x_3}. \tag{6.1.1}$$

Expressions such as (6.1.1) are called **Boolean expressions**.

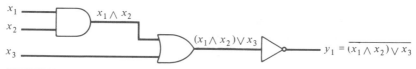

FIGURE 6.1.8

DEFINITION 6.1.9. *Boolean expressions* in the symbols $x_1, \ldots, x_n$ are defined recursively as follows.

$$0, 1, x_1, \ldots, x_n \text{ are Boolean expressions.} \tag{6.1.2}$$

If X_1 and X_2 are Boolean expressions, then

(a) (X_1) (b) $\overline{X}_1$

(c) $X_1 \vee X_2$ (d) $X_1 \wedge X_2$

$$\text{are Boolean expressions.} \tag{6.1.3}$$

If X is a Boolean expression in the symbols $x_1, \ldots, x_n$, we sometimes write

$$X = X(x_1, \ldots, x_n).$$

Either symbol x or $\overline{x}$ is called a *literal*.

EXAMPLE 6.1.10. Use Definition 6.1.9 to show that the right side of (6.1.1) is a Boolean expression in x_1, x_2, and x_3.

By (6.1.2), x_1 and x_2 are Boolean expressions. By (6.1.3d), $x_1 \wedge x_2$ is a Boolean expression. By (6.1.3a), $(x_1 \wedge x_2)$ is a Boolean expression. By (6.1.2), x_3 is a Boolean expression. Since $(x_1 \wedge x_2)$ and x_3 are Boolean expressions, by (6.1.3c), so is $(x_1 \wedge x_2) \vee x_3$. Finally, we may apply (6.1.3b), to conclude that

$$\overline{(x_1 \wedge x_2) \vee x_3}$$

is a Boolean expression.

If $X = X(x_1, \ldots, x_n)$ is a Boolean expression and $x_1, \ldots, x_n$ are assigned values $a_1, \ldots, a_n$ in $\{0, 1\}$, we may use Definitions 6.1.1–6.1.3 to compute a value for X. We denote this value $X(a_1, \ldots, a_n)$ or $X(x_i = a_i)$.

EXAMPLE 6.1.11. For $x_1 = 1$, $x_2 = 0$, and $x_3 = 0$, the Boolean expression $X(x_1, x_2, x_3) = \overline{(x_1 \wedge x_2) \vee x_3}$ of (6.1.1) becomes

$$X(1, 0, 0) = \overline{(1 \wedge 0) \vee 0}$$

$$= \overline{0 \vee 0} \qquad \text{since } 1 \wedge 0 = 0$$

$$= \overline{0} \qquad \text{since } 0 \vee 0 = 0$$

$$= 1 \qquad \text{since } \overline{0} = 1.$$

We have again computed the fourth row of the table in Example 6.1.5.

In a Boolean expression in which parentheses are not used to specify the order of operations, we assume that $\wedge$ is evaluated before $\vee$.

EXAMPLE 6.1.12. For $x_1 = 0$, $x_2 = 0$, and $x_3 = 1$, the value of the Boolean expression $x_1 \wedge x_2 \vee x_3$ is

$$x_1 \wedge x_2 \vee x_3 = 0 \wedge 0 \vee 1 = 0 \vee 1 = 1.$$

Example 6.1.8 showed how to represent a combinatorial circuit with one output as a Boolean expression. The following example shows how to construct a combinatorial circuit that represents a Boolean expression.

EXAMPLE 6.1.13. Find the combinatorial circuit corresponding to the Boolean expression

$$(x_1 \wedge (\overline{x}_2 \vee x_3)) \vee x_2$$

and write the logic table for the circuit obtained.

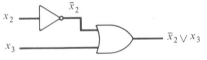

FIGURE 6.1.9

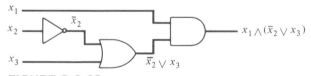

FIGURE 6.1.10

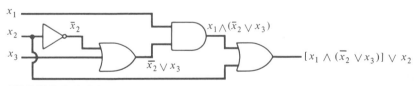

FIGURE 6.1.11

We begin with the expression $\bar{x}_2 \vee x_3$ in the innermost parentheses. This expression is converted to a combinatorial circuit as shown in Figure 6.1.9. The output of this circuit is ANDed with x_1 to produce the circuit drawn in Figure 6.1.10. Finally, the output of this circuit is ORed with x_2 to give the desired circuit drawn in Figure 6.1.11. The logic table is as follows:

x_1	x_2	x_3	$(x_1 \wedge (\bar{x}_2 \vee x_3)) \vee x_2$
1	1	1	1
1	1	0	1
1	0	1	1
1	0	0	1
0	1	1	1
0	1	0	1
0	0	1	0
0	0	0	0

EXERCISES

In Exercises 1–6, write the Boolean expression that represents the combinatorial circuit, write the logic table, and write the output of each gate symbolically as in Figure 6.1.8.

1H.

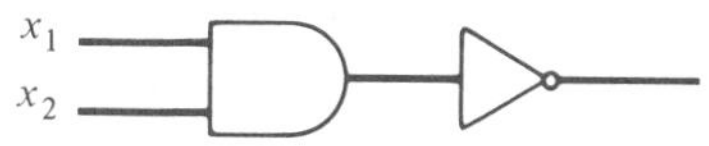

2.

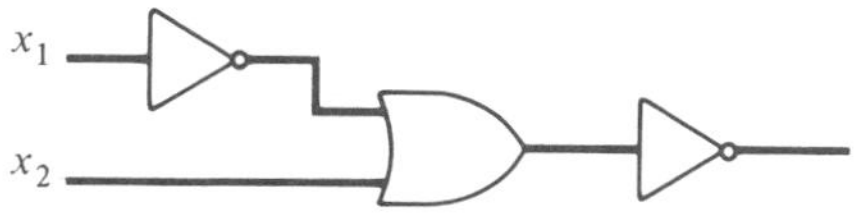

3H.

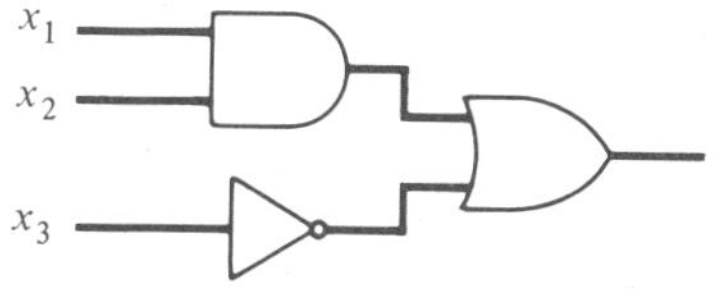

4.

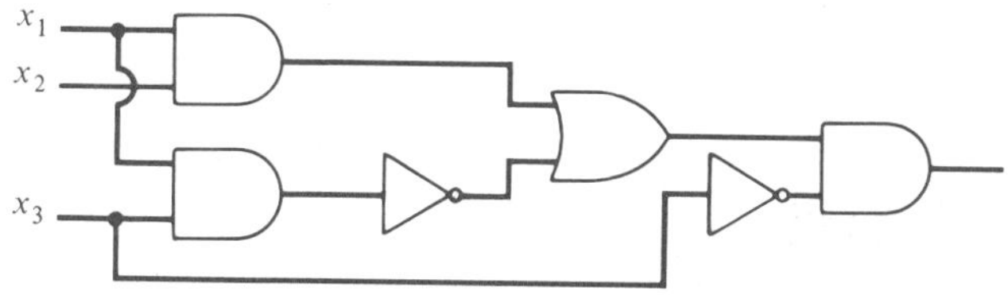

5H.

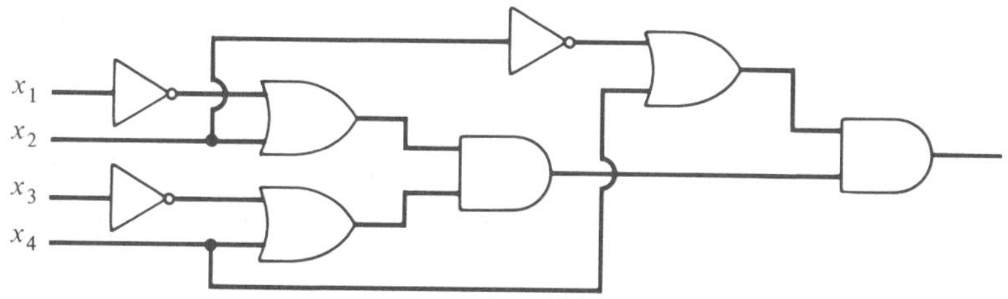

6.   The circuit at the bottom of Figure 6.1.7.

Exercises 7–9 refer to the circuit

7H.   Show that this circuit is not a combinatorial circuit.

8.   Show that if $x = 0$, the output y is uniquely determined.

9H.   Show that if $x = 1$, the output y is undetermined.

In Exercises 10–14, find the value of the Boolean expressions for

$$x_1 = 1, \quad x_2 = 1, \quad x_3 = 0, \quad x_4 = 1.$$

10. $\overline{x_1 \wedge x_2}$

11H. $x_1 \vee (\bar{x}_2 \wedge x_3)$

12. $(x_1 \wedge \bar{x}_2) \vee (x_1 \vee \bar{x}_3)$

13H. $(x_1 \wedge (x_2 \vee (x_1 \wedge \bar{x}_2))) \vee \overline{((x_1 \wedge \bar{x}_2) \vee (x_1 \wedge \bar{x}_3))}$

14. $(((x_1 \wedge x_2) \vee (x_3 \wedge \bar{x}_4)) \vee (\overline{(x_1 \vee x_3)} \wedge (\bar{x}_2 \vee x_3))) \vee (x_1 \wedge \bar{x}_3)$

15H.   Using Definition 6.1.9, show that each expression in Exercises 10–14 is a Boolean expression.

In Exercises 16–20, tell whether the given expression is a Boolean expression. If it is a Boolean expression, show that it is using Definition 6.1.9.

16. $x_1 \wedge (x_2 \vee x_3)$ **17H.** $x_1 \wedge \overline{x_2} \vee x_3$ **18.** (x_1)

19H. $((x_1 \wedge x_2) \vee \bar{x}_3$ **20.** $((x_1))$

21H.   Find the combinatorial circuit corresponding to each Boolean expression in Exercises 10–14 and write the logic table.

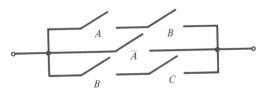

FIGURE 6.1.12

A **switching circuit** is an electrical network consisting of switches each of which is open or closed. An example is given in Figure 6.1.12. If switch X is open (closed), we write $X = 0$ ($X = 1$). Switches labeled with the same letter, such as B in Figure 6.1.12, are either all open or all closed. Switch X, like A in Figure 6.1.12, is open if and only if switch $\overline{X}$, like $\overline{A}$, is closed. If current can flow between the extreme left and right ends of the circuit, we say that the output of the circuit is 1; otherwise, we say that the output of the circuit is 0. A **switching table** gives the output of the circuit for all values of the switches. The switching table for Figure 6.1.12 is as follows:

A	B	C	Circuit Output
1	1	1	1
1	1	0	1
1	0	1	0
1	0	0	0
0	1	1	1
0	1	0	1
0	0	1	1
0	0	0	1

22. Draw a circuit with two switches A and B having the property that the circuit output is 1 precisely when both A and B are closed. This configuration is labeled $A \wedge B$ and is called a **series circuit**.

23H. Draw a circuit with two switches A and B having the property that the circuit output is 1 precisely when either A or B is closed. This configuration is labeled $A \vee B$ and is called a **parallel circuit**.

24. Show that the circuit of Figure 6.1.12 can be represented symbolically as

$$(A \wedge B) \vee \overline{A} \vee (B \wedge C).$$

Represent each circuit in Exercises 25–29 symbolically and give its switching table.

25H.

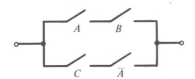

26.

27H.

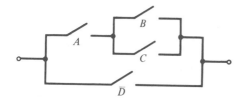

28.

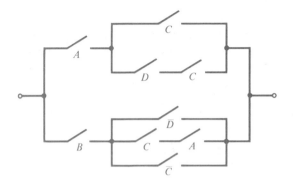

29H.

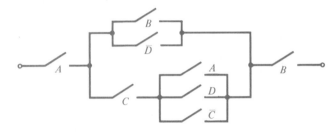

Represent the expressions in Exercises 30–34 as switching circuits and write the switching tables.

30. $(A \vee \bar{B}) \wedge A$ **31H.** $A \vee (\bar{B} \wedge C)$

32. $(\bar{A} \wedge B) \vee (C \wedge A)$ **33H.** $(A \wedge ((B \wedge \bar{C}) \vee (\bar{B} \wedge C)))$
                                                          $\vee (\bar{A} \wedge B \wedge C)$

34. $A \wedge ((B \wedge C \wedge \bar{D}) \vee ((\bar{B} \wedge C) \vee D) \vee (\bar{B} \wedge \bar{C} \wedge D)) \wedge (B \vee \bar{D})$

35H.   Does every Boolean expression represent a switching circuit? Why or why not?

6.2

Properties of Combinatorial Circuits

In the preceding section we defined two binary operators $\wedge$ and $\vee$ on $Z_2 = \{0, 1\}$ and a unary operator $\bar{\ }$ on Z_2. (Throughout the remainder of this chapter we let Z_2 denote the set $\{0, 1\}$.) We saw that these operators could be implemented in circuits as gates. In this section we discuss some properties of the system consisting of Z_2 and the operators $\wedge$, $\vee$, and $\bar{\ }$.

> **THEOREM 6.2.1.**   *If $\wedge$, $\vee$, and $\bar{\ }$ are as in Definitions 6.1.1, 6.1.2, and 6.1.3, then the following properties hold.*
> (a) *Associative laws:*
>
> $(a \vee b) \vee c = a \vee (b \vee c)$ $(a \wedge b) \wedge c = a \wedge (b \wedge c)$
> $$\textit{for all } a, b, c \in Z_2.$$

(b) *Commutative laws*:

$$a \vee b = b \vee a \qquad a \wedge b = b \wedge a \qquad \text{for all } a, b, \in Z_2.$$

(c) *Distributive laws*:

$$a \wedge (b \vee c) = (a \wedge b) \vee (a \wedge c)$$

$$a \vee (b \wedge c) = (a \vee b) \wedge (a \vee c) \qquad \text{for all } a, b, c, \in Z_2.$$

(d) *Identity laws*:

$$a \vee 0 = a \qquad a \wedge 1 = a \qquad \text{for all } a \in Z_2.$$

(e) *Complement laws*:

$$a \vee \bar{a} = 1, \quad a \wedge \bar{a} = 0 \qquad \text{for all } a \in Z_2.$$

Proof. The proofs are straightforward verifications. We shall prove the first Distributive Law only and leave the other equations as exercises (see Exercises 16 and 17).

We must show that

$$a \wedge (b \vee c) = (a \wedge b) \vee (a \wedge c) \qquad \text{for all } a, b, c \in Z_2. \qquad (6.2.1)$$

We simply evaluate both sides of (6.2.1) for all possible values of a, b, and c in Z_2 and verify that in each case we obtain the same result. The table gives the details.

a	b	c	$a \wedge (b \vee c)$	$(a \wedge b) \vee (a \wedge c)$
1	1	1	1	1
1	1	0	1	1
1	0	1	1	1
1	0	0	0	0
0	1	1	0	0
0	1	0	0	0
0	0	1	0	0
0	0	0	0	0

■

EXAMPLE 6.2.2. By using Theorem 6.2.1, show that the combinatorial circuits of Figure 6.2.1 have identical outputs for given identical inputs.

The Boolean expressions representing the circuits are, respectively,

$$x_1 \vee (x_2 \wedge x_3) \qquad (x_1 \vee x_2) \wedge (x_1 \vee x_3).$$

By Theorem 6.2.1c,

$$a \vee (b \wedge c) = (a \vee b) \wedge (a \vee c) \qquad \text{for all } a, b, c, \in Z_2. \qquad (6.2.2)$$

But (6.2.2) says that the combinatorial circuits of Figure 6.2.1 have identical outputs for given identical inputs.

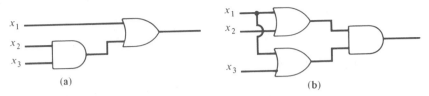

(a)   (b)

FIGURE 6.2.1

Arbitrary Boolean expressions are defined to be equal if they have the same values for all possible assignments of bits to the literals.

DEFINITION 6.2.3. Let

$$X_1 = X_1(x_1, \ldots, x_n) \quad \text{and} \quad X_2 = X_2(x_1, \ldots, x_n)$$

be Boolean expressions. We define X_1 to be *equal* to X_2 and write

$$X_1 = X_2$$

if

$$X_1(a_1, \ldots, a_n) = X_2(a_1, \ldots, a_n) \quad \text{for all } a_i \in Z_2.$$

EXAMPLE 6.2.4. Show that

$$\overline{(x \vee y)} = \bar{x} \wedge \bar{y}. \tag{6.2.3}$$

According to Definition 6.2.3, (6.2.3) holds if the equation is true for all choices of x and y in Z_2. Thus we may simply construct a table listing all possibilities to verify (6.2.3).

x	y	$\overline{(x \vee y)}$	$\bar{x} \wedge \bar{y}$
1	1	0	0
1	0	0	0
0	1	0	0
0	0	1	1

Because of the Associative Laws, Theorem 6.2.1a, we can unambiguously write

$$a_1 \vee a_2 \vee \cdots \vee a_n \tag{6.2.4}$$

or

$$a_1 \wedge a_2 \wedge \cdots \wedge a_n \tag{6.2.5}$$

for $a_i \in Z_2$. The combinatorial circuit corresponding to (6.2.4) is drawn as in Figure 6.2.2 and the combinatorial circuit corresponding to (6.2.5) is drawn as in Figure 6.2.3.

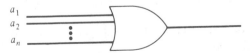

FIGURE 6.2.2

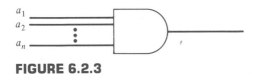

FIGURE 6.2.3

The properties listed in Theorem 6.2.1 hold for a variety of systems. Any system satisfying these properties is called a **Boolean algebra**. Abstract Boolean algebras are examined in Section 6.3.

Having defined equality of Boolean expressions, we define equivalence of combinatorial circuits.

DEFINITION 6.2.5. We say that two combinatorial circuits, each having inputs $x_1, \ldots, x_n$ and a single output, are *equivalent* if, whenever the circuits receive the same inputs, they produce the same outputs.

EXAMPLE 6.2.6. The combinatorial circuits of Figures 6.2.4 and 6.2.5 are equivalent since, as shown, they have identical logic tables.

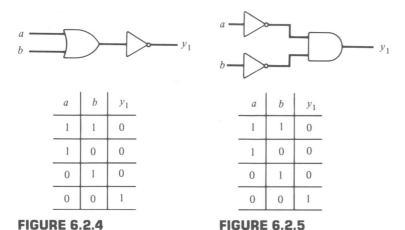

a	b	y_1
1	1	0
1	0	0
0	1	0
0	0	1

FIGURE 6.2.4

a	b	y_1
1	1	0
1	0	0
0	1	0
0	0	1

FIGURE 6.2.5

Example 6.2.6 shows that equivalent circuits may not have the same number of gates. In general, it is desirable to use as few gates as possible to minimize the cost of the components.

It follows immediately from the definitions that combinatorial circuits are equivalent if and only if the Boolean expressions that represent them are equal.

THEOREM 6.2.7. *Let C_1 and C_2 be combinatorial circuits represented, respectively, by the Boolean expressions $X_1 = X_1(x_1, \ldots, x_n)$ and $X_2 = X_2(x_1, \ldots, x_n)$. Then C_1 and C_2 are equivalent if and only if $X_1 = X_2$.*

Proof. The value $X_1(a_1, \ldots, a_n)$ [respectively, $X_2(a_1, \ldots, a_n)$] for $a_i \in Z_2$ is the output for circuit C_1 (respectively, C_2) for inputs $a_1, \ldots, a_n$.

According to Definition 6.2.5, circuits C_1 and C_2 are equivalent if and only if they have the same outputs $X_1(a_1, \ldots, a_n)$ and $X_2(a_1, \ldots, a_n)$ for all possible inputs $a_1, \ldots, a_n$. Thus circuits C_1 and C_2 are equivalent if and only if

$$X_1(a_1, \ldots, a_n) = X_2(a_1, \ldots, a_n) \qquad \text{for all values } a_i \in Z_2. \quad (6.2.6)$$

But by Definition 6.2.3, (6.2.6) holds if and only if $X_1 = X_2$. ∎

EXAMPLE 6.2.8. In Example 6.2.4 we showed that

$$\overline{(x \vee y)} = \bar{x} \wedge \bar{y}.$$

By Theorem 6.2.7, the combinatorial circuits (Figures 6.2.4 and 6.2.5) corresponding to these expressions are equivalent.

EXERCISES

Show that the combinatorial circuits of Exercises 1–5 are equivalent.

1H.

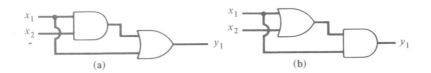

(a)   (b)

2.

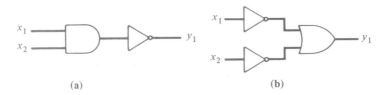

(a)   (b)

3H.

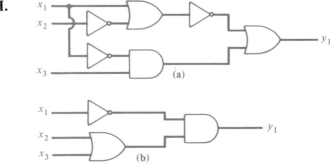

(a)

(b)

4.

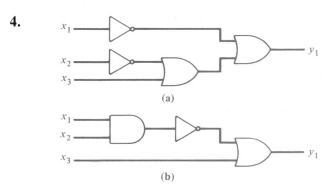

(a)

(b)

5H.

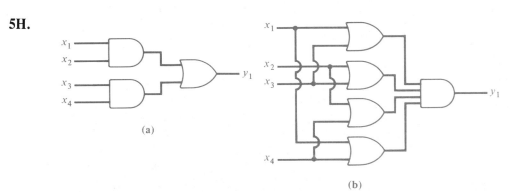

(a)

(b)

Verify the equations in Exercises 6–10.

6. $x_1 \vee x_1 = x_1$ **7H.** $x_1 \vee (x_1 \wedge x_2) = x_1$

8. $x_1 \wedge \bar{x}_2 = \overline{(\bar{x}_1 \vee x_2)}$ **9H.** $x_1 \wedge \overline{(x_2 \wedge x_3)} = (x_1 \wedge \bar{x}_2) \vee (x_1 \wedge \bar{x}_3)$

10. $(x_1 \vee x_2) \wedge (x_3 \vee x_4) = (x_3 \wedge x_1) \vee (x_3 \wedge x_2) \vee (x_4 \wedge x_1) \vee (x_4 \wedge x_2)$

Prove or disprove the equations in Exercises 11–15.

11H. $\bar{\bar{x}} = x$

12. $\bar{x}_1 \wedge \bar{x}_2 = x_1 \vee x_2$

13H. $\bar{x}_1 \wedge ((x_2 \wedge x_3) \vee (x_1 \wedge x_2 \wedge x_3)) = x_2 \wedge x_3$

14. $\overline{((\bar{x}_1 \wedge x_2) \vee (x_1 \wedge \bar{x}_3))} = (x_1 \vee \bar{x}_2) \wedge (x_1 \vee \bar{x}_3)$

15H. $(x_1 \vee x_2) \wedge (\bar{x}_3 \vee x_4) \wedge (x_3 \wedge \bar{x}_2) = 0$

16. Prove the second statement of Theorem 6.2.1c.

17. Prove Theorem 6.2.1, parts (a), (b), (d), and (e).

We say that two switching circuits are **equivalent** if the Boolean expressions that represent them are equal.

18. Show that the switching circuits shown are equivalent.

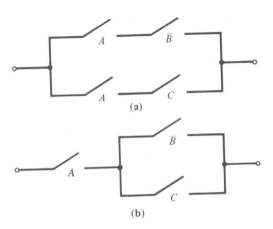

(a)

(b)

19H. For each switching circuit in Exercises 25–29, Section 6.1, find an equivalent switching circuit using parallel and series circuits having as few switches as you can.

20. For each Boolean expression in Exercises 30–34, Section 6.1, find a switching circuit using parallel and series circuits having as few switches as you can.

A **bridge circuit** is a switching circuit, like that shown, which uses nonparallel and nonseries circuits.

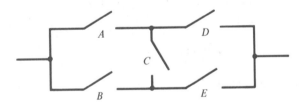

For each switching circuit find an equivalent switching circuit using bridge circuits having as few switches as you can.

21H.

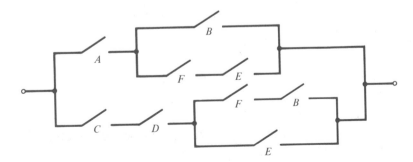

22.

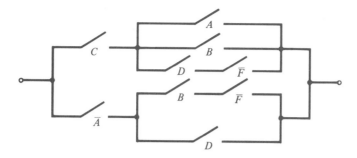

★23H.

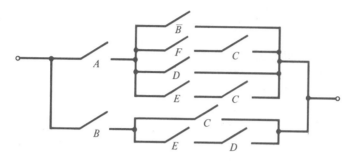

24. For each Boolean expression in Exercises 30–34, Section 6.1, find a switching circuit using bridge circuits having as few switches as you can.

6.3

Boolean Algebras

In this section we consider general systems that have properties like those given in Theorem 6.2.1. We will see that apparently diverse systems obey these same laws. We call such systems Boolean algebras.

DEFINITION 6.3.1. A *Boolean algebra B* consists of a set *S* containing distinct elements 0 and 1, binary operators + and · on *S*, and a unary operator ′ on *S* satisfying

(a) *Associative laws*:

$$(x + y) + z = x + (y + z) \qquad (x \cdot y) \cdot z = x \cdot (y \cdot z)$$
$$\text{for all } x, y, z \in S.$$

(b) *Commutative laws*:

$$x + y = y + x \qquad x \cdot y = y \cdot x \qquad \text{for all } x, y \in S.$$

(c) *Distributive laws*:

$$x \cdot (y + z) = (x \cdot y) + (x \cdot z)$$
$$x + (y \cdot z) = (x + y) \cdot (x + z) \qquad \text{for all } x, y, z \in S.$$

(d) *Identity laws*:

$$x + 0 = x \qquad x \cdot 1 = x \qquad \text{for all } x \in S.$$

(e) *Complement laws*:

$$x + x' = 1 \qquad x \cdot x' = 0 \qquad \text{for all } x \in S.$$

If B is a Boolean algebra, we write $B = (S, +, \cdot, ', 0, 1)$.

EXAMPLE 6.3.2. By Theorem 6.2.1, $(Z_2, \vee, \wedge, {}^{-}, 0, 1)$ is a Boolean algebra. (We are letting Z_2 denote the set $\{0, 1\}$.) The operators $+, \cdot, '$ in Definition 6.3.1 are $\vee, \wedge, {}^{-}$, respectively.

As is the standard custom, we will usually abbreviate $a \cdot b$ as ab. We also assume that $\cdot$ is evaluated before $+$. This allows us to eliminate some parentheses. For example, we can write $(xy) + z$ more simply as $xy + z$.

Several comments are in order concerning Definition 6.3.1. In the first place, 0 and 1 are merely symbolic names and, in general, have nothing to do with the numbers 0 and 1. This same comment applies to $+$ and $\cdot$, which merely denote binary operators and, in general, have nothing to do with ordinary addition and multiplication.

EXAMPLE 6.3.3. Let U be a universal set and let $S = \mathcal{P}(U)$, the power set of U. If we define the following operations

$$X + Y = X \cup Y$$

$$X \cdot Y = X \cap Y$$

$$X' = \overline{X}$$

on S, then $(S, \cup, \cap, {}^{-}, \varnothing, U)$ is a Boolean algebra. The empty set $\varnothing$ plays the role of 0 and the universal set U plays the role of 1. If we let X, Y, and Z be subsets of S, properties (a)–(e) of Definition 6.3.1 become the following well-known properties of sets:

(a') $(X \cup Y) \cup Z = X \cup (Y \cup Z) \qquad (X \cap Y) \cap Z = X \cap (Y \cap Z)$
     for all $X, Y, Z \in \mathcal{P}(U)$.
(b') $X \cup Y = Y \cup X \qquad X \cap Y = Y \cap X \qquad$ for all $X, Y \in \mathcal{P}(U)$.
(c') $X \cap (Y \cup Z) = (X \cap Y) \cup (X \cap Z)$
     $X \cup (Y \cap Z) = (X \cup Y) \cap (X \cup Z) \qquad$ for all $X, Y, Z \in \mathcal{P}(U)$.
(d') $X \cup \varnothing = X \qquad X \cap U = X \qquad$ for every $X \in \mathcal{P}(U)$.
(e') $X \cup \overline{X} = U \qquad X \cap \overline{X} = \varnothing \qquad$ for every $X \in \mathcal{P}(U)$.

At this point we will deduce several other properties of Boolean algebras. We begin by showing that the element x' in Definition 6.3.1e is unique.

THEOREM 6.3.4. *In a Boolean algebra, the element x' of Definition 6.3.1e is unique. Specifically, if $x + y = 1$ and $xy = 0$, then $y = x'$.*

Proof

$$y = y1 \qquad \text{Definition 6.3.1d}$$
$$= y(x + x') \qquad \text{Definition 6.3.1e}$$
$$= yx + yx' \qquad \text{Definition 6.3.1c}$$
$$= xy + yx' \qquad \text{Definition 6.3.1b}$$
$$= 0 + yx' \qquad \text{Given}$$
$$= xx' + yx' \qquad \text{Definition 6.3.1e}$$
$$= x'x + x'y \qquad \text{Definition 6.3.1b}$$
$$= x'(x + y) \qquad \text{Definition 6.3.1c}$$
$$= x'1 \qquad \text{Given}$$
$$= x' \qquad \text{Definition 6.3.1d} \qquad \blacksquare$$

DEFINITION 6.3.5. In a Boolean algebra, we call the element x' the *complement* of x.

We can now derive several additional properties of Boolean algebras.

THEOREM 6.3.6. *Let* $B = (S, +, \cdot, ', 0, 1)$ *be a Boolean algebra. The following properties hold.*
(a) *Idempotent laws*:

$$x + x = x \qquad xx = x \qquad \text{for all } x \in S.$$

(b) *Bound laws*:

$$x + 1 = 1 \qquad x0 = 0 \qquad \text{for all } x \in S.$$

(c) *Absorption laws*:

$$x + xy = x \qquad x(x + y) = x \qquad \text{for all } x, y \in S.$$

(d) *Involution law*:

$$(x')' = x \qquad \text{for all } x \in S.$$

(e) 0 *and* 1 *laws*:

$$0' = 1, \qquad 1' = 0.$$

(f) *DeMorgan's laws*:

$$(x + y)' = x'y' \qquad (xy)' = x' + y' \qquad \text{for all } x, y \in S.$$

Proof. We will prove (b) and the first statement of parts (a), (c), and (f) and leave the others as exercises (see Exercises 18–20).

(a)
$$x = x + 0 \qquad \text{Definition 6.3.1d}$$
$$= x + (xx') \qquad \text{Definition 6.3.1e}$$
$$= (x + x)(x + x') \qquad \text{Definition 6.3.1c}$$
$$= (x + x)1 \qquad \text{Definition 6.3.1e}$$
$$= x + x \qquad \text{Definition 6.3.1d}$$

(b)
$$x + 1 = (x + 1)1 \qquad \text{Definition 6.3.1d}$$
$$= (x + 1)(x + x') \qquad \text{Definition 6.3.1e}$$
$$= x + 1x' \qquad \text{Definition 6.3.1c}$$
$$= x + x'1 \qquad \text{Definition 6.3.1b}$$
$$= x + x' \qquad \text{Definition 6.3.1d}$$
$$= 1 \qquad \text{Definition 6.3.1e}$$
$$x0 = x0 + 0 \qquad \text{Definition 6.3.1d}$$
$$= x0 + xx' \qquad \text{Definition 6.3.1e}$$
$$= x(0 + x') \qquad \text{Definition 6.3.1c}$$
$$= x(x' + 0) \qquad \text{Definition 6.3.1b}$$
$$= xx' \qquad \text{Definition 6.3.1d}$$
$$= 0 \qquad \text{Definition 6.3.1e}$$

(c)
$$x + xy = x1 + xy \qquad \text{Definition 6.3.1d}$$
$$= x(1 + y) \qquad \text{Definition 6.3.1c}$$
$$= x(y + 1) \qquad \text{Definition 6.3.1b}$$
$$= x1 \qquad \text{Part (b)}$$
$$= x \qquad \text{Definition 6.3.1d}$$

(f) If we show that

$$(x + y)(x'y') = 0 \qquad (6.3.1)$$

and

$$(x + y) + x'y' = 1, \qquad (6.3.2)$$

it will follow from Theorem 6.3.4 that $x'y' = (x + y)'$. Now

$$
\begin{align*}
(x + y)(x'y') &= (x'y')(x + y) & \text{Definition 6.3.1b}\\
&= (x'y')x + (x'y')y & \text{Definition 6.3.1c}\\
&= x(x'y') + (x'y')y & \text{Definition 6.3.1b}\\
&= (xx')y' + x'(y'y) & \text{Definition 6.3.1a}\\
&= (xx')y' + x'(yy') & \text{Definition 6.3.1b}\\
&= 0y' + x'0 & \text{Definition 6.3.1e}\\
&= y'0 + x'0 & \text{Definition 6.3.1b}\\
&= 0 + 0 & \text{Part (b)}\\
&= 0 & \text{Definition 6.3.1d}
\end{align*}
$$

Therefore, (6.3.1) holds.

Next we verify (6.3.2).

$$
\begin{align*}
(x + y) + x'y' &= ((x + y) + x')((x + y) + y') & \text{Definition 6.3.1c}\\
&= ((y + x) + x')((x + y) + y') & \text{Definition 6.3.1b}\\
&= (y + (x + x'))(x + (y + y')) & \text{Definition 6.3.1a}\\
&= (y + 1)(x + 1) & \text{Definition 6.3.1e}\\
&= 1 \cdot 1 & \text{Part (b)}\\
&= 1 & \text{Definition 6.3.1d}
\end{align*}
$$

By Theorem 6.3.4, $x'y' = (x + y)'$. ∎

EXAMPLE 6.3.7. As explained in Example 6.3.3, if U is a set, $\mathcal{P}(U)$ can be considered a Boolean algebra. Therefore, DeMorgan's Laws, which for sets may be stated

$$(\overline{X \cup Y}) = \overline{X} \cap \overline{Y}, \quad (\overline{X \cap Y}) = \overline{X} \cup \overline{Y} \qquad \text{for all } X, Y \in \mathcal{P}(U),$$

hold. These equations may be verified directly, but Theorem 6.3.6 shows that they are a consequence of other laws.

The reader has surely noticed that equations involving elements of a Boolean algebra come in pairs. For example, the Identity Laws (Definition 6.3.1d) are

$$x + 0 = x, \qquad x1 = x.$$

Such pairs are said to be **dual**.

DEFINITION 6.3.8. The *dual* of a statement involving Boolean expressions is obtained by replacing 0 by 1, 1 by 0, + by ·, and · by + .

EXAMPLE 6.3.9.   The dual of

$$(x + y)' = x'y'$$

is

$$(xy)' = x' + y'.$$

Each condition in the definition of a Boolean algebra (Definition 6.3.1) includes its dual. Therefore, we have the following result.

THEOREM 6.3.10.   *The dual of a theorem about Boolean algebras is also a theorem.*

Proof.   Suppose that T is a theorem about Boolean algebras. Then there is a proof P of T involving only the definitions of a Boolean algebra (Definition 6.3.1). Let P' be the sequence of statements obtained by replacing every statement in P by its dual. Then P' is a proof of the dual of T.   ∎

EXAMPLE 6.3.11.   The dual of

$$x + x = x \tag{6.3.3}$$

is

$$xx = x. \tag{6.3.4}$$

We proved (6.3.3) earlier (see the proof of Theorem 6.3.6a). If we write the dual of each statement in the proof of (6.3.3), we obtain the following proof of (6.3.4):

$$
\begin{aligned}
x &= x1 \\
&= x(x + x') \\
&= xx + xx' \\
&= xx + 0 \\
&= xx.
\end{aligned}
$$

EXAMPLE 6.3.12.   The proofs given in Theorem 6.3.6 of the two statements of part (b) are dual to each other.

EXERCISES

1.   Verify properties (a′)–(e′) of Example 6.3.3.

2.   Let $S = \{1, 2, 3, 6\}$. Define

$$x + y = \text{lcm}\,(x, y), \qquad x \cdot y = \text{gcd}\,(x, y), \qquad x' = \frac{6}{x}$$

for $x, y \in S$. (lcm and gcd denote, respectively, the least common multiple and the greatest common divisor.) Show that $(S, +, \cdot, ', 1, 6)$ is a Boolean algebra.

3H. Let $S = \{1, 2, 4, 8\}$. Define $+$ and $\cdot$ as in Exercise 2 and define $x' = 8/x$. Show that $(S, +, \cdot, ', 1, 8)$ is not a Boolean algebra.

Let $S_n = \{1, 2, \ldots, n\}$. Define

$$x + y = \max \{x, y\}, \qquad x \cdot y = \min \{x, y\}.$$

4. Show that (a), (b), and (c) of Definition 6.3.1 hold for S_n.

5H. Show that it is possible to define 0, 1, and $'$ so that $(S_n, +, \cdot, ', 0, 1)$ is a Boolean algebra if and only if $n = 2$.

6. Rewrite the conditions of Theorem 6.3.6 for sets as in Example 6.3.3.

7H. Interpret Theorem 6.3.4 for sets as in Example 6.3.3.

Write the dual of each statement in Exercises 8–14.

8. $(x + y)(x + 1) = x + xy + y$

9H. $(x' + y')' = xy$

10. If $x + y = x + z$ and $x' + y = x' + z$, then $y = z$.

11H. $xy' = 0$ if and only if $xy = x$.

12. If $x + y = 0$, then $x = 0 = y$.

13H. $x = 0$ if and only if $y = xy' + x'y$ for all y.

14. $x + x(y + 1) = x$

15H. Prove the statements of Exercises 8–14.

16. Prove the duals of the statements of Exercises 8–14.

17H. Write the dual of Theorem 6.3.4. How does the dual relate to Theorem 6.3.4 itself?

18. Prove the second statements of parts (a), (c), and (f) of Theorem 6.3.6.

19H. Prove the second statements of parts (a), (c), and (f) of Theorem 6.3.6 by dualizing the proofs of the first statements given in the text.

20. Prove Theorem 6.3.6, parts (d) and (e).

★21H. Deduce part (a) of Definition 6.3.1 from parts (b)–(e) of Definition 6.3.1.

22. Let U be the set of positive integers. Let S be the collection of subsets X of U with either X or $\overline{X}$ finite. Show that $(S, \cup, \cap, \overline{}, \emptyset, U)$ is a Boolean algebra.

23H. Let $(S, +, \cdot, ', 0, 1)$ be a Boolean algebra and let A be a subset of S. Show that $(A, +, \cdot, ', 0, 1)$ is a Boolean algebra if and only if $1 \in A$ and, whenever $x, y \in A$, we have $x\overline{y} \in A$.

★24.   Let n be a positive integer. Let S be the set of all divisors of n, including 1 and n. Define $+$ and $\cdot$ as in Exercise 2 and define $x' = n/x$. What conditions must n satisfy so that $(S, +, \cdot, ', 1, n)$ is a Boolean algebra?

6.4

Boolean Functions and Synthesis of Circuits

A circuit is constructed to carry out a specified task. If we want to construct a combinatorial circuit, the problem can be given in terms of inputs and outputs. For example, suppose that we want to construct a combinatorial circuit to compute the **exclusive-OR** of x_1 and x_2. We can state the problem by listing the inputs and outputs that define the exclusive-OR. This is equivalent to giving the desired logic table.

> **DEFINITION 6.4.1.**   The *exclusive-OR* of x_1 and x_2, written $x_1 \oplus x_2$, is defined by Table 6.4.1

TABLE 6.4.1

x_1	x_2	$x_1 \oplus x_2$
1	1	0
1	0	1
0	1	1
0	0	0

A logic table, with one output, is a function. The domain is the set of inputs and the range is the set of outputs. For the exclusive-OR function given in Table 6.4.1, the domain is the set

$$\{(1, 1), (1, 0), (0, 1), (0, 0)\}$$

and the range is the set

$$Z_2 = \{0, 1\}.$$

If we could develop a formula for the exclusive-OR function of the form

$$x_1 \oplus x_2 = X(x_1, x_2),$$

where X is a Boolean expression, we could solve the problem of constructing the combinatorial circuit. We could merely construct the circuit corresponding to X.

Functions that can be represented by Boolean expressions are called **Boolean functions**.

DEFINITION 6.4.2. Let $X(x_1, \ldots, x_n)$ be a Boolean expression. A function f of the form

$$f(x_1, \ldots, x_n) = X(x_1, \ldots, x_n)$$

is called a *Boolean function*.

EXAMPLE 6.4.3. The function $f : Z_2^3 \to Z_2$ defined by

$$f(x_1, x_2, x_3) = x_1 \wedge (\bar{x}_2 \vee x_3)$$

is a Boolean function. The inputs and outputs are given in the following table.

x_1	x_2	x_3	$f(x_1, x_2, x_3)$
1	1	1	1
1	1	0	0
1	0	1	1
1	0	0	1
0	1	1	0
0	1	0	0
0	0	1	0
0	0	0	0

In the next example we show how an arbitrary function $f : Z_2^n \to Z_2$ can be realized as a Boolean function.

EXAMPLE 6.4.4. Show that the function f given by the table is a Boolean function.

x_1	x_2	x_3	$f(x_1, x_2, x_3)$
1	1	1	1
1	1	0	0
1	0	1	0
1	0	0	1
0	1	1	0
0	1	0	1
0	0	1	0
0	0	0	0

Consider the first row of the table and the combination

$$x_1 \wedge x_2 \wedge x_3. \tag{6.4.1}$$

Notice that if $x_1 = x_2 = x_3 = 1$, as indicated in the first row of the table, then (6.4.1) is 1. The values of x_i given by any other row of the table give

(6.4.1) the value 0. Similarly, for the fourth row of the table we may construct the combination

$$x_1 \wedge \bar{x}_2 \wedge \bar{x}_3. \tag{6.4.2}$$

Expression (6.4.2) has the value 1 for the values of x_i given by the fourth row of the table, whereas the values of x_i given by any other row of the table give (6.4.2) the value 0.

The procedure is clear. We consider a row R of the table where the output is 1. We then form the combination $x_1 \wedge x_2 \wedge x_3$ and place a bar over each x_i whose value is 0 in row R. The combination formed is 1 if and only if the x_i have the values given in row R. Thus, for row 6, we obtain the combination

$$\bar{x}_1 \wedge x_2 \wedge \bar{x}_3. \tag{6.4.3}$$

Next, we OR the terms (6.4.1)–(6.4.3) to obtain the Boolean expression

$$(x_1 \wedge x_2 (\wedge x_3) \vee (x_1 \wedge \bar{x}_2 \wedge \bar{x}_3) \vee (\bar{x}_1 \wedge x_2 \wedge \bar{x}_3). \tag{6.4.4}$$

We claim that $f(x_1, x_2, x_3)$ and (6.4.4) are equal. To verify this, first suppose that x_1, x_2, and x_3 have values given by a row of the table for which $f(x_1, x_2, x_3) = 1$. Then one of (6.4.1)–(6.4.3) is 1, so the value of (6.4.4) is 1. On the other hand, if x_1, x_2, x_3 have values given by a row of the table for which $f(x_1, x_2, x_3) = 0$, all of (6.4.1)–(6.4.3) are 0, so the value of (6.4.4) is 0. Thus f and the Boolean expression (6.4.4) agree on Z_2^3; therefore,

$$f(x_1, x_2, x_3) = (x_1 \wedge x_2 \wedge x_3) \vee (x_1 \wedge \bar{x}_2 \wedge \bar{x}_3) \vee (\bar{x}_1 \wedge x_2 \wedge \bar{x}_3),$$

as claimed.

After one more definition, we will show that the method of Example 6.4.4 can be used to represent any function $f : Z_2^n \to Z_2$.

DEFINITION 6.4.5. A *minterm* in the symbols $x_1, \ldots, x_n$ is a Boolean expression of the form

$$y_1 \wedge y_2 \wedge \cdots \wedge y_n,$$

where each y_i is either x_i or $\bar{x}_i$.

THEOREM 6.4.6. *If $f : Z_2^n \to Z_2$, then f is a Boolean function. If f is not identically zero, let $A_1, \ldots, A_k$ denote the elements A_i of Z_2^n for which $f(A_i) = 1$. For each $A_i = (a_1, \ldots, a_n)$, set*

$$m_i = y_1 \wedge \cdots \wedge y_n$$

where

$$y_i = \begin{cases} x_i & \text{if } a_i = 1 \\ \bar{x}_i & \text{if } a_i = 0. \end{cases}$$

Then

$$f(x_1, \ldots, x_n) = m_1 \vee m_2 \vee \cdots \vee m_k. \tag{6.4.5}$$

Proof. If $f(x_1, \ldots, x_n) = 0$ for all x_i, then f is a Boolean function, since 0 is a Boolean expression.

Suppose that f is not identically zero. Let $m_i(a_1, \ldots, a_n)$ denote the value obtained from m_i by replacing each x_i with a_i. It follows from the definition of m_i that

$$m_i(A) = \begin{cases} 1 & \text{if } A = A_i \\ 0 & \text{if } A \neq A_i. \end{cases}$$

Let $A \in Z_2^n$. If $A = A_i$ for some $i \in \{1, \ldots, k\}$, then $f(A) = 1$, $m_i(A) = 1$, and

$$m_1(A) \vee \cdots \vee m_k(A) = 1.$$

On the other hand, if $A \neq A_i$ for any $i \in \{1, \ldots, k\}$, then $f(A) = 0$, $m_i(A) = 0$ for $i = 1, \ldots, k$, and

$$m_1(A) \vee \cdots \vee m_k(A) = 0.$$

Therefore, (6.4.5) holds. ■

DEFINITION 6.4.7. The representation (6.4.5) of a Boolean function $f: Z_2^n \to Z_2$ is called the *disjunctive normal form* of the function f.

EXAMPLE 6.4.8. Design a combinatorial circuit that computes the exclusive-OR of x_1 and x_2.

The logic table for the exclusive-OR function $x_1 \oplus x_2$ is given in Table 6.4.1. The disjunctive normal form for this function is

$$x_1 \oplus x_2 = (x_1 \wedge \bar{x}_2) \vee (\bar{x}_1 \wedge x_2). \tag{6.4.6}$$

The combinatorial circuit corresponding to (6.4.6) is given in Figure 6.4.1.

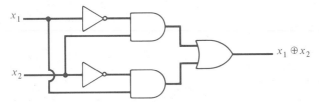

FIGURE 6.4.1

Suppose that a function is given by a Boolean expression such as

$$f(x_1, x_2, x_3) = (x_1 \vee x_2) \wedge x_3$$

and we wish to find the disjunctive normal form of f. We could write the logic table for f and then use Theorem 6.4.6. Alternatively, we can deal directly with the Boolean expression by using the definitions and results of Sections 6.2 and 6.3. We begin by distributing the term x_3 as follows:

$$(x_1 \vee x_2) \wedge x_3 = (x_1 \wedge x_3) \vee (x_2 \wedge x_3).$$

Although this represents the Boolean expression as a combination of terms of the form $y \wedge z$, it is not in disjunctive normal form, since each term does not contain all of the symbols x_1, x_2, and x_3. However, this is easily remedied, as follows:

$$(x_1 \wedge x_3) \vee (x_2 \wedge x_3) = (x_1 \wedge x_3 \wedge 1) \vee (x_2 \wedge x_3 \wedge 1)$$

$$= (x_1 \wedge x_3 \wedge (x_2 \vee \bar{x}_2)) \vee (x_2 \wedge x_3 \wedge (x_1 \vee \bar{x}_1))$$

$$= (x_1 \wedge x_2 \wedge x_3) \vee (x_1 \wedge \bar{x}_2 \wedge x_3) \vee (x_1 \wedge x_2 \wedge x_3)$$

$$\vee (\bar{x}_1 \wedge x_2 \wedge x_3)$$

$$= (x_1 \wedge x_2 \wedge x_3) \vee (x_1 \wedge \bar{x}_2 \wedge x_3) \vee (\bar{x}_1 \wedge x_2 \wedge x_3).$$

This expression is the disjunctive normal form of f.

Theorem 6.4.6 has a dual. In this case the function f is expressed as

$$f(x_1, \ldots, x_n) = M_1 \wedge M_2 \wedge \cdots \wedge M_k. \tag{6.4.7}$$

Each M_i is of the form

$$y_1 \vee \cdots \vee y_n \tag{6.4.8}$$

where y_i is either x_i or $\bar{x}_i$. A term of the form (6.4.8) is called a **maxterm** and the representation of f (6.4.7) is called the **conjunctive normal form**. Exercises 24–28 explore maxterms and the conjunctive normal form in more detail.

EXERCISES

In Exercises 1–10, find the disjunctive normal form of each function and draw the combinatorial circuit corresponding to the disjunctive normal form.

1H.

x	y	$f(x, y)$
1	1	1
1	0	0
0	1	1
0	0	1

2.

x	y	$f(x, y)$
1	1	0
1	0	1
0	1	0
0	0	1

3H.

x	y	z	$f(x, y, z)$
1	1	1	1
1	1	0	1
1	0	1	0
1	0	0	1
0	1	1	0
0	1	0	0
0	0	1	1
0	0	0	1

4.

x	y	z	$f(x, y, z)$
1	1	1	1
1	1	0	1
1	0	1	0
1	0	0	1
0	1	1	1
0	1	0	1
0	0	1	0
0	0	0	0

5H.

x	y	z	$f(x, y, z)$
1	1	1	1
1	1	0	1
1	0	1	1
1	0	0	0
0	1	1	0
0	1	0	1
0	0	1	1
0	0	0	1

6.

x	y	z	$f(x, y, z)$
1	1	1	0
1	1	0	1
1	0	1	1
1	0	0	1
0	1	1	1
0	1	0	1
0	0	1	1
0	0	0	0

7H.

x	y	z	$f(x, y, z)$
1	1	1	1
1	1	0	0
1	0	1	0
1	0	0	1
0	1	1	0
0	1	0	0
0	0	1	0
0	0	0	1

8.

x	y	z	$f(x, y, z)$
1	1	1	0
1	1	0	0
1	0	1	0
1	0	0	1
0	1	1	1
0	1	0	1
0	0	1	1
0	0	0	0

9H.

w	x	y	z	$f(w, x, y, z)$
1	1	1	1	1
1	1	1	0	0
1	1	0	1	1
1	1	0	0	0
1	0	1	1	0
1	0	1	0	0
1	0	0	1	0
1	0	0	0	1
0	1	1	1	1
0	1	1	0	0
0	1	0	1	0
0	1	0	0	0
0	0	1	1	1
0	0	1	0	0
0	0	0	1	0
0	0	0	0	0

10.

w	x	y	z	$f(w, x, y, z)$
1	1	1	1	0
1	1	1	0	0
1	1	0	1	1
1	1	0	0	1
1	0	1	1	1
1	0	1	0	1
1	0	0	1	0
1	0	0	0	1
0	1	1	1	0
0	1	1	0	1
0	1	0	1	1
0	1	0	0	1
0	0	1	1	0
0	0	1	0	1
0	0	0	1	0
0	0	0	0	1

In Exercises 11–20, find the disjunctive normal form of each function using algebraic techniques. (We abbreviate $a \wedge b$ as ab.)

11H. $f(x, y) = x \vee xy$

12. $f(x, y) = (x \vee y)(\bar{x} \vee \bar{y})$

13H. $f(x, y, z) = x \vee y(x \vee \bar{z})$

14. $f(x, y, z) = (yz \vee x\bar{z})(\overline{x\bar{y} \vee z})$

15H. $f(x, y, z) = (\bar{x}y \vee \overline{xz})(x \vee yz)$

16. $f(x, y, z) = x \vee (\bar{y} \vee (x\bar{y} \vee x\bar{z}))$

17H. $f(x, y, z) = (x \vee \bar{x}y \vee \overline{xy\bar{z}})(xy \vee \overline{xz})(y \vee xy\bar{z})$

18. $f(x, y, z) = (\bar{x}y \vee \overline{xz})(xyz \vee y\bar{z})(x\bar{y}z \vee x\bar{y} \vee x\bar{y}z \vee \overline{xyz})$

19H. $f(w, x, y, z) = wy \vee (w\bar{y} \vee z)(x \vee \overline{wz})$

20. $f(w, x, y, z) = (\overline{wx}\bar{y}z \vee x\bar{y}\bar{z})(\overline{wy}z \vee xy\bar{z} \vee yxz)$
$(\overline{wz} \vee xy \vee \overline{w}\bar{y}z \vee xy\bar{z} \vee \bar{x}yz)$

21H. How many Boolean functions are there from Z_2^n into Z_2?

Let F denote the set of all functions from Z_2^n into Z_2. Define

$$(f \vee g)(x) = f(x) \vee g(x) \qquad x \in Z_2^n$$

$$(f \wedge g)(x) = f(x) \wedge g(x) \qquad x \in Z_2^n$$

$$\bar{f}(x) = \overline{f(x)} \qquad x \in Z_2^n$$

$$0(x) = 0 \qquad x \in Z_2^n$$

$$1(x) = 1 \qquad x \in Z_2^n.$$

22. How many elements does F have?

23. Show that $(F, \vee, \wedge, {}^-, 0, 1)$ is a Boolean algebra.

24. By dualizing the procedure of Example 6.4.4, explain how to find the conjunctive normal form of a Boolean function from Z_2^n into Z_2.

25H. Find the conjunctive normal form of each function in Exercises 1–10.

26. By using algebraic methods, find the conjunctive normal form of each function in Exercises 11–20.

27H. Show that if $m_1 \vee \cdots \vee m_k$ is the disjunctive normal form of $f(x_1, \ldots, x_n)$, then $\bar{m}_1 \wedge \cdots \wedge \bar{m}_k$ is the conjunctive normal form of $\bar{f}(x_1, \ldots, x_n)$.

28. Using the method of Exercise 27, find the conjunctive normal form of $\bar{f}$ for each function f of Exercises 1–10.

29H. Show that the disjunctive normal form (6.4.5) is unique; that is, show that if we have a Boolean function

$$f(x_1, \ldots, x_n) = m_1 \vee \cdots \vee m_k = m_1' \vee \cdots \vee m_j'$$

where each m_i, m_i' is a minterm, then $k = j$ and the subscripts on the m_i' may be permuted so that $m_i = m_i'$ for $i = 1, \ldots, k$.

6.5

Applications

In the preceding section we showed how to design a combinatorial circuit using AND, OR, and NOT gates that would compute an arbitrary function from Z_2^n into Z_2, where $Z_2 = \{0, 1\}$. In this section we consider using other kinds of gates to implement a circuit. We also consider the problem of efficient design. We will conclude by looking at several useful circuits, having multiple outputs. Throughout this section, we write ab for $a \wedge b$.

Before considering alternatives to AND, OR, and NOT gates, we must give a precise definition of gate.

DEFINITION 6.5.1. A *gate* is a function from Z_2^n into Z_2.

EXAMPLE 6.5.2. The AND gate is the function $\wedge$ from Z_2^2 into Z_2 defined as in Definition 6.1.1. The NOT gate is the function $^-$ from Z_2 into Z_2 defined as in Definition 6.1.3.

We are interested in gates that allow us to construct arbitrary combinatorial circuits.

DEFINITION 6.5.3. A set of gates $\{g_1, \ldots, g_k\}$ is said to be *functionally complete* if, given any positive integer n and a function f from Z_2^n into Z_2, it is possible to construct a combinatorial circuit that computes f using only the gates $g_1, \ldots, g_k$.

EXAMPLE 6.5.4. Theorem 6.4.6 shows that the set of gates $\{AND, OR, NOT\}$ is functionally complete.

It is an interesting fact that we can eliminate either AND or OR from the set $\{AND, OR, NOT\}$ and still obtain a functionally complete set of gates.

THEOREM 6.5.5. *The sets of gates*

$$\{AND, NOT\} \qquad \{OR, NOT\}$$

are functionally complete.

Proof. We will show that the set of gates $\{AND, NOT\}$ is functionally complete and leave the problem of showing that the other set is functionally complete for the exercises (see Exercise 1).

We have

$$x \vee y = \overline{\overline{x}} \vee \overline{\overline{y}} \qquad \text{Involution Law}$$

$$= \overline{\overline{x}\,\overline{y}} \qquad \text{DeMorgan's Law.}$$

Therefore, an OR gate can be replaced by one AND gate and three NOT gates. (The combinatorial circuit is shown in Figure 6.5.1.)

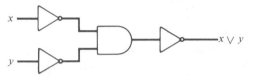

FIGURE 6.5.1

Given any function $f : Z_2^n \to Z_2$, by Theorem 6.4.6 we can construct a combinatorial circuit C using AND, OR, and NOT gates that computes f. But Figure 6.5.1 shows that each OR gate can be replaced by AND and NOT gates. Therefore, the circuit C can be modified so that it consists only of AND and NOT gates. Thus the set of gates {AND, NOT} is functionally complete. ∎

Although none of AND, OR, or NOT singly forms a functionally complete set (see Exercises 2–4), it is possible to define a new gate that by itself forms a functionally complete set.

DEFINITION 6.5.6. A *NAND gate* receives inputs x_1 and x_2, where x_1 and x_2 are bits, and produces output denoted $x_1 \uparrow x_2$, where

$$x_1 \uparrow x_2 = \begin{cases} 0 & \text{if } x_1 = 1 \text{ and } x_2 = 1 \\ 1 & \text{otherwise.} \end{cases}$$

A NAND gate is drawn as shown in Figure 6.5.2.

FIGURE 6.5.2

Many basic circuits used in digital computers today are built from NAND gates (see [D'Angelo, p. 176]).

THEOREM 6.5.7. *The set* {NAND} *is a functionally complete set of gates.*

Proof. First we observe that

$$x \uparrow y = \overline{xy}.$$

Therefore,

$$\bar{x} = \overline{xx} = x \uparrow x \qquad (6.5.1)$$

$$x \vee y = \overline{\overline{x}\overline{y}} = \bar{x} \uparrow \bar{y} = (x \uparrow x) \uparrow (y \uparrow y). \qquad (6.5.2)$$

Equations (6.5.1) and (6.5.2) show that both OR and NOT can be written in terms of NAND. By Theorem 6.5.5, the set {OR, NOT} is functionally complete. It follows that the set {NAND} is also functionally complete. ∎

EXAMPLE 6.5.8. Design combinatorial circuits using NAND gates to compute the functions $f_1(x) = \bar{x}$ and $f_2(x, y) = x \vee y$.

The combinatorial circuits, derived from equations (6.5.1) and (6.5.2), are shown in Figure 6.5.3.

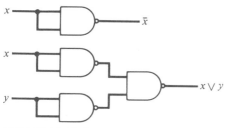

FIGURE 6.5.3

Consider the problem of designing a combinatorial circuit using AND, OR, and NOT gates to compute the function f.

x	y	z	$f(x, y, z)$
1	1	1	1
1	1	0	1
1	0	1	0
1	0	0	1
0	1	1	0
0	1	0	0
0	0	1	0
0	0	0	0

The disjunctive normal form of f is

$$f(x, y, z) = xyz \vee xy\bar{z} \vee x\bar{y}\bar{z}. \qquad (6.5.3)$$

The combinatorial circuit corresponding to (6.5.3) is shown in Figure 6.5.4.

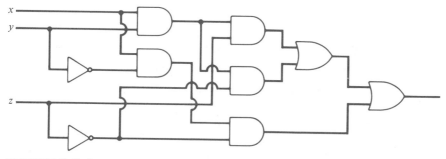

FIGURE 6.5.4

The circuit in Figure 6.5.4 has nine gates. As we will show, it is possible to design a circuit having fewer gates. The problem of finding the best circuit is called the **minimization problem**. There are many definitions of "best."

To find a simpler combinatorial circuit equivalent to that in Figure 6.5.4, we attempt to simplify the Boolean expression (6.5.3) that represents it. The equations

$$Ea \lor E\bar{a} = E \qquad\qquad (6.5.4)$$

$$E = E \lor Ea, \qquad\qquad (6.5.5)$$

where E represents an arbitrary Boolean expression, are useful in simplifying Boolean expressions.

Equation (6.5.4) may be derived as follows:

$$Ea \lor E\bar{a} = E(a \lor \bar{a}) = E1 = E$$

using the properties of Boolean algebras. Equation (6.5.5) is essentially the Absorption Law (Theorem 6.3.6c).

Using (6.5.4) and (6.5.5), we may simplify (6.5.3) as follows:

$$
\begin{aligned}
xyz \lor xy\bar{z} \lor x\bar{y}\bar{z} &= xy \lor x\bar{y}\bar{z} &&\text{by (6.5.4)}\\
&= xy \lor xy\bar{z} \lor x\bar{y}\bar{z} &&\text{by (6.5.5)}\\
&= xy \lor x\bar{z} &&\text{by (6.5.4).}
\end{aligned}
$$

A further simplification,

$$xy \lor x\bar{z} = x(y \lor \bar{z}), \qquad\qquad (6.5.6)$$

is possible using the Distributive Law (Definition 6.3.1c). The combinatorial circuit corresponding to (6.5.6), which requires only three gates, is shown in Figure 6.5.5.

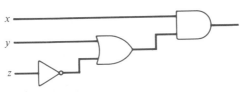

FIGURE 6.5.5

EXAMPLE 6.5.9. The combinatorial circuit in Figure 6.4.1 uses five AND, OR, and NOT gates to compute the exclusive-OR $x \oplus y$ of x and y. Design a circuit that computes $x \oplus y$ using fewer AND, OR, and NOT gates.

Unfortunately, (6.5.4) and (6.5.5) do not help us simplify the disjunctive normal form $x\bar{y} \lor \bar{x}y$ of $x \oplus y$. Thus we must experiment with various Boolean rules until we produce an expression that requires fewer than five gates. One solution is provided by the expression

$$(x \lor y)\overline{xy}$$

whose implementation requires only four gates. The combinatorial circuit is drawn in Figure 6.5.6.

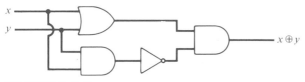

FIGURE 6.5.6

The set of gates available determines the minimization problem. Since the state of technology determines the available gates, the minimization problem changes through time. In the 1950s, the typical problem was to minimize circuits consisting of AND, OR, and NOT gates. Solutions such as the Quine–McCluskey method and the method of Karnaugh maps were provided. The reader is referred to [Berztiss] and [Mendelson] for the details of these methods.

Advances in solid-state technology have made it possible to manufacture very small components, called **integrated circuits**, which are themselves entire circuits. Thus circuit design today consists of combining basic gates such as AND, OR, NOT, and NAND gates and integrated circuits to compute the desired functions. Boolean algebra remains an essential tool, as a glance at a current book on logic design such as [D'Angelo] will show.

We conclude this section by considering several useful combinatorial circuits having multiple outputs. A circuit with n outputs can be characterized by n Boolean expressions, as the next example shows.

EXAMPLE 6.5.10. Write two Boolean expressions to describe the combinatorial circuit of Figure 6.5.7.

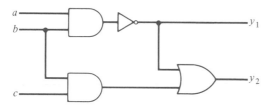

FIGURE 6.5.7

The output y_1 is described by the expression

$$y_1 = \overline{ab}$$

and y_2 is described by the expression

$$y_2 = bc \vee \overline{ab}.$$

Our first circuit is called a **half-adder**.

DEFINITION 6.5.11. A *half-adder* accepts as input two bits x and y and produces as output the binary sum cs of x and y. The term cs is a two-bit binary number. We call s the *sum bit* and c the *carry bit*.

EXAMPLE 6.5.12 Half-Adder Circuit. Design a half-adder combinatorial circuit.

The table for the half-adder circuit is as follows:

x	y	c	s
1	1	1	0
1	0	0	1
0	1	0	1
0	0	0	0

This function has two outputs c and s. We observe that $c = xy$ and $s = x \oplus y$. Thus we obtain the half-adder circuit of Figure 6.5.8. We used the circuit of Figure 6.5.6 to realize the exclusive-OR.

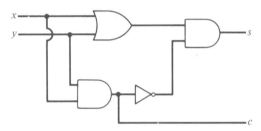

FIGURE 6.5.8

A **full-adder** sums three bits and is useful for adding two bits and a third carry bit from a previous addition.

DEFINITION 6.5.13. A *full-adder* accepts as input three bits x, y, and z and produces as output the binary sum cs of x, y, and z. The term cs is a two-bit binary number.

EXAMPLE 6.5.14 Full-Adder Circuit. Design a full-adder combinatorial circuit.

The table for the full-adder circuit is as follows:

x	y	z	c	s
1	1	1	1	1
1	1	0	1	0
1	0	1	1	0
1	0	0	0	1
0	1	1	1	0
0	1	0	0	1
0	0	1	0	1
0	0	0	0	0

Checking the eight possibilities, we see that

$$s = x \oplus y \oplus z;$$

hence we can use two exclusive-OR circuits to compute s.

To compute c, we first find the disjunctive normal form

$$c = xyz \vee xy\bar{z} \vee x\bar{y}z \vee \bar{x}yz \qquad (6.5.7)$$

of c. Next, we use (6.5.4) and (6.5.5) to simplify (6.5.7) as follows:

$$\begin{aligned}
xyz \vee xy\bar{z} \vee x\bar{y}z \vee \bar{x}yz &= xy \vee x\bar{y}z \vee \bar{x}yz \\
&= xy \vee xyz \vee x\bar{y}z \vee \bar{x}yz \\
&= xy \vee xz \vee \bar{x}yz \\
&= xy \vee xz \vee xyz \vee \bar{x}yz \\
&= xy \vee xz \vee yz.
\end{aligned}$$

Additional gates can be eliminated by writing

$$c = xy \vee z(x \vee y).$$

We obtain the full-adder circuit given in Figure 6.5.9.

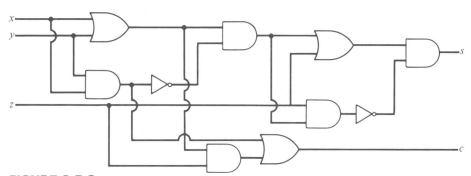

FIGURE 6.5.9

Our last example shows how we may use half-adder and full-adder circuits to construct a circuit to add binary numbers.

EXAMPLE 6.5.15. Using half-adder and full-adder circuits, design a combinatorial circuit which computes the sum of two three-bit numbers.

We will let $M = x_3x_2x_1$ and $N = y_3y_2y_1$ denote the numbers to be added and let $z_4z_3z_2z_1$ denote the sum. The circuit that computes the sum of M and N is drawn in Figure 6.5.10. It is an implementation of the standard algorithm for adding numbers, inasmuch as the "carry bit" is indeed *carried* into the next binary addition.

If we were using three-bit registers for addition, so that the sum of two three-bit numbers would have to be no more than three bits, we could use the z_4 bit

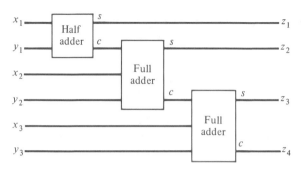

FIGURE 6.5.10

in Example 6.5.15 as an overflow flag. If $z_4 = 1$, overflow occurred; if $z_4 = 0$, there was no overflow.

In the next chapter (Example 7.1.3), we will discuss a sequential circuit that makes use of a primitive, internal memory to add binary numbers.

EXERCISES

1H.   Show that the set of gates {OR, NOT} is functionally complete.

Show that each set of gates below is not functionally complete.

2.   {AND}   **3H.** {OR}   **4.** {NOT}   **5H.** {AND, OR}

6.   Draw a circuit using only NAND gates that computes xy.

7H.   Write xy using only $\uparrow$.

8.   Prove or disprove: $x \uparrow (y \uparrow z) = (x \uparrow y) \uparrow z$, for all $x, y, z \in Z_2$.

Write Boolean expressions to describe the multiple output circuits in Exercises 9–11.

9H.

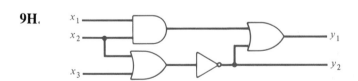

10.

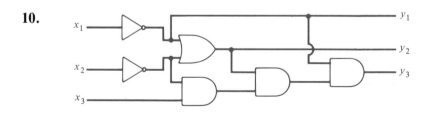

11.

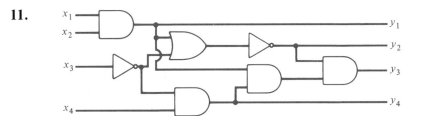

12H. Design circuits using only NAND gates to compute the functions of Exercises 1–10, Section 6.4.

13. Can you reduce the number of NAND gates used in any of your circuits for Exercise 12?

14. Design circuits using as few AND, OR, and NOT gates as you can to compute the functions of Exercises 1–10, Section 6.4.

15H. Design a half-adder circuit using only NAND gates.

★16. Design a half-adder circuit using five NAND gates.

A **NOR gate** receives inputs x_1 and x_2, where x_1 and x_2 are bits, and produces output denoted $x_1 \downarrow x_2$, where

$$x_1 \downarrow x_2 = \begin{cases} 0 & \text{if } x_1 = 1 \text{ or } x_2 = 1 \\ 1 & \text{otherwise.} \end{cases}$$

17H. Write xy, $x \vee y$, $\bar{x}$, and $x \uparrow y$ in terms of $\downarrow$.

18. Write $x \downarrow y$ in terms of $\uparrow$.

19H. Write the logic table for the NOR function.

20. Show that the set of gates {NOR} is functionally complete.

21H. Design circuits using only NOR gates to compute the functions of Exercises 1–10, Section 6.4.

22. Can you reduce the number of NOR gates used in any of your circuits for Exercise 21?

23H. Design a half-adder circuit using only NOR gates.

★24. Design a half-adder circuit using five NOR gates.

25H. Design a circuit with three inputs that outputs 1 precisely when two or three inputs have value 1.

26. Design a circuit that multiplies the binary numbers x_2x_1 and y_2y_1. The output will be of the form $z_4z_3z_2z_1$.

27H. A **2's module** is a circuit that accepts as input two bits b and FLAGIN and outputs bits c and FLAGOUT. If FLAGIN $= 1$, then $c = \bar{b}$ and FLAGOUT $= 1$. If FLAGIN $= 0$ and $b = 1$, then FLAGOUT $= 1$. If FLAGIN $= 0$ and $b = 0$, then FLAGOUT $= 0$. If FLAGIN $= 0$, then $c = b$. Design a circuit to implement a 2's module.

The **2's complement** of a binary number can be computed by using the following algorithm.

ALGORITHM 6.5.16 Finding the 2's Complement. This algorithm computes the 2's complement $C_N C_{N-1} \cdots C_2 C_1$ of the binary number $M = B_N B_{N-1} \cdots B_2 B_1$. The number M is scanned from right to left and the bits are copied until 1 is found. Thereafter, if $B_I = 0$, we set $C_I = 1$ and if $B_I = 1$, we set $C_I = 0$. The flag F indicates whether a 1 has been found ($F = 1$) or not ($F = 0$).

    1. [Initialization.] Set $F := 0$ (1 has not been found); $I := 1$ (the Ith bit is being scanned).

    2. [Copy bit.] $C_I := B_I$. If $B_I = 1$, set $F := 1$; go to step 4; otherwise, go to step 4.

    3. [Copy complement.] $C_I := B_I \oplus 1$.

    4. [Done?] If $I = N$, stop; otherwise, $I := I + 1$.

    5. [Flag set?] If $F = 0$, go to step 2; otherwise, go to step 3.

Find the 2's complement of the numbers in Exercises 28–30 using Algorithm 6.5.16.

28.   101100   **29H.**   11011   **30.**   011010110

31H.   Using 2's modules, design a circuit that computes the 2's complement $y_3 y_2 y_1$ of the three-bit binary number $x_3 x_2 x_1$.

★32.   Let $*$ be a binary operator on a set S containing 0 and 1. Write a set of axioms for $*$, modeled after rules that NAND satisfies, so that if we define

$$\bar{x} = x * x$$

$$x \vee y = (x * x) * (y * y)$$

$$x \wedge y = (x * y) * (x * y),$$

then $(S, \vee, \wedge, ^-, 0, 1)$ is a Boolean algebra.

★33.   Let $*$ be a binary operator on a set S containing 0 and 1. Write a set of axioms for $*$, modeled after rules which NOR satisfies, and definitions for $^-$, $\vee$, and $\wedge$ so that $(S, \vee, \wedge, ^-, 0, 1)$ is a Boolean algebra.

6.6

Notes

General references on Boolean algebras are [Halmos, 1967; Hohn; and Mendelson]. [Hohn] has many applications, and [Mendelson] contains over 150 references on Boolean algebras and combinatorial circuits. In addition, the following books contain information on Boolean algebras and related topics: [Berztiss; Fisher; Gill; Lipschutz, 1976; Liu, 1985; Sahni; Stone; and Tremblay] with [Gill; Stone; and Tremblay] having the most extensive treatments. Books on logic design include [D'Angelo; Hill; and Kohavi]. [D'Angelo] is a very nice introduction to circuit hardware. The introduction to Chapter 3 contains a brief but enlightening discussion of the current state of logic design.

For a history of logic see [Kline]. For a brief history of the computer and its relationship to logic, see Part One of [Goldstine]. This book also includes a very interesting and highly personal account of recent computer history. [Hailperin] gives a technical discussion of Boole's mathematics. Additional references are also provided. Boole's book, *The Laws of Thought*, has been reprinted (see [Boole]).

Because of our interest in applications of Boolean algebra, most of our discussion was limited to the Boolean algebra $(Z_2, \vee, \wedge, ^-, 0, 1)$. However, versions of most of our results remain valid for arbitrary, finite Boolean algebras.

Boolean expressions in the symbols $x_1, \ldots, x_n$ over an arbitrary Boolean algebra $(S, \vee, \wedge, ^-, 0, 1)$ are defined recursively as

$$\{x \mid x \in S\}, x_1, \ldots, x_n \text{ are Boolean expressions.}$$

If X_1 and X_2 are Boolean expressions, so are

$$(X_1) \quad \overline{X}_1 \quad X_1 \vee X_2 \quad X_1 \wedge X_2.$$

A **Boolean function** over S is defined as a function from S^n to S of the form

$$f(x_1, \ldots, x_n) = X(x_1, \ldots, x_n),$$

where X is a Boolean expression in the symbols $x_1, \ldots, x_n$ over S. A disjunctive normal form can be defined for f. Another result is that if X and Y are Boolean expressions over S and

$$X(x_1, \ldots, x_n) = Y(x_1, \ldots, x_n)$$

for all $x_i \in S$, then Y is derivable from X using the definition (Definition 6.3.1) of a Boolean algebra. Other results are that any finite Boolean algebra has 2^n elements and that if two Boolean algebras both have 2^n elements they are essentially the same. It follows that any finite Boolean algebra is essentially Example 6.3.3, the Boolean algebra of subsets of a finite, universal set U. The proofs of these results can be found in [Mendelson].

COMPUTER EXERCISES

1. Write a program that inputs a Boolean expression in X and Y and prints the logic table of the expression. Use a higher-level language that has the capability of evaluating a Boolean expression.

2. Write a program that inputs a Boolean expression in X, Y, and Z and prints the logic table of the expression.

3. Write a program that outputs the disjunctive normal form of a Boolean expression $P(X, Y)$.

4. Write a program that outputs the conjunctive normal form of a Boolean expression $P(X, Y)$.

5. Write a program that outputs the disjunctive normal form of a Boolean expression $P(X, Y, Z)$.

6. Write a program that outputs the conjunctive normal form of a Boolean expression $P(X, Y, Z)$.

7. Write a program that computes the 2's complement of an n-bit binary number.

CHAPTER 7

Automata, Grammars, and Languages

In Chapter 6 we discussed combinatorial circuits in which the output depended only on the input. These circuits have no memory. In this chapter we begin by discussing circuits in which the output depends not only on the input, but also on the state of the system at the time the input is introduced. The state of the system is determined by previous processing. In this sense, these circuits have memory. Such circuits are called sequential circuits and are obviously important in computer design.

Finite-state machines are abstract models of machines with a primitive, internal memory. A finite-state automaton is a special kind of finite-state machine that is closely linked to a particular type of language. In the latter part of this chapter, we will discuss finite-state machines, finite-state automata, and languages in some detail.

7.1

Sequential Circuits and Finite-State Machines

Operations within a digital computer are carried out at discrete intervals of time. Output depends on the state of the system as well as on the input. We will assume that the state of the system changes only at time $t = 0, 1, \ldots$. A simple way to introduce sequencing in circuits is to introduce a **unit time delay**.

DEFINITION 7.1.1. A *unit time delay* accepts as input a bit x_t at time t and outputs the bit x_{t-1}, the bit received as input at time $t - 1$. The unit time delay is drawn as shown in Figure 7.1.1.

x_t ———— | Delay | ———— x_{t-1}

FIGURE 7.1.1

As an example of the use of the unit time delay, we discuss the **serial adder**.

DEFINITION 7.1.2. A *serial adder* accepts as input two binary numbers

$$x = 0x_N x_{N-1} \cdots x_0 \quad \text{and} \quad y = 0y_N y_{N-1} \cdots y_0$$

and outputs the sum $z_{N+1} z_N \cdots z_0$ of x and y. The numbers x and y are input sequentially in pairs, $x_0, y_0; \ldots; x_N, y_N; 0, 0$. The sum is output $z_0, z_1, \ldots, z_{N+1}$.

EXAMPLE 7.1.3 Serial-Adder Circuit. A circuit, using a unit time delay, which implements a serial adder is shown in Figure 7.1.2.

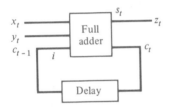

FIGURE 7.1.2

Let us show how the serial adder computes the sum of

$$x = 010 \quad \text{and} \quad y = 011.$$

We begin by setting $x_0 = 0$ and $y_0 = 1$. (We assume that at this instant $i = 0$. This can be arranged by first setting $x = y = 0$.) The state of the system is shown in Figure 7.1.3. Next, we set $x_1 = y_1 = 1$. The unit time delay sends $i = 0$ as the third bit to the full adder. The state of the system is shown in Figure 7.1.3. Finally, we set $x_2 = y_2 = 0$. This time the unit time delay sends $i = 1$ as the third bit to the full adder. The state of the system is shown in Figure 7.1.3. We obtain the sum $z = 101$.

A **finite-state machine** is an abstract model of a machine with a primitive, internal memory.

DEFINITION 7.1.4. A *finite-state machine* M consists of
(a) A finite set $\mathscr{I}$ of *input symbols*.
(b) A finite set $\mathscr{O}$ of *output symbols*.
(c) A finite set $\mathscr{S}$ of *states*.

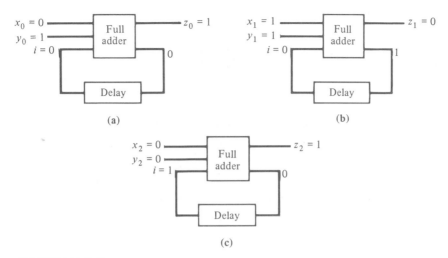

FIGURE 7.1.3

(d) A *next-state function f* from $\mathscr{S} \times \mathscr{I}$ into $\mathscr{S}$.
(e) An *output function g* from $\mathscr{S} \times \mathscr{I}$ into $\mathbb{O}$.
(f) An *initial state* $\sigma^* \in \mathscr{S}$.
We write $M = (\mathscr{I}, \mathbb{O}, \mathscr{S}, f, g, \sigma^*)$.

EXAMPLE 7.1.5. Let $\mathscr{I} = \{a, b\}$, $\mathbb{O} = \{0, 1\}$, and $\mathscr{S} = \{\sigma_0, \sigma_1\}$. Define the functions $f : \mathscr{S} \times \mathscr{I} \to \mathscr{S}$ and $g : \mathscr{S} \times \mathscr{I} \to \mathbb{O}$ by the rules given in Table 7.1.1.

TABLE 7.1.1

$\mathscr{S}$ \ $\mathscr{I}$	f		g	
	a	b	a	b
σ_0	σ_0	σ_1	0	1
σ_1	σ_1	σ_1	1	0

Then $M = (\mathscr{I}, \mathbb{O}, \mathscr{S}, f, g, \sigma_0)$ is a finite-state machine.
Table 7.1.1 is interpreted to mean

$$f(\sigma_0, a) = \sigma_0 \qquad g(\sigma_0, a) = 0$$

$$f(\sigma_0, b) = \sigma_1 \qquad g(\sigma_0, b) = 1$$

$$f(\sigma_1, a) = \sigma_1 \qquad g(\sigma_1, a) = 1$$

$$f(\sigma_1, b) = \sigma_1 \qquad g(\sigma_1, b) = 0.$$

The next-state and output functions can also be defined by a **transition diagram**. Before formally defining a transition diagram, we will illustrate how a transition diagram is constructed.

EXAMPLE 7.1.6.   Draw the transition diagram for the finite-state machine of Example 7.1.5.

The transition diagram is a digraph. The vertices are the states (see Figure 7.1.4). The initial state is indicated by an arrow as shown. If we are in state σ and inputting i causes output o and moves us to state σ', we draw a directed edge from vertex σ to vertex σ' and label it i/o. For example, if we are in state σ_0, and we input a, Table 7.1.1 tells us that we output 0 and remain in state σ_0. Thus we draw a directed loop on vertex σ_0 and label it $a/0$ (see Figure 7.1.4). On the other hand, if we are in state σ_0 and we input b, we output 1 and move to state σ_1. Thus we draw a directed edge from σ_0 to σ_1 and label it $b/1$. By considering all such possibilities, we obtain the transition diagram of Figure 7.1.4.

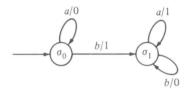

FIGURE 7.1.4

DEFINITION 7.1.7.   Let $M = (\mathscr{I}, \mathbb{O}, \mathscr{S}, f, g, \sigma^*)$ be a finite-state machine. The *transition diagram* of M is a digraph G whose vertices are the members of $\mathscr{S}$. An arrow designates the initial state σ^*. A directed edge (σ, σ') exists in G if there exists an input i with $f(\sigma, i) = \sigma'$. In this case, if $g(\sigma, i) = o$, the edge (σ, σ') is labeled i/o.

We can regard the finite-state machine $M = (\mathscr{I}, \mathbb{O}, \mathscr{S}, f, g, \sigma^*)$ as a simple computer. We begin in state σ^*, input a string over $\mathscr{I}$, and produce a string of output.

DEFINITION 7.1.8.   Let $M = (\mathscr{I}, \mathbb{O}, \mathscr{S}, f, g, \sigma^*)$ be a finite-state machine. An *input string* for M is a string over $\mathscr{I}$. The string

$$y_1 \cdots y_n$$

is the *output string* for M corresponding to the input string

$$\alpha = x_1 \cdots x_n$$

if there exist states $\sigma_0, \ldots, \sigma_n \in \mathscr{S}$ with

$$\sigma_0 = \sigma^*;$$

$$\sigma_i = f(\sigma_{i-1}, x_i) \qquad \text{for } i = 1, \ldots, n;$$

$$y_i = g(\sigma_{i-1}, x_i) \qquad \text{for } i = 1, \ldots, n.$$

EXAMPLE 7.1.9.   Find the output string corresponding to the input string

$$aababba \qquad (7.1.1)$$

for the finite-state machine of Example 7.1.5.

Initially, we are in state σ_0. The first symbol input is a. We locate the outgoing edge in the transition diagram of M (Figure 7.1.4) from σ_0 labeled a/x, which tells us that if a is input, x is output. In our case, 0 is output. The edge points to the next state σ_0. Next, a is input again. As before, we output 0 and remain in state σ_0. Next, b is input. In this case, we output 1 and change to state σ_1. Continuing in this way, we find that the output string is

$$0011001. \qquad (7.1.2)$$

EXAMPLE 7.1.10.   Design a finite-state machine that performs serial addition.

We will represent the finite-state machine by its transition diagram. Since the serial adder accepts pairs of bits, the input set will be

$$\{00, 01, 10, 11\}.$$

The output set is

$$\{0, 1\}.$$

Given an input xy, we take one of two actions: either we add x and y, or we add x, y, and 1, depending on whether the carry bit was 0 or 1. Thus there are two states, which we will call C (carry) and NC (no carry). The initial state is NC. At this point, we can draw the vertices and designate the initial state in our transition diagram (see Figure 7.1.5).

FIGURE 7.1.5

Next, we consider the possible inputs at each vertex. For example, if 00 is input to NC, we should output 0 and remain in state NC. Thus NC has a loop labeled 00/0. As another example, if 11 is input to C, we compute $1 + 1 + 1 = 11$. In this case we output 1 and remain in state C. Thus C has a loop labeled 11/1. As a final example, if we are in state NC and 11 is input, we should output 0 and move to state C. By considering all possibilities, we arrive at the transition diagram of Figure 7.1.6.

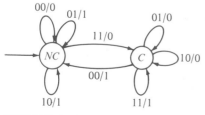

FIGURE 7.1.6

EXERCISES

In Exercises 1–5, draw the transition diagram of the finite-state machine $(\mathcal{I}, \mathcal{O}, \mathcal{S}, f, g, \sigma_0)$.

1H. $\quad \mathcal{I} = \{a, b\}, \mathcal{O} = \{0, 1\}, \mathcal{S} = \{\sigma_0, \sigma_1\}$

$\mathcal{S}$ $\diagdown$ $\mathcal{I}$	f		g	
	a	b	a	b
σ_0	σ_1	σ_1	1	1
σ_1	σ_0	σ_1	0	1

2. $\quad \mathcal{I} = \{a, b\}, \mathcal{O} = \{0, 1\}, \mathcal{S} = \{\sigma_0, \sigma_1\}$

$\mathcal{S}$ $\diagdown$ $\mathcal{I}$	f		g	
	a	b	a	b
σ_0	σ_1	σ_0	0	0
σ_1	σ_0	σ_0	1	1

3H. $\quad \mathcal{I} = \{a, b\}, \mathcal{O} = \{0, 1\}, \mathcal{S} = \{\sigma_0, \sigma_1, \sigma_2\}$

$\mathcal{S}$ $\diagdown$ $\mathcal{I}$	f		g	
	a	b	a	b
σ_0	σ_1	σ_1	0	1
σ_1	σ_2	σ_1	1	1
σ_2	σ_0	σ_0	0	0

4. $\quad \mathcal{I} = \{a, b, c\}, \mathcal{O} = \{0, 1\}, \mathcal{S} = \{\sigma_0, \sigma_1, \sigma_2\}$

$\mathcal{S}$ $\diagdown$ $\mathcal{I}$	f			g		
	a	b	c	a	b	c
σ_0	σ_0	σ_1	σ_2	0	1	0
σ_1	σ_1	σ_1	σ_0	1	1	1
σ_2	σ_2	σ_1	σ_0	1	0	0

5H. $\mathscr{I} = \{a, b, c\}$, $\mathbb{O} = \{0, 1, 2\}$, $\mathscr{S} = \{\sigma_0, \sigma_1, \sigma_2, \sigma_3\}$

$\mathscr{S}$ \ $\mathscr{I}$	f			g		
	a	b	c	a	b	c
σ_0	σ_1	σ_0	σ_2	1	1	2
σ_1	σ_0	σ_2	σ_2	2	0	0
σ_2	σ_3	σ_3	σ_0	1	0	1
σ_3	σ_1	σ_1	σ_0	2	0	2

In Exercises 6–10, find the sets $\mathscr{I}$, $\mathbb{O}$, and $\mathscr{S}$, the initial state, and the table defining the next-state and output functions for each finite-state machine.

6.

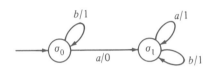

7H.

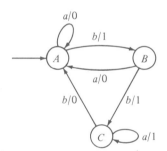

8.

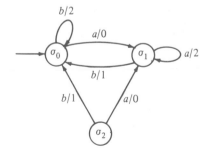

9H.

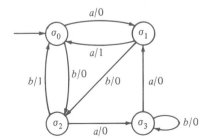

10.

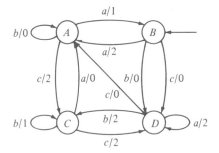

In Exercises 11–20, find the output string for the given input string and finite-state machine.

11H.   *abba*; Exercise 1   **12.**   *abba*; Exercise 2

13H.   *aabbaba*; Exercise 3   **14.**   *aabbcc*; Exercise 4

15H.   *aabaab*; Exercise 5   **16.**   *aaa*; Exercise 6

17H.   *aabbabaab*; Exercise 7   **18.**   *baaba*; Exercise 8

19H.   *bbababbabaaa*; Exercise 9   **20.**   *cacbccbaabac*; Exercise 10

In Exercises 21–26, design a finite-state machine having the given properties. The input is always a bit string.

21H.   Outputs 1 if an even number of 1's have been input; otherwise, outputs 0

22.   Outputs 1 if k 1's have been input, where k is a multiple of 3; otherwise, outputs 0

23H.   Outputs 1 if two or more 1's are input; otherwise, outputs 0

24.   Outputs 1 whenever it sees 101; otherwise, outputs 0

25H.   Outputs 1 when it sees 101 and thereafter; otherwise, outputs 0

26.   Outputs 1 when it sees the first 0 and until it sees another 0; thereafter it outputs 0; in all other cases, outputs 0

27H.   Let $\alpha = x_1 \cdots x_n$ be a bit string. Let $\beta = y_1 \cdots y_n$ where

$$y_i = \begin{cases} a & \text{if } x_i = 0 \\ b & \text{if } x_i = 1, \end{cases}$$

for $i = 1, \ldots, n$. Let $\gamma = y_n \cdots y_1$.

Show that if γ is input to the finite-state machine of Figure 7.1.4, the output is the 2's complement of α (see Algorithm 6.5.16 for a description of 2's complement).

The sequential circuit shown is called a **flip-flop**.

28.   Assume that in the flip-flop circuit, initially $x_1 = 1$, $x_2 = 1$, and $y = 0$. Determine the values of a, b, and c.

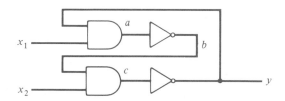

29H. Assume that the circuit of Exercise 28 is changed so that $x_2 = 0$. Determine the values of a, b, c, and y.

30. Assume that the circuit of Exercise 29 is changed so that $x_2 = 1$. Show that the values of a, b, c, and y do not change.

31H. Assume that the circuit of Exercise 30 is changed by setting $x_1 = 0$ and then $x_1 = 1$. Determine the values of a, b, c, and y.

32. By examining the results of Exercises 28–31, explain how a flip-flop circuit can be used to store one bit.

★33. Show that there is no finite-state machine that receives a bit string and outputs 1 whenever the number of 1's input equals the number of 0's input and outputs 0 otherwise.

★34H. Show that there is no finite-state machine that performs serial multiplication. Specifically, show that there is no finite-state machine that inputs binary numbers $X = x_1 \cdots x_n$, $Y = y_1 \cdots y_n$, as the sequence of two-bit numbers

$$x_n y_n, \ x_{n-1} y_{n-1}, \ \ldots, \ x_1 y_1, \ 00, \ \ldots, \ 00,$$

where there are n 00's, and outputs $z_{2n}, \ \ldots, \ z_1$, where $Z = z_1 \cdots z_{2n} = XY$.

EXAMPLE: If there is such a machine, to multiply 101×1001, we would input $11,00,10,01,00,00,00,00$. The first pair 11 is the pair of rightmost bits ($10\underline{1}$, $100\underline{1}$); the second pair 00 is the next pair of bits ($1\underline{0}1$, $10\underline{0}1$); and so on. We pad the input string with four pairs of 00's—the length of the longest number 1001 to be multiplied. Since $101 \times 1001 = 101101$, it is alleged that we obtain

Input	Output
11	1
00	0
10	1
01	1
00	0
00	1
00	0
00	0

Finite-State Automata

A **finite-state automaton** is a special kind of finite-state machine. Finite-state automata are of special interest because of their relationship to languages as we shall see in Section 7.5.

DEFINITION 7.2.1. A *finite-state automaton* $A = (\mathcal{S}, \mathbb{O}, \mathcal{I}, f, g, \sigma*)$ is a finite-state machine in which the set of output symbols is $\{0, 1\}$ and where the current state determines the last output. Those states for which the last output was 1 are called *accepting states*.

EXAMPLE 7.2.2. Draw the transition diagram of the finite-state machine A defined by the table. The initial state is σ_0. Show that A is a finite-state automaton and determine the set of accepting states.

$\mathcal{S}$ \ $\mathcal{I}$	f		g	
	a	b	a	b
σ_0	σ_1	σ_0	1	0
σ_1	σ_2	σ_0	1	0
σ_2	σ_2	σ_0	1	0

The transition diagram is shown in Figure 7.2.1. If we are in state σ_0, the last output was 0. If we are in either state σ_1 or σ_2, the last output was 1; thus A is a finite-state automaton. The accepting states are σ_1 and σ_2.

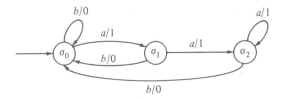

FIGURE 7.2.1

Example 7.2.2 shows that the finite-state machine defined by a transition diagram will be a finite-state automaton if the set of output symbols is $\{0, 1\}$ and if, for each state σ, all incoming edges to σ have the same output label.

The transition diagram of a finite-state automaton is usually drawn with the accepting states in double circles and the output symbols omitted. When the transition diagram of Figure 7.2.1 is redrawn in this way, we obtain the transition diagram of Figure 7.2.2.

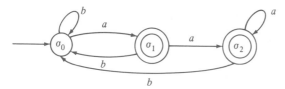

FIGURE 7.2.2

EXAMPLE 7.2.3. Draw the transition diagram of the finite-state autr aton of Figure 7.2.3 as a transition diagram of a finite-state machine.

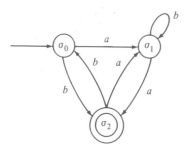

FIGURE 7.2.3

Since σ_2 is an accepting state, we label all its incoming edges with output 1 (see Figure 7.2.4.). The states σ_0 and σ_1 are not accepting, so we label all their incoming edges with output 0. We obtain the transition diagram of Figure 7.2.4.

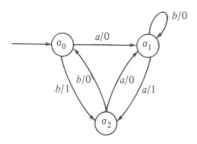

FIGURE 7.2.4

As an alternative to Definition 7.2.1, we can regard a finite-state automaton A as consisting of

1. A finite set $\mathcal{I}$ of *input symbols*.
2. A finite set $\mathcal{S}$ of *states*.
3. A *next-state function* f from $\mathcal{I} \times \mathcal{S}$ into $\mathcal{S}$.
4. A subset $\mathcal{A}$ of $\mathcal{S}$ of *accepting states*.
5. An *initial state* $\sigma^* \in \mathcal{S}$.

If we use this characterization, we write $A = (\mathcal{I}, \mathcal{S}, f, \mathcal{A}, \sigma^*)$.

EXAMPLE 7.2.4. The transition diagram of the finite-state automaton
$A = (\mathcal{I}, \mathcal{S}, f, \mathcal{A}, \sigma^*)$, where

$$\mathcal{I} = \{a, b\}$$

$$\mathcal{S} = \{\sigma_0, \sigma_1, \sigma_2\};$$

$$\mathcal{A} = \{\sigma_2\};$$

$$\sigma^* = \sigma_0;$$

and f is given by the following table:

	f	
$\mathcal{S}$	a	b
σ_0	σ_0	σ_1
σ_1	σ_0	σ_2
σ_2	σ_0	σ_2

is shown in Figure 7.2.5.

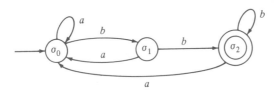

FIGURE 7.2.5

If a string is input to a finite-state automaton, we will end at either an accepting
or a nonaccepting state. The status of this final state determines whether the
string is **accepted** by the finite-state automaton.

DEFINITION 7.2.5.   Let $A = (\mathcal{I}, \mathcal{S}, f, \mathcal{A}, \sigma^*)$ be a finite-state automaton. Let $\alpha = x_1 \cdots x_n$ be a nonnull string over $\mathcal{I}$. If there exist states
$\sigma_0, \ldots, \sigma_n$ satisfying
   (a) $\sigma_0 = \sigma^*$;
   (b) $f(\sigma_{i-1}, x_i) = \sigma_i$ for $i = 1, \ldots, n$;
   (c) $\sigma_n \in \mathcal{A}$;
we say that α is *accepted by* A. We let $Ac(A)$ denote the set of strings accepted
by A and we say that A *accepts* $Ac(A)$.
   Let $\alpha = x_1 \cdots x_n$ be a string over $\mathcal{I}$. Define states $\sigma_0, \ldots, \sigma_n$ by (a)
and (b) above. We call the (directed) path $(\sigma_0, \ldots, \sigma_n)$ the path *representing* α in A.

It follows from Definition 7.2.5 that if the path P represents the string α in a finite-state automaton A, then A accepts α if and only if P ends at an accepting state.

EXAMPLE 7.2.6. Is the string $abaa$ accepted by the finite-state automaton of Figure 7.2.2?

We begin at state σ_0. When a is input, we move to state σ_1. When b is input, we move to state σ_0. When a is input, we move to state σ_1. Finally, when the last symbol a is input, we move to state σ_2. The path $(\sigma_0, \sigma_1, \sigma_0, \sigma_1, \sigma_2)$ represents the string $abaa$. Since the final state σ_2 is an accepting state, the string $abaa$ is accepted by the finite-state automaton of Figure 7.2.2.

EXAMPLE 7.2.7. Is the string $\alpha = abbabba$ accepted by the finite-state automaton of Figure 7.2.3?

The path representing α terminates at σ_1. Since σ_1 is not an accepting state, the string α is not accepted by the finite-state automaton of Figure 7.2.3.

We next give two examples illustrating design problems.

EXAMPLE 7.2.8. Design a finite-state automaton that accepts precisely those nonnull strings over $\{a, b\}$ that contain no a's.

The idea is to use two states:

A: An a was found.
NA: No a's were found.

The state NA is the initial state and the only accepting state. It is now a simple matter to draw the edges (see Figure 7.2.6).

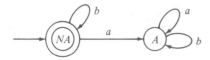

FIGURE 7.2.6

EXAMPLE 7.2.9. Design a finite-state automaton that accepts precisely those strings over $\{a, b\}$ that contain an odd number of a's.

This time the two states are

E: An even number of a's was found.
O: An odd number of a's was found.

The initial state is E and the accepting state is O. We obtain the transition diagram shown in Figure 7.2.7.

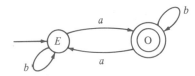

FIGURE 7.2.7

A finite-state automaton is essentially an algorithm to decide whether or not a given string is accepted. As an example, we convert the transition diagram of Figure 7.2.7 to an algorithm.

ALGORITHM 7.2.10.   This algorithm determines whether a string over $\{a, b\}$ is accepted by the finite-state automaton whose transition diagram is given in Figure 7.2.7. The string is held in the array $S(1), \ldots, S(N)$, one character per array element. At termination, ACCEPT is 'YES' if the string is accepted; otherwise, ACCEPT is 'NO'.
  1. [Initialization.] $I := 0$. (I is the current character being examined.)
  2. [State E.] $I := I + 1$. If $I > N$, set ACCEPT = 'NO'; stop.
  3. [State E.] If $S(I) = 'a'$, go to step 5.
  4. [State E.] If $S(I) = 'b'$, go to step 2.
  5. [State O.] $I := I + 1$. If $I > N$, set ACCEPT = 'YES'; stop.
  6. [State O.] If $S(I) = 'a'$, go to step 2.
  7. [State O.] If $S(I) = 'b'$, go to step 5.

If two finite-state automata accept precisely the same strings, we say that the automata are **equivalent**.

DEFINITION 7.2.11.   The finite-state automata A and A' are *equivalent* if $Ac(A) = Ac(A')$.

EXAMPLE 7.2.12.   It can be verified that the finite-state automata of Figures 7.2.6 and 7.2.8 are equivalent (see Exercise 33).

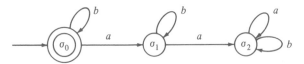

FIGURE 7.2.8

EXERCISES

In Exercises 1–3, show that each finite-state machine is a finite-state automaton and redraw the transition diagram as the diagram of a finite-state automaton.

1H.

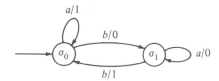

2.

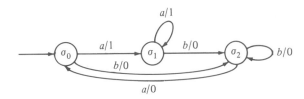

3H.

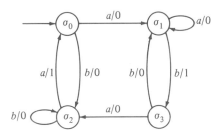

In Exercises 4–6, redraw the transition diagram of the finite-state automaton as the transition diagram of a finite-state machine.

4.

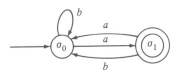

5H.

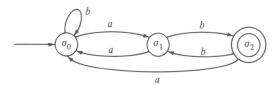

6.

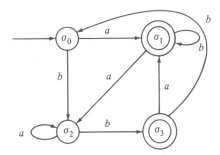

In Exercises 7–9, draw the transition diagram of the finite-state automaton $(\mathcal{I}, \mathcal{S}, f, \mathcal{A}, \sigma_0)$.

7H. $\mathcal{I} = \{a, b\}$, $\mathcal{S} = \{\sigma_0, \sigma_1, \sigma_2\}$, $\mathcal{A} = \{\sigma_0\}$

$\mathcal{S}$ ＼ $\mathcal{I}$	f	
	a	b
σ_0	σ_1	σ_0
σ_1	σ_2	σ_0
σ_2	σ_0	σ_2

8. $\mathcal{I} = \{a, b\}$, $\mathcal{S} = \{\sigma_0, \sigma_1, \sigma_2\}$, $\mathcal{A} = \{\sigma_0, \sigma_2\}$

$\mathcal{S}$ ＼ $\mathcal{I}$	f	
	a	b
σ_0	σ_1	σ_1
σ_1	σ_0	σ_2
σ_2	σ_0	σ_1

9H. $\mathcal{I} = \{a, b, c\}$, $\mathcal{S} = \{\sigma_0, \sigma_1, \sigma_2, \sigma_3\}$, $\mathcal{A} = \{\sigma_1, \sigma_2\}$

$\mathcal{S}$ ＼ $\mathcal{I}$	f		
	a	b	c
σ_0	σ_1	σ_0	σ_2
σ_1	σ_0	σ_3	σ_0
σ_2	σ_3	σ_2	σ_0
σ_3	σ_1	σ_0	σ_1

10. For each finite-state automaton in Exercises 1–6, find the sets $\mathcal{I}$, $\mathcal{S}$, and $\mathcal{A}$, the initial state, and the table defining the next-state function.

11H. Which of the finite-state machines of Exercises 1–10, Section 7.1, are finite-state automata?

12. What must the table of a finite-state machine M look like in order for M to be a finite-state automaton?

In Exercises 13–17, determine whether the given string is accepted by the given finite-state automaton.

13H. *abbaa*; Figure 7.2.2   **14.** *abbaa*; Figure 7.2.3

15H. *aabaabb*; Figure 7.2.5   **16.** *aaabbbaab*; Exercise 5

17H. *aaababbab*; Exercise 6

18. Show that a string α over $\{a, b\}$ is accepted by the finite-state automaton of Figure 7.2.2 if and only if α ends with *a*.

19H. Show that a string α over $\{a, b\}$ is accepted by the finite-state automaton of Figure 7.2.5 if and only if α ends with *bb*.

★20. Characterize the strings accepted by the finite-state automata of Exercises 1–9.

In Exercises 21–31, draw the transition diagram of a finite-state automaton that accepts the given set of nonnull strings over $\{a, b\}$.

21H. Even number of *a*'s

22. Exactly one *b*

23H. At least one *b*

24. Exactly two *a*'s

25H. At least two *a*'s

26. Contains *m* *a*'s, where *m* is a multiple of 3

27H. Starts with *baa*

★28. Contains *abba*

29H. Every *b* is followed by *a*

★30. Ends with *aba*

★31H. Starts with *ab* and ends with *baa*

32. Write algorithms, similar to Algorithm 7.2.10, that decide whether or not a given string is accepted by the finite-state automata of Exercises 1–9.

33H. Give a formal argument to show that the finite-state automata of Figures 7.2.6 and 7.2.8 are equivalent.

34. Let L be a finite set of nonnull strings over $\{a, b\}$. Show that there is a finite-state automaton that accepts L.

35H. Let L be the set of strings accepted by the finite-state automaton of Exercise 6. Let S denote the set of all nonnull strings over $\{a,b\}$. Design a finite-state automaton that accepts $S - L$.

36. Let L_i be the set of strings accepted by the finite-state automaton $A_i = (\mathcal{I}, \mathcal{S}_i, f_i, \mathcal{A}_i, \sigma_i^*)$, $i = 1, 2$. Let

$$A = (\mathcal{I}, \mathcal{S}_1 \times \mathcal{S}_2, f, \mathcal{A}, \sigma^*)$$

where

$$f((\sigma, \sigma'), x) = (f_1(\sigma, x), f_2(\sigma', x)),$$

$$\mathscr{A} = \{(\sigma, \sigma') \mid \sigma \in \mathscr{A}_1 \text{ and } \sigma' \in \mathscr{A}_2\},$$

$$\sigma^* = (\sigma_1^*, \sigma_2^*).$$

Show that $Ac(A) = L_1 \cap L_2$.

37H. Let L_i be the set of strings accepted by the finite-state automaton $A_i = (\mathscr{I}, \mathscr{S}_i, f_i, \mathscr{A}_i, \sigma_i^*)$, $i = 1, 2$. Let

$$A = (\mathscr{I}, \mathscr{S}_1 \times \mathscr{S}_2, f, \mathscr{A}, \sigma^*)$$

where.

$$f((\sigma, \sigma'), x) = (f_1(\sigma, x), f_2(\sigma', x)),$$

$$\mathscr{A} = \{(\sigma, \sigma') \mid \sigma \in \mathscr{A}_1 \text{ or } \sigma' \in \mathscr{A}_2\},$$

$$\sigma^* = (\sigma_1^*, \sigma_2^*).$$

Show that $Ac(A) = L_1 \cup L_2$.

In Exercises 38–42, let $L_i = Ac(A_i)$, $i = 1, 2$. Draw the transition diagrams of the finite-state automata that accept $L_1 \cap L_2$ and $L_1 \cup L_2$.

38. A_1 given by Exercise 4; A_2 given by Exercise 5

39H. A_1 given by Exercise 4; A_2 given by Exercise 6

40. A_1 given by Exercise 5; A_2 given by Exercise 6

41H. A_1 given by Exercise 6; A_2 given by Exercise 6

42. A_1 given by Figure 7.5.7, Section 7.5; A_2 given by Exercise 6

7.3

Languages and Grammars

According to *Webster's New Collegiate Dictionary*, language is a "body of words and methods of combining words used and understood by a considerable community." Such languages are often called **natural languages** to distinguish them from **formal languages**, which are used to model natural languages and to communicate with computers. The rules of a natural language are very complex and difficult to completely characterize. On the other hand, it is possible to specify completely the rules by which certain formal languages are constructed. We begin with the definition of a formal language.

DEFINITION 7.3.1.   Let A be a finite set. A *(formal) language L* over A is a subset of A^*, the set of all strings over A.

EXAMPLE 7.3.2. Let $A = \{a, b\}$. The set L of all strings over A containing an odd number of a's is a language over A. As we saw in Example 7.2.9, L is precisely the set of strings over A accepted by the finite-state automaton of Figure 7.2.7.

One way to define a language is to give a list of rules that the language is assumed to obey.

DEFINITION 7.3.3. A *phrase-structure grammar* (or, simply, *grammar*) G consists of
   (a) A finite set N of *nonterminal symbols*.
   (b) A finite set T of *terminal symbols* where $N \cap T = \varnothing$.
   (c) A finite subset P of $[(N \cup T)^* - T^*] \times (N \cup T)^*$, called the set of *productions*.
   (d) A *starting symbol* $\sigma^* \in N$.
We write $G = (N, T, P, \sigma^*)$.

A production $(A, B) \in P$ is usually written

$$A \rightarrow B.$$

Definition 7.3.3c states that in the production $A \rightarrow B$, $A \in (N \cup T)^* - T^*$ and $B \in (N \cup T)^*$; thus A must include at least one nonterminal symbol, whereas B can consist of any combination of nonterminal and terminal symbols.

EXAMPLE 7.3.4. Let

$$N = \{\sigma^*, S\}$$

$$T = \{a, b\}$$

$$P = \{\sigma^* \rightarrow b\sigma^*, \sigma^* \rightarrow aS, S \rightarrow bS, S \rightarrow b\}.$$

Then $G = (N, T, P, \sigma^*)$ is a grammar.

Given a grammar G, we can construct a language $L(G)$ from G by using the productions to derive the strings that make up $L(G)$.

DEFINITION 7.3.5. Let $G = (N, T, P, \sigma^*)$ be a grammar.
   If $\alpha \rightarrow \beta$ is a production and $x\alpha y \in (N \cup T)^*$, we say that $x\beta y$ is *directly derivable* from $x\alpha y$ and write

$$x\alpha y \Rightarrow x\beta y.$$

If $\alpha_i \in (N \cup T)^*$ for $i = 1, \ldots, n$, and α_{i+1} is directly derivable from α_i for $i = 1, \ldots, n - 1$, we say that α_n is *derivable from* α_1 and write

$$\alpha_1 \overset{*}{\Rightarrow} \alpha_n.$$

We call

$$\alpha_1 \Rightarrow \alpha_2 \Rightarrow \cdots \Rightarrow \alpha_n$$

the *derivation of* α_n *(from* α_1). By convention, any element of $(N \cup T)^*$, is derivable from itself.

The *language generated by* G, written $L(G)$, consists of all strings over T derivable from σ^*.

EXAMPLE 7.3.6.   Let G be the grammar of Example 7.3.4.

The string *abSbb* is directly derivable from *aSbb*, written

$$aSbb \Rightarrow abSbb,$$

by using the production $S \to bS$.

The string *bbab* is derivable from σ^*, written

$$\sigma^* \Rightarrow bbab.$$

The derivation is

$$\sigma^* \Rightarrow b\sigma^* \Rightarrow bb\sigma^* \Rightarrow bbaS \Rightarrow bbab.$$

The only derivations of σ^* are

$$\sigma^* \Rightarrow b\sigma^*$$
$$\vdots$$
$$\Rightarrow b^n\sigma^* \qquad n \geq 0$$
$$\Rightarrow b^n aS$$
$$\vdots$$
$$\Rightarrow b^n ab^{m-1}S$$
$$\Rightarrow b^n ab^m \qquad n \geq 0, \quad m \geq 1.$$

Thus $L(G)$ consists of the strings over $\{a, b\}$ containing precisely one a that end with b.

An alternative way to state the productions of a grammar is by using **Backus normal form** (or **Backus–Naur form** or **BNF**). In BNF the nonterminal symbols begin with "<" and end with ">." The production $S \to T$ is written $S ::= T$. Productions of the form

$$S ::= T_1, \ S ::= T_2, \ \ldots, \ S ::= T_n$$

may be combined as

$$S ::= T_1 \mid T_2 \mid \cdots \mid T_n.$$

The bar "|" is read "or."

EXAMPLE 7.3.7.   An integer is defined as a string consisting of an optional sign (+ or −) followed by a string of digits (0 through 9). The following grammar generates all integers.

<digit> ::= 0|1|2|3|4|5|6|7|8|9

<integer> ::= <signed integer> | <unsigned integer>

<signed integer> ::= +<unsigned integer> | −<unsigned integer>

<unsigned integer> ::= <digit> | <digit><unsigned integer>

The starting symbol is <integer>.

For example, the derivation of the integer −901 is

$$<integer> \Rightarrow <signed\ integer>$$
$$\Rightarrow -<unsigned\ integer>$$
$$\Rightarrow -<digit><unsigned\ integer>$$
$$\Rightarrow -<digit><digit><unsigned\ integer>$$
$$\Rightarrow -<digit><digit><digit>$$
$$\Rightarrow -9<digit><digit>$$
$$\Rightarrow -90<digit>$$
$$\Rightarrow -901.$$

In the notation of Definition 7.3.3, this language consists of

1. The set N = {<digit>, <integer>, <signed integer>, <unsigned integer>} of nonterminal symbols.

2. The set T = {0, 1, 2, 3, 4, 5, 6, 7, 8, 9, +, −} of terminal symbols.

3. The productions

<digit> → 0, . . . , <digit> → 9,

<integer> → <signed integer>,

<integer> → <unsigned integer>,

<signed integer> → +<unsigned integer>,

<signed integer> → −<unsigned integer>,

<unsigned integer> → <digit>,

<unsigned integer> → <digit><unsigned integer>.

4. The starting symbol <integer>.

Parts of higher-level computer languages, such as FORTRAN, Pascal, and ALGOL, can be written in BNF. (See [Jensen, pp. 110–114] for the syntax of Pascal in BNF.) Example 7.3.7 shows how an integer constant in a higher-level computer language might be specified in BNF.

Grammars are classified according to the types of productions that define the grammars.

DEFINITION 7.3.8. Let G be a grammar and let λ denote the null string.
(a) If every production is of the form

$$\alpha A \beta \rightarrow \alpha \delta \beta, \quad \text{where } \alpha, \beta \in (N \cup T)^*, \quad A \in N,$$
$$\delta \in (N \cup T)^* - \{\lambda\}, \tag{7.3.1}$$

we call G a *context-sensitive grammar*.
(b) If every production is of the form

$$A \rightarrow \delta, \quad \text{where } A \in N, \quad \delta \in (N \cup T)^* - \{\lambda\}, \tag{7.3.2}$$

we call G a *context-free* (or *type 2*) *grammar*.
(c) If every production is of the form

$$A \rightarrow \alpha \text{ or } A \rightarrow \alpha B, \quad \text{where } A, B \in N, \quad \alpha \in T - \{\lambda\},$$

we call G a *regular* (or *type 3*) *grammar*.

According to (7.3.1), in a context-sensitive grammar, we may replace A by δ if A is in the context of α and β. In a context-free grammar, (7.3.2) states that we may replace A by δ anytime. A regular grammar has especially simple substitution rules: We replace a nonterminal symbol by either a terminal symbol or a terminal symbol followed by a nonterminal symbol.

Notice that a regular grammar is a context-free grammar and that a context-free grammar is a context-sensitive grammar.

Some definitions allow $\alpha \in T^* - \{\lambda\}$ in Definition 7.3.8c; however, it can be shown (see Exercise 25) that the two definitions produce the same languages.

There are type 1 grammars, although we will not be discussing them. A grammar is of **type 1** if every production is of the form $\alpha \rightarrow \beta$, where $|\alpha| \leq |\beta|$. A context-sensitive grammar is a type 1 grammar.

EXAMPLE 7.3.9. The grammar G defined by

$$T = \{a, b, c\}, \quad N = \{\sigma^*, A, B, C, D, E\},$$

with productions

$$\sigma^* \rightarrow aAB, \quad \sigma^* \rightarrow aB, \quad A \rightarrow aAC, \quad A \rightarrow aC, \quad B \rightarrow Dc,$$

$$D \rightarrow b, \quad CD \rightarrow CE, \quad CE \rightarrow DE, \quad DE \rightarrow DC, \quad Cc \rightarrow Dcc,$$

and starting symbol σ^* is context-sensitive. For example, the production $CE \rightarrow DE$ says that we can replace C by D if C is followed by E and the production $Cc \rightarrow Dcc$ says that we can replace C by Dc if C is followed by c.

We can derive DC from CD since

$$CD \Rightarrow CE \Rightarrow DE \Rightarrow DC.$$

The string $a^3 b^3 c^3$ is in $L(G)$, since we have

$$\sigma^* \Rightarrow aAB \Rightarrow aaACB \Rightarrow aaaCCDc \Rightarrow aaaDCCc \Rightarrow aaaDCDcc$$

$$\Rightarrow aaaDDCcc \Rightarrow aaaDDDccc \Rightarrow aaabbbccc.$$

It can be shown (see Exercise 26) that

$$L(G) = \{a^n b^n c^n \mid n = 1, 2, \ldots\}.$$

It is natural to allow the language $L(G)$ to inherit a property of a grammar G. The next definition makes this concept precise.

DEFINITION 7.3.10. A language L is *context-sensitive* (respectively, *context-free, regular*) if there is a context-sensitive (respectively, context-free, regular) grammar G with $L = L(G)$.

EXAMPLE 7.3.11. According to Example 7.3.9, the language

$$L = \{a^n b^n c^n \mid n = 1, 2, \ldots\}$$

is context-sensitive. It follows from the so-called *Pumping Lemma* (see [Bobrow, p. 243]) that there is no context-free grammar G with $L = L(G)$; hence L is not a context-free language.

EXAMPLE 7.3.12. The grammar G defined by

$$T = \{a, b\}, N = \{\sigma*\},$$

with productions

$$\sigma* \to a\sigma*b, \qquad \sigma* \to ab$$

and starting symbol $\sigma*$, is context-free. The only derivations of $\sigma*$ are

$$\sigma* \Rightarrow a\sigma*b$$
$$\vdots$$
$$\Rightarrow a^{n-1}\sigma*b^{n-1}$$
$$\Rightarrow a^{n-1}abb^{n-1} = a^n b^n.$$

Thus $L(G)$ consists of the strings over $\{a, b\}$ of the form $a^n b^n$, $n = 1$, $2, \ldots$. This language is context-free. In Section 7.5 (see Example 7.5.6), we will show that $L(G)$ is not regular.

It follows from Examples 7.3.11 and 7.3.12 that the set of context-free languages is a proper subset of the set of context-sensitive languages and that the set of regular languages is a proper subset of the set of context-free languages. It can also be shown that there are languages that are not context-sensitive.

EXAMPLE 7.3.13. The grammar G defined in Example 7.3.4 is regular. Thus the language

$$L(G) = \{b^n ab^m \mid n = 0, 1, \ldots; m = 1, 2, \ldots\}$$

it generates is regular.

EXAMPLE 7.3.14. The grammar of Example 7.3.7 is context-free, but not regular. However, if we change the productions to

<digit> ::= 0|1|2|3|4|5|6|7|8|9

<integer> ::= <signed integer> | <unsigned integer>

<signed integer> ::= +<unsigned integer> | −<unsigned integer>

<unsigned integer> ::= <digit> | 0<unsigned integer> |

$\qquad\qquad$ 1<unsigned integer> | · · · | 9<unsigned integer>,

the resulting grammar is regular. Since the language generated is unchanged, it follows that the set of strings representing integers is a regular language.

Example 7.3.14 motivates the following definition.

DEFINITION 7.3.15. Grammars G and G' are *equivalent* if $L(G) = L(G')$.

EXAMPLE 7.3.16. The grammars of Examples 7.3.7 and 7.3.14 are equivalent.

EXERCISES

In Exercises 1–6, determine whether the given grammar is context-sensitive, context-free, regular, or none of these. Give all characterizations that apply.

1H. $T = \{a, b\}$, $N = \{\sigma^*, A\}$, with productions

$$\sigma^* \to b\sigma^*, \ \sigma^* \to aA, \ A \to a\sigma^*,$$

$$A \to bA, \ A \to a, \ \sigma^* \to b,$$

and starting symbol σ^*.

2. $T = \{a, b, c\}$, $N = \{\sigma^*, A, B\}$, with productions

$$\sigma^* \to AB, \ AB \to BA, \ A \to aA,$$

$$B \to Bb, \ A \to a, \ B \to b,$$

and starting symbol σ^*.

3H. $T = \{a, b\}$, $N = \{\sigma^*, A, B\}$, with productions

$$\sigma^* \to A, \ \sigma^* \to AAB, \ Aa \to ABa, \ A \to aa,$$

$$Bb \to ABb, \ AB \to ABB, \ B \to b,$$

and starting symbol σ^*.

4. $T = \{a, b, c\}$, $N = \{\sigma^*, A, B\}$, with productions

$$\sigma^* \rightarrow BAB, \ \sigma^* \rightarrow ABA, \ A \rightarrow AB, \ B \rightarrow BA,$$

$$A \rightarrow aA, \ A \rightarrow ab, \ B \rightarrow b,$$

and starting symbol σ^*.

5H. $<S> ::= b<S> \mid a<A> \mid a$
$<A> ::= a<S> \mid b<B>$
$<B> ::= b<A> \mid a<S> \mid b$
with starting symbol $<S>$.

6. $T = \{a, b\}$, $N = \{\sigma^*, A, B\}$, with productions

$$\sigma^* \rightarrow AA\sigma^*, \ AA \rightarrow B, \ B \rightarrow bB, \ A \rightarrow a,$$

and starting symbol σ^*.

In Exercises 7–11, show that the given string α is in $L(G)$ for the given grammar G by giving a derivation of α.

7H.   *bbabbab*, Exercise 1   **8.**   *abab*, Exercise 2

9H.   *aabbaab*, Exercise 3   **10.**   *abbbaabab*, Exercise 4

11H.   *abaabbabba*, Exercise 5

12.   Write the grammars of Examples 7.3.4 and 7.3.9 and Exercises 1–4 and 6 in BNF.

★13H.   Let G be the grammar of Exercise 1. Show that $\alpha \in L(G)$ if and only if α is nonnull and contains an even number of a's.

★14.   Let G be the grammar of Exercise 5. Characterize $L(G)$.

In Exercises 15–24, write a grammar that generates the nonnull strings having the given property.

15H.   Strings over $\{a, b\}$ starting with a

16.   Strings over $\{a, b\}$ ending with ba

17H.   Strings over $\{a, b\}$ containing ba

★18.   Strings over $\{a, b\}$ not ending with ab

19H.   Integers with no leading 0's

20.   Floating-point numbers (numbers like .294, 89., 67.284)

21H.   Exponential numbers (numbers including floating-point numbers and numbers like 6.9E3, 8E12, 9.6E-4, 9E-10)

22.   Boolean expressions in $X_1, \ldots, X_n$

23H.   All nonnull strings over $\{a, b\}$

24.   Strings $x_1 \cdots x_n$ over $\{a, b\}$ with $x_1 \cdots x_n = x_n \cdots x_1$

★25H. Let G be a grammar and let λ denote the null string. Show that if every production is of the form

$$A \rightarrow \alpha \text{ or } A \rightarrow \alpha B, \text{ where } A, B \in N, \quad \alpha \in T^* - \{\lambda\},$$

there is a regular grammar G' with $L(G) = L(G')$.

★26. Let G be the grammar of Example 7.3.9. Show that

$$L(G) = \{a^n b^n c^n \mid n = 1, 2, \ldots\}.$$

27H. Show that the language

$$\{a^n b^n c^k \mid n, k \in \{1, 2, \ldots\}\}$$

is a context-free language.

7.4

Nondeterministic Finite-State Automata

In this section and the next, we show that regular grammars and finite-state automata are essentially the same in that either is a specification of a regular language. We begin with an example that illustrates how we can convert a finite-state automaton to a regular grammar.

EXAMPLE 7.4.1. Write the regular grammar given by the finite-state automaton of Figure 7.2.7.

The terminal symbols are the output symbols $\{a, b\}$. The states E and O become the nonterminal symbols. The initial state E becomes the starting symbol. The productions correspond to the directed edges. If there is an edge labeled x from S to S', we write the production

$$S \rightarrow xS'.$$

In our case, we obtain the productions

$$E \rightarrow bE$$

$$E \rightarrow aO$$

$$O \rightarrow aE$$

$$O \rightarrow bO. \tag{7.4.1}$$

In addition, if there is an edge labeled x from state S to an accepting state, we include the production

$$S \rightarrow x.$$

In our case, we obtain the additional productions

$$O \rightarrow b$$

$$E \rightarrow a. \tag{7.4.2}$$

Then the grammar $G = (N, T, P, E)$, with $N = \{O, E\}$, $T = \{a, b\}$, and P consisting of the productions (7.4.1) and (7.4.2), generates the language $L(G)$, which is the same as the set of strings accepted by the finite-state automaton of Figure 7.2.7.

THEOREM 7.4.2. *Let A be a finite-state automaton given as a transition diagram. Let σ^* be the initial state. Let T be the set of output symbols and let N be the set of states. Define productions*

$$S \rightarrow xS'$$

if there is an edge labeled x from S to S', and

$$S \rightarrow x$$

if there is an edge labeled x from S to an accepting state. Let G be the regular grammar

$$G = (N, T, P, \sigma^*).$$

Then the set of strings accepted by A is equal to $L(G)$.

Proof. Suppose that the string $x_1 \cdots x_n$ is accepted by A. Then there is a path $(\sigma^*, S_1, \ldots, S_n)$, where S_n is an accepting state, with edges successively labeled $x_1, \ldots, x_n$. It follows that there are productions

$$\sigma^* \rightarrow x_1 S_1$$

$$S_{i-1} \rightarrow x_i S_i \qquad \text{for } i = 2, \ldots, n-1$$

$$S_{n-1} \rightarrow x_n.$$

The derivation

$$\sigma^* \Rightarrow x_1 S_1$$
$$\Rightarrow x_1 x_2 S_2$$
$$\vdots$$
$$\Rightarrow x_1 \cdots x_{n-1} S_{n-1}$$
$$\Rightarrow x_1 \cdots x_n \qquad (7.4.3)$$

shows that $x_1 \cdots x_n \in L(G)$.

Conversely, suppose that $x_1 \cdots x_n \in L(G)$. Then there is a derivation of the form (7.4.3). If, in the transition diagram, we begin at σ^* and trace the path $(\sigma^*, S_1, \ldots, S_n)$, we can generate the string $x_1 \cdots x_n$. The last production used in (7.4.3) is $S_{n-1} \rightarrow x_n$; thus the last state reached is an accepting state. Therefore, $x_1 \cdots x_n$ is accepted by A. The proof is complete. ■

Next, we consider the reverse situation. Given a regular grammar G, we want to construct a finite-state automaton A so that $L(G)$ is precisely the set of strings accepted by A. It might seem, at first glance, that we can simply reverse the procedure of Theorem 7.4.2. However, the next example shows that the situation is a bit more complex.

EXAMPLE 7.4.3. Consider the regular grammar defined by

$$T = \{a, b\}, \qquad N = \{\sigma^*, C\}$$

with productions

$$\sigma^* \rightarrow b\sigma^*, \quad \sigma^* \rightarrow aC, \quad C \rightarrow bC, \quad C \rightarrow b$$

and starting symbol σ^*.

We represent each state as a vertex with σ^* the initial state and we add an additional vertex F to serve as an accepting state (see Figure 7.4.1). We can now reverse the procedure of Theorem 7.4.2. That is, if there is a production of the form

$$S \rightarrow xS'$$

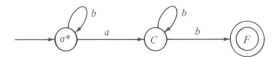

FIGURE 7.4.1

we draw an edge from state S to state S' and label it x. If there is a production of the form

$$S \rightarrow x$$

we draw an edge from state S to state F and label it x. We obtain the graph of Figure 7.4.1.

Unfortunately, the graph of Figure 7.4.1 is not a finite-state automaton. There are several problems. Vertex C has no outgoing edge labeled a and vertex F has no outgoing edges at all. Also, vertex C has two outgoing edges labeled b. A diagram like that of Figure 7.4.1 defines another kind of automaton called a **nondeterministic finite-state automaton**. The reason for the word "nondeterministic" is that when we are in a state where there are multiple outgoing edges all having the same label x, if x is input the situation is nondeterministic—we have a choice of next states. For example, if in Figure 7.4.1, we are in state C and b is input, we have a choice of next states—we can either remain in state C or go to state F.

DEFINITION 7.4.4. A *nondeterministic finite-state automaton* A consists of

(a) A finite set $\mathscr{I}$ of *input symbols.*
(b) A finite set $\mathscr{S}$ of *states.*
(c) A *next-state function* f from $\mathscr{S} \times \mathscr{I}$ into $\mathscr{P}(\mathscr{S})$.

(d) A subset $\mathcal{A}$ of $\mathcal{S}$ of *accepting states*.

(e) An *initial state* $\sigma^* \in \mathcal{S}$.

We write $A = (\mathcal{I}, \mathcal{S}, f, \mathcal{A}, \sigma^*)$.

The only difference between a nondeterministic finite-state automaton and a finite-state automaton is that in a finite-state automaton, the next-state function takes us to a uniquely defined state, whereas in a nondeterministic finite-state automaton the next-state function takes us to a set of states.

EXAMPLE 7.4.5.   For the nondeterministic finite-state automaton of Figure 7.4.1, we have

$$\mathcal{I} = \{a, b\}$$

$$\mathcal{S} = \{\sigma^*, C, F\}$$

$$\mathcal{A} = \{F\}.$$

The initial state is σ^* and the next-state function f is given by

$\mathcal{S}$ \ $\mathcal{I}$	a	b
σ^*	$\{C\}$	$\{\sigma^*\}$
C	$\varnothing$	$\{C, F\}$
F	$\varnothing$	$\varnothing$

We draw the transition diagram of a nondeterministic finite-state automaton similarly to that of a finite-state automaton. We draw an edge from state S to each state in the set $f(S, x)$ and label each x.

EXAMPLE 7.4.6.   The transition diagram of the nondeterministic finite-state automaton

$$\mathcal{I} = \{a, b\}$$

$$\mathcal{S} = \{\sigma^*, C, D\}$$

$$\mathcal{A} = \{C, D\}$$

with initial state σ^* and next-state function

$\mathcal{S}$ \ $\mathcal{I}$	a	b
σ^*	$\{\sigma^*, C\}$	$\{D\}$
C	$\varnothing$	$\{C\}$
D	$\{C, D\}$	$\varnothing$

is shown in Figure 7.4.2.

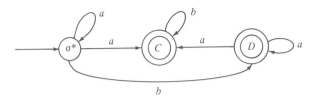

FIGURE 7.4.2

A nonnull string α is accepted by a nondeterministic finite-state automaton A if there is some path representing α in the transition diagram of A beginning at the initial state and ending in an accepting state. The formal definition follows.

DEFINITION 7.4.7. Let $A = (\mathscr{I}, \mathscr{S}, f, \mathscr{A}, \sigma^*)$ be a nondeterministic finite-state automaton. Let $\alpha = x_1 \cdots x_n$ be a nonnull string over $\mathscr{I}$. If there exist states $\sigma_0, \ldots, \sigma_n$ satisfying

   (a) $\sigma_0 = \sigma^*$;
   (b) $\sigma_i \in f(\sigma_{i-1}, x_i)$ for $i = 1, \ldots, n$;
   (c) $\sigma_n \in \mathscr{A}$;

we say that α is *accepted* by A. We let $\mathrm{Ac}(A)$ denote the set of strings accepted by A and we say that A *accepts* $\mathrm{Ac}(A)$.

If A and A' are nondeterministic finite-state automata and $\mathrm{Ac}(A) = \mathrm{Ac}(A')$, we say that A and A' are *equivalent*.

If $\alpha = x_1 \cdots x_n$ is a string over $\mathscr{I}$ and there exist states $\sigma_0, \ldots, \sigma_n$ satisfying (a) and (b) above, we call the path $(\sigma_0, \ldots, \sigma_n)$ a *path representing* α in A.

EXAMPLE 7.4.8. The string

$$\alpha = bbabb$$

is accepted by the nondeterministic finite-state automaton of Figure 7.4.1, since the path $(\sigma^*, \sigma^*, \sigma^*, C, C, F)$, which ends in an accepting state, represents α. Notice that the path $P = (\sigma^*, \sigma^*, \sigma^*, C, C, C)$ also represents α, but that P does not end in an accepting state. Nevertheless, the string α is accepted because there is at least one path representing α that ends at an accepting state. A string β will fail to be accepted if no path represents β or every path representing β ends at a nonaccepting state.

EXAMPLE 7.4.9. The string $\alpha = aabaabbb$ is accepted by the nondeterministic finite-state automaton of Figure 7.4.2. The reader should locate the path representing α, which ends at state C.

EXAMPLE 7.4.10. The string $\alpha = abba$ is not accepted by the nondeterministic finite-state automaton of Figure 7.4.2. Starting at σ^*, when we input a, there are two choices: Go to C or remain at σ^*. If we go to C, when we input two b's, our moves are determined and we remain at C. But now when we input the final a, there is no edge along which to move. On the other hand, suppose that when we input the first a, we remain at σ^*. Then, when we input b, we move to D. But now when we input the next b, there is no edge along which to move. Since there is no path representing α in Figure 7.4.2, the string α is not accepted by the nondeterministic finite-state automaton of Figure 7.4.2.

We formulate the construction of Example 7.4.3 as a theorem.

THEOREM 7.4.11. *Let* $G = (N, T, P, \sigma^*)$ *be a regular grammar. Let*

$$\mathcal{I} = T$$
$$\mathcal{S} = N \cup \{F\}, \text{ where } F \notin N \cup T$$
$$f(\sigma, x) = \{\sigma' \mid \sigma \to x\sigma' \in P\} \cup \{F \mid \sigma \to x \in P\}$$
$$\mathcal{A} = \{F\}.$$

Then the nondeterministic finite-state automaton $A = (\mathcal{I}, \mathcal{S}, f, \mathcal{A}, \sigma^*)$ *accepts precisely the strings* $L(G)$.

    Proof.   The proof is essentially the same as the proof of Theorem 7.4.2 and is, therefore, omitted.   ■

It may seem that a nondeterministic finite-state automaton is a more general concept than a finite-state automaton; however, in the next section we will show that given a nondeterministic finite-state automaton A, we can construct a finite-state automaton that is equivalent to A.

EXERCISES

In Exercises 1–5, draw the transition diagram of the nondeterministic finite-state automaton $(\mathcal{I}, \mathcal{S}, f, \mathcal{A}, \sigma_0)$.

1H. $\mathcal{I} = \{a, b\}, \mathcal{S} = \{\sigma_0, \sigma_1, \sigma_2\}, \mathcal{A} = \{\sigma_0\}$

$\mathcal{S}$	$\mathcal{I}$	
	a	b
σ_0	$\varnothing$	$\{\sigma_1, \sigma_2\}$
σ_1	$\{\sigma_2\}$	$\{\sigma_0, \sigma_1\}$
σ_2	$\{\sigma_0\}$	$\varnothing$

2. $\mathcal{I} = \{a, b\}, \mathcal{S} = \{\sigma_0, \sigma_1, \sigma_2\}, \mathcal{A} = \{\sigma_0, \sigma_1\}$

$\mathcal{S}$	$\mathcal{I}$	
	a	b
σ_0	$\{\sigma_1\}$	$\{\sigma_0, \sigma_2\}$
σ_1	$\varnothing$	$\{\sigma_2\}$
σ_2	$\{\sigma_1\}$	$\varnothing$

3H. $\mathscr{I} = \{a, b\}$, $\mathscr{S} = \{\sigma_0, \sigma_1, \sigma_2, \sigma_3\}$, $\mathscr{A} = \{\sigma_1\}$

$\mathscr{S}$ \ $\mathscr{I}$	a	b
σ_0	$\varnothing$	$\{\sigma_3\}$
σ_1	$\{\sigma_1, \sigma_2\}$	$\{\sigma_3\}$
σ_2	$\varnothing$	$\{\sigma_0, \sigma_1, \sigma_3\}$
σ_3	$\varnothing$	$\varnothing$

4. $\mathscr{I} = \{a, b, c\}$, $\mathscr{S} = \{\sigma_0, \sigma_1, \sigma_2\}$, $\mathscr{A} = \{\sigma_0\}$

$\mathscr{S}$ \ $\mathscr{I}$	a	b	c
σ_0	$\{\sigma_1\}$	$\varnothing$	$\varnothing$
σ_1	$\{\sigma_0\}$	$\{\sigma_2\}$	$\{\sigma_0, \sigma_2\}$
σ_2	$\{\sigma_0, \sigma_1, \sigma_2\}$	$\{\sigma_0\}$	$\{\sigma_0\}$

5H. $\mathscr{I} = \{a, b, c\}$, $\mathscr{S} = \{\sigma_0, \sigma_1, \sigma_2, \sigma_3\}$, $\mathscr{A} = \{\sigma_0, \sigma_3\}$

$\mathscr{S}$ \ $\mathscr{I}$	a	b	c
σ_0	$\{\sigma_1\}$	$\{\sigma_0, \sigma_1, \sigma_3\}$	$\varnothing$
σ_1	$\{\sigma_2, \sigma_3\}$	$\varnothing$	$\varnothing$
σ_2	$\varnothing$	$\{\sigma_0, \sigma_3\}$	$\{\sigma_1, \sigma_2\}$
σ_3	$\varnothing$	$\varnothing$	$\{\sigma_0\}$

For each nondeterministic finite-state automaton in Exercises 6–10, find the sets $\mathscr{I}$, $\mathscr{S}$, and $\mathscr{A}$, the initial state, and the table defining the next-state function.

6.

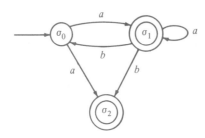

7H.

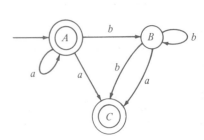

8.

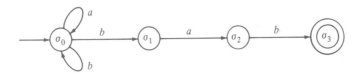

9H.

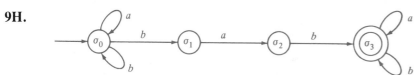

10.

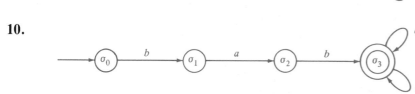

11H. Write the regular grammars given by the finite-state automata of Exercises 4–9, Section 7.2.

12. Represent the grammars of Exercises 1 and 5, Section 7.3, and Example 7.3.14 by nondeterministic finite-state automata.

13H. Is the string *bbabbb* accepted by the nondeterministic finite-state automaton of Figure 7.4.1? Prove your answer.

14. Is the string *bbabab* accepted by the nondeterministic finite-state automaton of Figure 7.4.1? Prove your answer.

15H. Show that a string α over $\{a, b\}$ is accepted by the nondeterministic finite-state automaton of Figure 7.4.1 if and only if α contains exactly one *a* and ends with *b*.

16. Is the string *aaabba* accepted by the nondeterministic finite-state automaton of Figure 7.4.2? Prove your answer.

17H. Is the string *aaaab* accepted by the nondeterministic finite-state automaton of Figure 7.4.2? Prove your answer.

18. Characterize the strings accepted by the nondeterministic finite-state automaton of Figure 7.4.2.

19H. Show that the strings accepted by the nondeterministic finite-state automaton of Exercise 8 are precisely those strings over $\{a, b\}$ that end *bab*.

★20. Characterize the strings accepted by the nondeterministic finite-state automata of Exercises 1–7, 9, and 10.

Design nondeterministic finite-state automata that accept the nonnull strings over $\{a, b\}$ having the properties specified in Exercises 21–29.

21H. Starting either *abb* or *ba*

22. Ending either *abb* or *ba*

23H. Containing either *abb* or *ba*

★24. Containing *bab* and *bb*

25H. Having each *b* preceded and followed by an *a*

26. Starting with *abb* and ending with *ab*

★27H. Starting with *ab* but not ending with *ab*

28.   Not containing *ba* or *bbb*

★29H.   Not containing *abba* or *bbb*

30.   Write regular grammars that generate the strings of Exercises 21–29.

7.5

Relationships Between Languages and Automata

In the preceding section we showed (Theorem 7.4.2) that if A is a finite-state automaton, there exists a regular grammar G, with $L(G) = \text{Ac}(A)$. As a partial converse, we showed (Theorem 7.4.11) that if G is a regular grammar, there exists a nondeterministic finite-state automaton A with $L(G) = \text{Ac}(A)$. In this section we show (Theorem 7.5.4) that if G is a regular grammar, there exists a finite-state automaton A with $L(G) = \text{Ac}(A)$. This result will be deduced from Theorem 7.4.11 by showing that any nondeterministic finite-state automaton can be converted to an equivalent finite-state automaton (Theorem 7.5.3). We will first illustrate the method by an example.

EXAMPLE 7.5.1.   Find a finite-state automaton equivalent to the nondeterministic finite-state automaton of Figure 7.4.1.

    The set of input symbols is unchanged. The states consist of all subsets

$$\varnothing, \{\sigma^*\}, \{C\}, \{F\}, \{\sigma^*, C\}, \{\sigma^*, F\}, \{C, F\}, \{\sigma^*, C, F\}$$

of the original set $\mathscr{S} = \{\sigma^*, C, F\}$ of states. The initial state is $\{\sigma^*\}$. The accepting states are all subsets

$$\{F\}, \{\sigma^*, F\}, \{C, F\}, \{\sigma^*, C, F\}$$

of $\mathscr{S}$ that contain an accepting state of the original nondeterministic finite-state automaton.

    An edge is drawn from X to Y and labeled x if $X = \varnothing = Y$ or if

$$\bigcup_{S \in X} f(S, x) = Y.$$

We obtain the finite-state automaton of Figure 7.5.1. The states

$$\{\sigma^*, F\}, \{\sigma^*, C\}, \{\sigma^*, C, F\}, \{F\},$$

which can never be reached, can be deleted. Thus we obtain the simplified, equivalent finite-state automaton of Figure 7.5.2.

EXAMPLE 7.5.2.   The finite-state automaton equivalent to the nondeterministic finite-state automaton of Example 7.4.6 is shown in Figure 7.5.3.

We now formally justify the method of Examples 7.5.1 and 7.5.2.

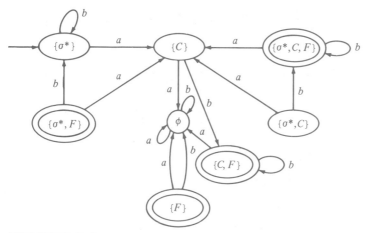

FIGURE 7.5.1

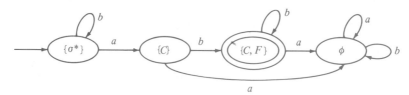

FIGURE 7.5.2

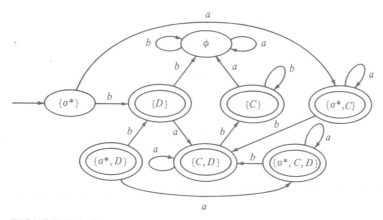

FIGURE 7.5.3

THEOREM 7.5.3. *Let* $A = (\mathcal{I}, \mathcal{S}, f, \mathcal{A}, \sigma^*)$ *be a nondeterministic finite-state automaton. Let*

(a) $\mathcal{S}' = \mathcal{P}(\mathcal{S})$.

(b) $\mathcal{I}' = \mathcal{I}$.

(c) $\sigma^{*\prime} = \{\sigma^*\}$.

(d) $\mathcal{A}' = \{X \subseteq \mathcal{S} \mid X \cap \mathcal{A} \neq \varnothing\}$.

(e) $f'(X, x) = \begin{cases} \varnothing & \text{if } X = \varnothing \\ \displaystyle\bigcup_{S \in X} f(S, x) & \text{if } X \neq \varnothing. \end{cases}$

Then the finite-state automaton $A' = (\mathscr{I}', \mathscr{S}', f', \mathscr{A}', \sigma^{*'})$ is equivalent to A.

Proof. Suppose that the string $\alpha = x_1 \cdots x_n$ is accepted by A. Then there exist states $\sigma_0, \ldots, \sigma_n \in \mathscr{S}$ with

$$\sigma_0 = \sigma^*;$$

$$\sigma_i \in f(\sigma_{i-1}, x_i) \qquad \text{for } i = 1, \ldots, n;$$

$$\sigma_n \in \mathscr{A}.$$

Set $Y_0 = \{\sigma_0\}$ and

$$Y_i = f'(Y_{i-1}, x_i) \qquad \text{for } i = 1, \ldots, n.$$

Since

$$Y_1 = f'(Y_0, x_1) = f'(\{\sigma_0\}, x_1) = f(\sigma_0, x_1),$$

it follows that $\sigma_1 \in Y_1$. Now

$$\sigma_2 \in f(\sigma_1, x_2) \subseteq \bigcup_{S \in Y_1} f(S, x_2) = f'(Y_1, x_2) = Y_2.$$

Again,

$$\sigma_3 \in f(\sigma_2, x_3) \subseteq \bigcup_{S \in Y_2} f(S, x_3) = f'(Y_2, x_3) = Y_3.$$

The argument may be continued (formally, we would use induction) to show that $\sigma_n \in Y_n$. Since σ_n is an accepting state in A, Y_n is an accepting state in A'. Thus, in A', we have

$$f'(\sigma^{*'}, x_1) = f'(Y_0, x_1) = Y_1$$

$$f'(Y_1, x_2) = Y_2$$

$$\vdots$$

$$f'(Y_{n-1}, x_n) = Y_n.$$

Therefore, α is accepted by A'.

Now suppose that the string $\alpha = x_1 \cdots x_n$ is accepted by A'. Then there exist subsets $Y_0, \ldots, Y_n$ of $\mathscr{S}$ such that

$$Y_0 = \sigma^{*'} = \{\sigma^*\};$$

$$f'(Y_{i-1}, x_i) = Y_i \qquad \text{for } i = 1, \ldots, n;$$

there exists a state $\sigma_n \in Y_n \cap \mathscr{A}$.

Since

$$\sigma_n \in Y_n = f'(Y_{n-1}, x_n) = \bigcup_{S \in Y_{n-1}} f(S, x_n),$$

there exists $\sigma_{n-1} \in Y_{n-1}$ with $\sigma_n \in f(\sigma_{n-1}, x_n)$. Similarly, since

$$\sigma_{n-1} \in Y_{n-1} = f'(Y_{n-2}, x_{n-1}) = \bigcup_{S \in Y_{n-2}} f(S, x_{n-1}),$$

there exists $\sigma_{n-2} \in Y_{n-2}$ with $\sigma_{n-1} \in f(\sigma_{n-2}, x_{n-1})$. Continuing, we obtain

$$\sigma_i \in Y_i \qquad \text{for } i = 0, \ldots, n,$$

with

$$\sigma_i \in f(\sigma_{i-1}, x_i) \qquad \text{for } i = 1, \ldots, n.$$

In particular,

$$\sigma_0 \in Y_0 = \{\sigma^*\}.$$

Thus $\sigma_0 = \sigma^*$, the initial state in A. Since σ_n is an accepting state in A, the string α is accepted by A. ∎

The next theorem summarizes these results and those of the preceding section.

THEOREM 7.5.4. *A language L is regular if and only if there exists a finite-state automaton that accepts precisely the strings in L.*

   Proof. This theorem restates Theorems 7.4.2, 7.4.11, and 7.5.3. ∎

EXAMPLE 7.5.5. Find a finite-state automaton A that accepts precisely the strings generated by the regular grammar G having productions

$$\sigma^* \to b\sigma^*, \quad \sigma^* \to aC, \quad C \to bC, \quad C \to b.$$

The starting symbol is σ^*, the set of terminal symbols is $\{a, b\}$, and the set of nonterminal symbols is $\{\sigma^*, C\}$.

   The nondeterministic finite-state automaton A' that accepts $L(G)$ is shown in Figure 7.4.1. A finite-state automaton equivalent to A' is shown in Figure 7.5.1 and an equivalent simplified finite-state automaton A is shown in Figure 7.5.2. The finite-state automaton A accepts precisely the strings generated by G.

   We close this section by giving some applications of the methods and theory we have developed.

EXAMPLE 7.5.6. Show that the language

$$L = \{a^n b^n \mid n = 1, 2, \ldots\}$$

is not regular.

   If L is regular, there exists a finite-state automaton A such that $Ac(A) = L$. Suppose that A has k states. The string $\alpha = a^{k+1} b^{k+1}$ is accepted by A. Consider the path P, which represents α. Since there are k states, some state σ is revisited on the part of the path representing a^{k+1}. Thus there is a circuit

C, all of whose edges are labeled a, which contains σ. We change the path P to obtain a path P' as follows. When we arrive at σ in P, we follow C. After returning to σ on C, we continue on P to the end. If the length of C is j, the path P' represents the string $\alpha' = a^{j+k+1}b^{k+1}$. Since P and P' end at the same state σ' and σ' is an accepting state, α' is accepted by A. This is a contradiction, since α' is not of the form $a^n b^n$. Therefore, L is not regular.

EXAMPLE 7.5.7. Let L be the set of strings accepted by the finite-state automaton A of Figure 7.5.4. Construct a finite-state automaton that accepts the strings

$$L^R = \{x_n \cdots x_1 \mid x_1 \cdots x_n \in L\}.$$

We want to convert A to a finite-state automaton that accepts L^R. The string $\alpha = x_1 \cdots x_n$ is accepted by A if there is a path P in A representing α that starts at σ_1 and ends at σ_3. If we start at σ_3 and trace P in reverse, we end at σ_1 and process the edges in the order $x_n, \ldots, x_1$. Thus we need only reverse all arrows in Figure 7.5.4 and make σ_3 the starting state and σ_1 the accepting state (see Figure 7.5.5). The result is a nondeterministic finite-state automaton that accepts L^R.

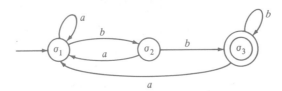

FIGURE 7.5.4

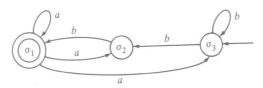

FIGURE 7.5.5

After finding an equivalent finite-state automaton and eliminating the unreachable states, we obtain the equivalent finite-state automaton of Figure 7.5.6.

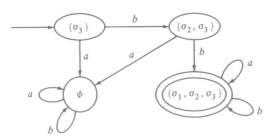

FIGURE 7.5.6

EXAMPLE 7.5.8. Let L be the set of strings accepted by the finite-state automaton A of Figure 7.5.7. Construct a nondeterministic finite-state automaton that accepts the strings

$$L^R = \{x_n \cdots x_1 \mid x_1 \cdots x_n \in L\}.$$

If A had only one accepting state, we could use the procedure of Example 7.5.7 to construct the desired nondeterministic finite-state automaton. Thus we first construct a nondeterministic finite-state automaton equivalent to A with one accepting state. To do this we introduce an additional state σ_5. Then we arrange for paths terminating at σ_3 or σ_4 to optionally terminate at σ_5 (see Figure 7.5.8). The desired nondeterministic finite-state automaton is obtained from Figure 7.5.8 by the method of Example 7.5.7 (see Figure 7.5.9). Of course, if desired, we could construct an equivalent finite-state automaton.

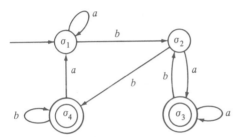

FIGURE 7.5.7

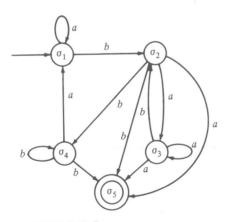

FIGURE 7.5.8

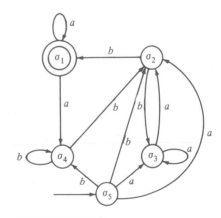

FIGURE 7.5.9

The methods of Examples 7.5.7 and 7.5.8 can be used to deduce the following result.

THEOREM 7.5.9. *If L is a regular language, the language*

$$L^R = \{x_n \cdots x_1 \mid x_1 \cdots x_n \in L\}$$

is also regular.

Proof. Let A be a finite-state automaton that accepts L. Construct an equivalent nondeterministic finite-state automaton A' with one accepting state as outlined in Example 7.5.8. Finally, construct an equivalent nondeterministic finite-state automaton A'' that accepts L^R as outlined in Example 7.5.7. By Theorems 7.5.3 and 7.5.4, L^R is a regular language. ∎

In the definition of regular grammar (Definition 7.3.8c), the allowable productions were

$$A \to \alpha, \quad A \to \alpha B, \qquad \text{where } A, B \in N, \quad \alpha \in T - \{\lambda\}.$$

It follows from Theorem 7.5.9 that if we allow productions of the form

$$A \to \alpha, \quad A \to B\alpha, \qquad \text{where } A, B \in N, \quad \alpha \in T - \{\lambda\},$$

we still obtain a regular language.

THEOREM 7.5.10. *Let G be a grammar and let λ denote the null string. If every production is of the form*

$$A \to \alpha, \quad A \to B\alpha, \qquad \text{where } A, B \in N, \quad \alpha \in T - \{\lambda\},$$

then $L(G)$ is a regular language.

Proof. Let G' be the grammar with the productions

$$\{A \to \alpha \mid A \to \alpha \text{ is a production in } G\} \cup$$

$$\{A \to \alpha B \mid A \to B\alpha \text{ is a production in } G\}.$$

Now

$$L(G) = L(G')^R$$

where

$$L(G')^R = \{x_n \cdots x_1 \mid x_1 \cdots x_n \in L(G')\}.$$

By Theorem 7.5.9, $L(G)$ is a regular language. ∎

EXERCISES

1H. Find finite-state automata equivalent to the nondeterministic finite-state automata of Exercises 1–10, Section 7.4.

In Exercises 2–6, find finite-state automata that accept the strings generated by the regular grammars.

2. Grammar of Exercise 1, Section 7.3

3H. Grammar of Exercise 5, Section 7.3

4. $<S> ::= a<A> \mid a<B>$
$<A> ::= a<B> \mid b<S> \mid b$
$<B> ::= b<S> \mid b$
with starting symbol $<S>$

5H. $<S> ::= a<S> \mid a<A> \mid b<C> \mid a$
$<A> ::= b<A> \mid a<C>$
$<B> ::= a<S> \mid a$
$<C> ::= a<B> \mid a<C>$
with starting symbol $<S>$

6. $<S> ::= a<A> \mid a<B>$
$<A> ::= b<S> \mid b$
$<B> ::= a<B> \mid a<C>$
$<C> ::= a<S> \mid b<A> \mid a<C> \mid a$

7H.   Find finite-state automata that accept the nonnull strings of Exercises 21–29, Section 7.4.

8.   By eliminating unreachable states from the finite-state automaton of Figure 7.5.3, find a simpler, equivalent finite-state automaton.

9H.   Show that the nondeterministic finite-state automaton of Figure 7.5.5 accepts a string α over $\{a, b\}$ if and only if α begins bb.

★10.   Characterize the strings accepted by the nondeterministic finite-state automata of Figures 7.5.7 and 7.5.9.

In Exercises 11–21, find a nondeterministic finite-state automaton that accepts the given set of strings. If S_1 and S_2 are sets of strings, we let

$$S_1^+ = \{u_1 u_2 \cdots u_n \mid u_i \in S_1, n \in \{1, 2, \ldots\}\};$$
$$S_1 S_2 = \{uv \mid u \in S_1, v \in S_2\}.$$

11H. $Ac(A)^R$, where A is the automaton of Exercise 4, Section 7.2

12. $Ac(A)^R$, where A is the automaton of Exercise 5, Section 7.2

13H. $Ac(A)^R$, where A is the automaton of Exercise 6, Section 7.2

14. $Ac(A)^+$, where A is the automaton of Exercise 4, Section 7.2

15H. $Ac(A)^+$, where A is the automaton of Exercise 5, Section 7.2

16. $Ac(A)^+$, where A is the automaton of Exercise 6, Section 7.2

17H. $Ac(A)^+$, where A is the automaton of Figure 7.5.7

18. $Ac(A_1)Ac(A_2)$, where A_1 is the automaton of Exercise 4, Section 7.2, and A_2 is the automaton of Exercise 5, Section 7.2

19H. $Ac(A_1)Ac(A_2)$, where A_1 is the automaton of Exercise 5, Section 7.2, and A_2 is the automaton of Exercise 6, Section 7.2

20. $Ac(A_1)Ac(A_1)$, where A_1 is the automaton of Exercise 6, Section 7.2

21H. $Ac(A_1)Ac(A_2)$, where A_1 is the automaton of Figure 7.5.7 and A_2 is the automaton of Exercise 5, Section 7.2

22. Find a regular grammar that generates the language L^R, where L is the language generated by the grammar of Exercise 5, Section 7.3.

23H. Find a regular grammar that generates the language L^+, where L is the language generated by the grammar of Exercise 5, Section 7.3.

24. Let L_1 (respectively, L_2) be the language generated by the grammar of Exercise 5, Section 7.3 (respectively, Example 7.5.5). Find a regular grammar that generates the language L_1L_2.

★25H. Show that the set

$$L = \{x_1 \cdots x_n \mid x_1 \cdots x_n = x_n \cdots x_1\}$$

of strings over $\{a, b\}$ is not a regular language.

26. Show that if L_1 and L_2 are regular languages over $\mathcal{I}$ and S is the set of all nonnull strings over $\mathcal{I}$, then any of $S - L_1$, $L_1 \cup L_2$, $L_1 \cap L_2$, L_1^+, and L_1L_2 is either empty or is a regular language.

★27H. Show, by example, that there are context-free languages L_1 and L_2 such that $L_1 \cap L_2$ is not context-free.

★28. Prove or disprove: If L is a regular language, so is

$$\{u^n \mid u \in L, n \in \{1, 2, \ldots\}\}.$$

7.6

Notes

General references on automata, grammars, and languages are [Arbib, 1969; Hopcroft, 1969, 1979; Kain; Kohavi; McNaughton; Minsky; and Nelson]. In addition, the following books on discrete mathematics contain information about automata, grammars, and languages: [Berztiss; Birkhoff; Bobrow; Fisher; Gersting; Gill; Levy; Lipschutz, 1976; Sahni; and Tremblay] with [Bobrow] having the most extensive treatment. See [D'Angelo] for the details of hardware that implements sequential circuits.

A finite-state machine has a primitive, internal memory in the sense that it remembers which state it is in. By permitting an external memory on which the machine can read and write data, we can define more powerful machines. Other enhancements are achieved by allowing the machine to scan the input string in either direction and by allowing the machine to alter the input string. It is then possible to characterize the classes of machines that accept context-free languages, context-sensitive languages, and languages generated by phrase structure grammars.

Turing machines form a particularly important class of machines. Like a finite-state machine, a Turing machine is always in a particular state. The input string to a Turing machine is assumed to reside on a tape that is infinite in both directions. A Turing machine scans one character at a time and after scanning a character, the machine either halts or does some, none, or all of: alter the character, move one position left or right, change states. In particular, the input string can be changed. A Turing machine T accepts a string α if, when α is input to T, T halts in an accepting state. It can be shown that a language L is generated by a phrase-structure grammar if and only if there is a Turing machine that accepts L.

The real importance of Turing machines results from the widely held belief that any function that can be computed by some, perhaps hypothetical, digital computer can be computed by some Turing machine. This last assertion is known as **Turing's hypothesis** or **Church's thesis**. Church's thesis implies that a Turing machine is the correct abstract model of a digital computer. These ideas also yield the following formal definition of algorithm. An **algorithm** is a Turing machine that, given an input string, eventually stops.

COMPUTER EXERCISES

In Exercises 1–4, implement the finite-state machines as programs. Provide sample runs for various input strings.

1.	Example 7.1.5	**2.**	Example 7.1.10
3.	Exercises 1–10, Section 7.1	**4.**	Exercises 21–26, Section 7.1

In Exercises 5–11, implement the finite-state automata as programs similarly to Algorithm 7.2.10. Provide sample runs for various input strings.

5.	Figure 7.2.2	**6.**	Figure 7.2.3
7.	Example 7.2.4	**8.**	Example 7.2.8
9.	Example 7.2.9	**10.**	Exercises 1–9, Section 7.2
11.	Exercises 21–31, Section 7.2		

12. Write programs that determine whether a given string is in the regular languages of Exercises 2–6, Section 7.5.

13. Write a program that simulates an arbitrary finite-state machine. The program should initially receive as input, the next-state function, the output function, and the initial state. The program should then be able to accept strings, simulate the action of the finite-state machine, and output the strings produced by the finite-state machine.

Appendix A: Logic

This appendix examines some techniques of **logic**. Logic studies methods of reasoning, specifically, methods to separate valid reasoning from invalid reasoning. Results in many disciplines are established by logical reasoning. In mathematics, correct arguments must be supplied to provide proofs of theorems. In computer science, arguments must be given to show that programs do what they claim to do.

A.1

Propositions

An assertion such as

$$1 + 1 = 3$$

which is either true or false, but not both, is called a **proposition**. We will use lowercase letters, such as p, q, or r, to represent propositions. We will also use the notation

$$p: 1 + 1 = 3$$

to define p to be the proposition $1 + 1 = 3$.

EXAMPLE A.1.1.   We define

> p: There is a Nobel prize in computer science.
> q: Earth is the only planet in the universe that has life.
> r: Type control Z to exit insert mode.

Then p and q are propositions. p is false and q is either true or false, but no one knows which at this time. r is not a proposition, since r is neither true nor false (r is a command).

Propositions may be combined, as the next example illustrates.

EXAMPLE A.1.2.   If p and q are as in Example A.1.1, we may form the propositions

> p and q: There is a Nobel prize in computer science and Earth is the only planet in the universe that has life.
> p or q: There is a Nobel prize in computer science or Earth is the only planet in the universe that has life.

DEFINITION A.1.3.   Let p and q be propositions. The *conjunction* of p and q, denoted $p \wedge q$, is the proposition

$$p \text{ and } q.$$

The *disjunction* of p and q, denoted $p \vee q$, is the proposition

$$p \text{ or } q.$$

Propositions such as $p \wedge q$ and $p \vee q$ that result from combining propositions are called **compound propositions**.

EXAMPLE A.1.4.   Let

$$p: 1 + 1 = 3.$$
$$q: \text{A decade is 10 years.}$$

The conjunction of p and q is

$$p \wedge q: 1 + 1 = 3 \text{ and a decade is 10 years.}$$

The disjunction of p and q is

$$p \vee q: 1 + 1 = 3 \text{ or a decade is 10 years.}$$

If the propositions $p_1, \ldots, p_n$ are combined to form the compound proposition P, we will sometimes write

$$P = P(p_1, \ldots, p_n).$$

The truth values of compound propositions can be described by **truth tables**. The truth table of the compound proposition $P(p_1, \ldots, p_n)$ lists all possible combinations of truth values for $p_1, \ldots, p_n$, T denoting true and F denoting false, and for each such combination lists the truth value of $P(p_1, \ldots, p_n)$.

DEFINITION A.1.5. The truth values of the compound propositions $p \wedge q$ and $p \vee q$ are defined by the following truth tables.

p	q	$p \wedge q$	p	q	$p \vee q$
T	T	T	T	T	T
T	F	F	T	F	T
F	T	F	F	T	T
F	F	F	F	F	F

Notice that in the truth tables in Definition A.1.5 all four possible combinations of truth assignments for p and q are given.

According to Definition A.1.5, the conjunction $p \wedge q$ is true when both p and q are true and is false otherwise. The disjunction $p \vee q$ is false when both p and q are false and is true otherwise. The *or* in the disjunction, $p \vee q$, is used in the *inclusive* sense, that is, $p \vee q$ is true if either p or q or *both* are true. There is also an **exclusive-or** (see Exercise 14) in which p *exor* q is true if either p or q but *not* both is true.

EXAMPLE A.1.6. Using Definition A.1.5, we find that for p and q as in Example A.1.4, since p is false and q is true, $p \wedge q$ is false and $p \vee q$ is true.

If p is a proposition, we may form the **negation** $\bar{p}$ of p by reversing its truth values. $\bar{p}$ is sometimes read "not p."

DEFINITION A.1.7. The *negation* $\bar{p}$ of p is defined by the following truth table.

p	$\bar{p}$
T	F
F	T

EXAMPLE A.1.8. Let

> p: Cary Grant starred in "Rear Window."

The negation of p is the proposition

> $\bar{p}$: Cary Grant did not star in "Rear Window."

EXAMPLE A.1.9. Let

> p: Beethoven lived in the eighteenth century.
> q: The first all-electronic digital computer was constructed in the twentieth century.
> r: The cow jumped over the moon.

Represent the proposition

> Either Beethoven lived in the eighteenth century and it is not true that the first all-electronic digital computer was constructed in the twentieth century; or the cow jumped over the moon.

symbolically and determine whether it is true or false.

The proposition may be written symbolically as

$$(p \wedge \bar{q}) \vee r.$$

If we replace each symbol by its truth value, we find

$$\begin{aligned} (p \wedge \bar{q}) \vee r &= (T \wedge \bar{T}) \vee F \\ &= (T \wedge F) \vee F \\ &= F \vee F \\ &= F. \end{aligned}$$

Therefore, the given proposition is false.

It is possible for a compound proposition $P(p_1, \ldots, p_n)$ to be true no matter what truth assignments are made for the propositions $p_1, \ldots, p_n$. Our next definition makes this notion, as well as the situation in which a compound proposition is always false, precise.

DEFINITION A.1.10. Let $P = P(p_1, \ldots, p_n)$ be a proposition.

The proposition P is a *tautology* if P is true for every truth assignment for $p_1, \ldots, p_n$.

The proposition P is a *contradiction* if P is false for every truth assignment for $p_1, \ldots, p_n$.

Notice that the negation of a tautology is a contradiction and that the negation of a contradiction is a tautology.

EXAMPLE A.1.11. The truth table shows that the proposition $p \vee \bar{p}$ is a tautology and that the proposition $p \wedge \bar{p}$ is a contradiction.

p	$p \vee \bar{p}$	$p \wedge \bar{p}$
T	T	F
F	T	F

The execution of statements within a structure in a computer program is controlled by evaluating the truth of propositions.

For example, in the **if-then-else** structure

> **if** p **then**
>> *action* 1
>
> **else**
>> *action* 2

if the proposition p is true, *action* 1 (but not *action* 2) is executed and if the proposition p is false, *action* 2 (but not *action* 1) is executed.

In the **while loop** structure

$$\textbf{while } p \textbf{ do}$$
$$action$$

action will be repeatedly executed as long as p is true. In the while loop or the if-then-else structure, we will use indentation to indicate the extent of *action*.

EXAMPLE A.1.12. In the following code,

```
i := 1
j := 1
while (i < 2 and j < 5) or i + j = 5 do
        begin
        i := i + 2
        j := j + 1
        end
```

the while loop will be executed twice. If we let

$$p{:}\ i < 2 \qquad q{:}\ j < 5 \qquad r{:}\ i + j = 5$$

the proposition can be written $(p \wedge q) \vee r$. Initially, p is true, q is true, and r is false. In this case, $(p \wedge q) \vee r$ is true and the while loop is executed. At this point, i is 3 and j is 2, so this time p is false, q is true, and r is true. Therefore, $(p \wedge q) \vee r$ is again true and the while loop is executed a second time. At this point i is 5 and j is 3, so this time p is false, q is true, and r is false. Therefore, $(p \wedge q) \vee r$ is false and the while loop ceases to execute.

EXERCISES

Evaluate each proposition in Exercises 1–5 for the truth values

$$p = \text{F}, \qquad q = \text{T}, \qquad r = \text{F}.$$

1H. $p \vee q$ **2.** $\overline{p} \vee \overline{q}$

3H. $\overline{p} \vee q$ **4.** $p \vee \overline{(q \wedge r)}$

5H. $(p \vee q) \wedge (\overline{p} \vee r)$

Write the truth table of each proposition in Exercises 6–12.

6. $p \wedge \overline{q}$ **7H.** $(\overline{p} \vee \overline{q}) \vee p$ **8.** $(p \vee q) \wedge \overline{p}$

9H. $(p \wedge q) \wedge \overline{p}$ **10.** $(p \wedge q) \vee (\overline{p} \vee q)$ **11H.** $\overline{(p \wedge q)} \vee (r \wedge \overline{p})$

12. $(p \vee q) \wedge (\overline{p} \vee q) \wedge (p \vee \overline{q}) \wedge (\overline{p} \vee \overline{q})$

13H.   Determine whether each proposition in Exercises 1–12 is a tautology, a contradiction, or neither.

14.   Give the truth table for the exclusive-or of p and q in which p *exor* q is true if either p or q but not both is true.

Assume that a, b, and c are real numbers. In Exercises 15–19, represent the given statement symbolically by letting

$$p: a < b, \qquad q: b < c, \qquad r: a < c.$$

15H. $a < b$ and $b < c$.

16. $(a \geq b$ and $b < c)$ or $a \geq c$.

17H.   It is not true that $(a < b$ and $b < c)$.

18. $a < b$ or it is not true that $(b < c$ and $a < c)$.

19H.   (It is not true that $(a < b$ and $(a < c$ or $b < c)))$ or $(a \geq b$ and $a < c)$.

In Exercises 20–24, formulate the symbolic expression in words using

p: Today is Monday.
q: It is raining.
r: It is hot.

20. $p \vee q$ **21H.** $\bar{p} \wedge (q \vee r)$ **22.** $\overline{p \vee q} \wedge r$

23H. $(p \wedge (q \vee r)) \wedge (r \vee (q \vee p))$

24. $(p \vee (\bar{p} \wedge (\overline{q \vee r}))) \wedge (p \vee (r \vee q))$

For each program segment in Exercises 25–29, determine the number of times the statement $x := x + 1$ will be executed.

25H. $i := 1$
         if $i < 2$ **or** $i > 0$ **then**
             $x := x + 1$
         else
             $x := x + 2$

26. $i := 2$
         if $(i < 0$ **and** $i > 1)$ **or** $i = 3$ **then**
             $x := x + 1$
         else
             $x := x + 2$

27H. $i := 1$
         while $i < 3$ **do**
             begin
             $x := x + 1$
             $i := i + 1$
             end

28. $i := 1$
   while $(i > 0$ **and** $i < 3)$ **or** $i = 3$ **do**
      begin
      $x := x + 1$
      $i := i + 1$
      end

29H. $i := 1$
   while $i > 0$ **and** $i < 4$ **do**
      begin
      if $i < 2$ **then**
         $i := i + 1$
      else
         $i := i + 2$
      $x := x + 1$
      end

A.2

Conditional Propositions and Logical Equivalence

Consider the statement

> If the Mathematics Department gets an additional \$20,000,
>   then it will hire one new faculty member.   (A.2.1)

Statement (A.2.1) states that on the condition that the Mathematics Department gets an additional \$20,000, then the Mathematics Department will hire one new faculty member. A proposition such as (A.2.1) is called a **conditional proposition**.

DEFINITION A.2.1. If p and q are propositions, the compound proposition

$$\text{if } p, \text{ then } q \qquad (A.2.2)$$

is called a *conditional proposition* and is denoted

$$p \rightarrow q.$$

The proposition p is called the *hypothesis* (or *antecedent*) and the proposition q is called the *conclusion* (or *consequent*).

EXAMPLE A.2.2. If we define

p: the Mathematics Department gets an additional \$20,000
q: the Mathematics Department hires one new faculty member.

then statement (A.2.1) assumes the form (A.2.2). The hypothesis is the statement, "the Mathematics Department gets an additional $20,000," and the conclusion is the statement, "the Mathematics Department hires one new faculty member."

Some statements not of the form (A.2.2) may be restated as conditional propositions, as the next example illustrates.

EXAMPLE A.2.3.   Restate each proposition in the form (A.2.2) of a conditional proposition.
   (a) Mary will be a good student if she studies hard.
   (b) John may take calculus only if he has sophomore, junior, or senior standing.
   (c) When you sing, my ears hurt.
   (d) A necessary condition for the triangle t to be equilateral is that t be equiangular.
   (e) A sufficient condition for the function f to be continuous is that f be differentiable.

   (a) The hypothesis is the clause following *if*; thus, an equivalent formulation is

   If Mary studies hard, then she will be a good student.

   (b) The *only if* clause is the conclusion; that is,

   if p, then q

is considered logically the same as

   p only if q.

An equivalent formulation is

If John takes calculus, then he has sophomore, junior, or senior standing.

   The "if p, then q" formulation emphasizes the hypothesis, whereas the "p only if q" formulation emphasizes the conclusion; the difference is only stylistic.
   (c) *When* means the same as *if*; thus an equivalent formulation is

   If you sing, then my ears hurt.

   (d) A **necessary condition** is another name for the conclusion; thus an equivalent formulation is

   If the triangle t is equilateral, then t is equiangular.

   (e) A **sufficient condition** is another name for the hypothesis; thus an equivalent formulation is

   If the function f is differentiable, then f is continuous.

Consider the problem of assigning a truth value to the conditional proposition (A.2.2). Certainly if the hypothesis p is true and the conclusion q is also true (i.e., the hypothesis and the conclusion are both true), then the conditional proposition (A.2.2) should be true. On the other hand, if the hypothesis p is true and the conclusion q is false, then (A.2.2) should be false. (You should not be able to deduce a false conclusion from a true hypothesis!) The standard definition declares (A.2.2) to be true in case the hypothesis p is false regardless of the truth value of the conclusion q. (Exercises 45 and 46 illustrate reasons for adopting this rule.) The preceding discussion is summarized in our next definition.

DEFINITION A.2.4. The truth value of the conditional proposition $p \rightarrow q$ is defined by the following truth table:

p	q	$p \rightarrow q$
T	T	T
T	F	F
F	T	T
F	F	T

EXAMPLE A.2.5. Let

$$p: 1 > 2 \qquad q: 4 < 8.$$

Then p is false and q is true. Therefore,

$$p \rightarrow q \text{ is true}, \qquad q \rightarrow p \text{ is false}.$$

EXAMPLE A.2.6. Assuming that p is true, q is false, and r is true, find the truth value of each proposition.
  (a) $(p \wedge q) \rightarrow r$
  (b) $(p \vee q) \rightarrow \bar{r}$
  (c) $p \wedge (q \rightarrow r)$
  (d) $p \rightarrow (q \rightarrow r)$

We replace each symbol p, q, and r by its truth value to obtain the truth value of the proposition:
  (a) $(T \wedge F) \rightarrow T = F \rightarrow T = \text{True}$
  (b) $(T \vee F) \rightarrow \bar{T} = T \rightarrow F = \text{False}$
  (c) $T \wedge (F \rightarrow T) = T \wedge T = \text{True}$
  (d) $T \rightarrow (F \rightarrow T) = T \rightarrow T = \text{True}$

In ordinary language, the hypothesis and conclusion in a conditional proposition are normally related, but in logic, the hypothesis and conclusion in a conditional proposition are not required to refer to the same subject matter. For example, in logic, we permit propositions such as

If $5 < 3$, then Alfred Hitchcock directed ''The Third Man.''

(In fact, since the hypothesis is false, this proposition is true.) Logic is concerned with the form of propositions and the relations of propositions to each other and not with the subject matter itself.

Example A.2.5 shows that the proposition $p \rightarrow q$ can be true while the proposition $q \rightarrow p$ is false. We call the proposition $q \rightarrow p$ the **converse** of the proposition $p \rightarrow q$. Thus a conditional proposition can be true while its converse is false.

EXAMPLE A.2.7. Write each conditional proposition symbolically. Write the converse of each statement symbolically and in words. Also, find the truth value of each conditional proposition and its converse.
(a) If $1 < 2$, then $3 < 6$.
(b) If $1 > 2$, then $3 < 6$.

(a) Let

$$p: 1 < 2, \qquad q: 3 < 6.$$

The given statement may be written symbolically as

$$p \rightarrow q.$$

Since p and q are both true, this statement is true. The converse may be written symbolically as

$$q \rightarrow p$$

and in words as

$$\text{If } 3 < 6, \text{ then } 1 < 2.$$

Since p and q are both true, the converse $q \rightarrow p$ is true.
(b) Let

$$p: 1 > 2 \qquad q: 3 < 6.$$

The given statement may be written symbolically as

$$p \rightarrow q.$$

Since p is false and q is true, this statement is true. The converse may be written symbolically as

$$q \rightarrow p$$

and in words as

$$\text{If } 3 < 6, \text{ then } 1 > 2.$$

Since q is true and p is false, the converse $q \rightarrow p$ is false.

Of particular interest are true conditional propositions. Mathematical theorems are often phrased as conditional propositions. A proof of a theorem of this form is a verification that the conditional proposition is true.

Let $P = P(p_1, \ldots, p_n)$ and $Q = Q(p_1, \ldots, p_n)$ be compound propositions and suppose that $P \rightarrow Q$ is true. We know that if P is false, $P \rightarrow Q$ is true regardless of whether Q is true or false. On the other hand, if P is true, Q must also be true or otherwise $P \rightarrow Q$ would be false. This discussion motivates our next definition.

DEFINITION A.2.8. The compound proposition $P = P(p_1, \ldots, p_n)$ *logically implies* the compound proposition $Q = Q(p_1, \ldots, p_n)$, written

$$P \Rightarrow Q,$$

provided that given any truth values of $p_1, \ldots, p_n$, if P is true, Q is also true.

Notice that $P(p_1, \ldots, p_n) \Rightarrow Q(p_1, \ldots, p_n)$ holds precisely when

$$P(p_1, \ldots, p_n) \rightarrow Q(p_1, \ldots, p_n)$$

is a tautology.

EXAMPLE A.2.9. Show that

$$\overline{p \vee q} \Rightarrow \overline{p}. \tag{A.2.3}$$

The proposition $\overline{p \vee q}$ is true precisely when $p \vee q$ is false. For $p \vee q$ to be false, both p and q must be false. Therefore, $\overline{p}$ is true. We have shown that whenever $\overline{p \vee q}$ is true, so is $\overline{p}$. Therefore, (A.2.3) holds.

Another useful compound proposition is

$$p \text{ if and only if } q. \tag{A.2.4}$$

Such a statement is interpreted to mean

$$\text{(if } p \text{, then } q\text{) } and \text{ (if } q \text{, then } p\text{).} \tag{A.2.5}$$

Let us determine the truth value of the proposition (A.2.4). Suppose that p and q are both true. Then both conditional propositions in (A.2.5) are true. *And*ing true values produces a true value; hence (A.2.5) is true. Since (A.2.4) is interpreted as (A.2.5), we regard (A.2.4) as true if both p and q are true. If both p and q are false, again both conditional propositions in (A.2.5) are true, so (A.2.5) is true. Therefore, if both p and q are false, we regard (A.2.4) as true. If p is false and q is true, then the second conditional proposition in (A.2.5) is false. When one value in an *and* operation is false, the result is false. Thus we regard (A.2.4) as false if p is false and q is true. Similarly, if p is true and q is false, we regard (A.2.4) as false. This motivates the following definition.

DEFINITION A.2.10. If p and q are propositions, the compound proposition

$$p \text{ if and only if } q$$

is called a *biconditional proposition* and is denoted

$$p \leftrightarrow q.$$

The truth value of the proposition $p \leftrightarrow q$ is defined by the following truth table:

p	q	$p \leftrightarrow q$
T	T	T
T	F	F
F	T	F
F	F	T

Notice that $p \leftrightarrow q$ is true precisely when either p and q are both true or p and q are both false. An alternative way to state "p if and only if q" is "p is a necessary and sufficient condition for q." "p if and only if q" is sometimes written "p iff q."

EXAMPLE A.2.11. Let a, b, and c be the lengths of the sides of the triangle T with c the longest length. The statement

$$T \text{ is a right triangle if and only if } a^2 + b^2 = c^2 \qquad \text{(A.2.6)}$$

can be written symbolically as

$$p \leftrightarrow q$$

if we define

$$p: T \text{ is a right triangle} \qquad q: a^2 + b^2 = c^2.$$

Statement (A.2.6) asserts that

$$\text{If } T \text{ is a right triangle, then } a^2 + b^2 = c^2. \qquad \text{(A.2.7)}$$

and

$$\text{If } a^2 + b^2 = c^2, \text{ then } T \text{ is a right triangle.} \qquad \text{(A.2.8)}$$

An alternative way to state (A.2.6) is: A necessary and sufficient condition for the triangle T to be a right triangle is that the sides satisfy $a^2 + b^2 = c^2$.

Of particular interest are true statements of the form: P if and only if Q. Let $P = P(p_1, \ldots, p_n)$ and $Q = Q(p_1, \ldots, p_n)$ be compound propositions and suppose that $P \leftrightarrow Q$ is true. According to Definition A.2.10 either P and Q are both true or P and Q are both false. This discussion motivates our next definition.

DEFINITION A.2.12. The compound propositions $P = P(p_1, \ldots, p_n)$ and $Q = Q(p_1, \ldots, p_n)$ are *logically equivalent* and we write

$$P \equiv Q,$$

provided that given any truth values of $p_1, \ldots, p_n$, either P and Q are both true or P and Q are both false.

Notice that $P \equiv Q$ holds precisely when

$$P(p_1, \ldots, p_n) \leftrightarrow Q(p_1, \ldots, p_n)$$

is a tautology.

EXAMPLE A.2.13.   We will verify the first of **DeMorgan's Laws**

$$\overline{p \vee q} \equiv \overline{p} \wedge \overline{q}, \qquad \overline{p \wedge q} \equiv \overline{p} \vee \overline{q}$$

leaving the second as an exercise (see Exercise 47).

By writing the truth tables for $P = \overline{p \vee q}$ and $Q = \overline{p} \wedge \overline{q}$, we can verify that given any truth values of p and q, either P and Q are both true or P and Q are both false:

p	q	$\overline{p \vee q}$	$\overline{p} \wedge \overline{q}$
T	T	F	F
T	F	F	F
F	T	F	F
F	F	T	T

Thus P and Q are logically equivalent.

Our next example gives a logically equivalent form of the negation of $p \rightarrow q$. This characterization is often useful in proofs.

EXAMPLE A.2.14.   Show that the negation of $p \rightarrow q$ is logically equivalent to $p \wedge \overline{q}$.

We must show that

$$\overline{p \rightarrow q} \equiv p \wedge \overline{q}.$$

By writing the truth tables for $P = \overline{p \rightarrow q}$ and $Q = p \wedge \overline{q}$, we can verify that given any truth values of p and q, either P and Q are both true or P and Q are both false:

p	q	$\overline{p \rightarrow q}$	$p \wedge \overline{q}$
T	T	F	F
T	F	T	T
F	T	F	F
F	F	F	F

Thus P and Q are logically equivalent.

We will later need the following result, which shows that $p \rightarrow q$ is logically equivalent to $p \wedge \bar{q} \rightarrow c$, where c is a contradiction.

EXAMPLE A.2.15.   The truth table shows that

$$p \rightarrow q \equiv (p \wedge \bar{q}) \rightarrow (r \wedge \bar{r}).$$

p	q	r	$p \rightarrow q$	$p \wedge \bar{q}$	$r \wedge \bar{r}$	$(p \wedge \bar{q}) \rightarrow (r \wedge \bar{r})$
T	T	T	T	F	F	T
T	T	F	T	F	F	T
T	F	T	F	T	F	F
T	F	F	F	T	F	F
F	T	T	T	F	F	T
F	T	F	T	F	F	T
F	F	T	T	F	F	T
F	F	F	T	F	F	T

We defined $p \leftrightarrow q$ so that it would mean the same as $p \rightarrow q$ and $q \rightarrow p$. We can now show that according to our definitions, $p \leftrightarrow q$ is logically equivalent to $p \rightarrow q$ and $q \rightarrow p$.

EXAMPLE A.2.16.   The truth table shows that

$$p \leftrightarrow q \equiv (p \rightarrow q) \wedge (q \rightarrow p).$$

p	q	$p \leftrightarrow q$	$p \rightarrow q$	$q \rightarrow p$	$(p \rightarrow q) \wedge (q \rightarrow p)$
T	T	T	T	T	T
T	F	F	F	T	F
F	T	F	T	F	F
F	F	T	T	T	T

We conclude this section by defining the **contrapositive** of a conditional proposition. We will see (Theorem A.2.19) that the contrapositive is an alternative, logically equivalent form of the conditional proposition. Exercise 48 gives another logically equivalent form of the conditional proposition.

DEFINITION A.2.17.   The *contrapositive* (or *transposition*) of the conditional proposition $p \rightarrow q$ is the proposition $\bar{q} \rightarrow \bar{p}$.

Notice the difference between the contrapositive and the converse. The converse of a conditional proposition merely reverses the roles of p and q, whereas the contrapositive reverses the roles of p and q *and* negates each of them.

EXAMPLE A.2.18.   Write the proposition

If $1 < 4$, then $5 > 8$.

symbolically. Write the converse and the contrapositive both symbolically and in words. Find the truth value of each proposition.

If we define

$$p: 1 < 4, \qquad q: 5 > 8$$

then the given proposition may be written symbolically as

$$p \to q.$$

The converse is

$$q \to p$$

or, in words,

$$\text{If } 5 > 8, \text{ then } 1 < 4.$$

The contrapositive is

$$\bar{q} \to \bar{p}$$

or, in words,

$$\text{If } 5 \leq 8, \text{ then } 1 \geq 4.$$

We see that $p \to q$ is false, $q \to p$ is true, and $\bar{q} \to \bar{p}$ is false.

An important fact is that a conditional proposition and its contrapositive are logically equivalent.

THEOREM A.2.19. *The conditional proposition $p \to q$ and its contrapositive $\bar{q} \to \bar{p}$ are logically equivalent.*

Proof. The truth table

p	q	$p \to q$	$\bar{q} \to \bar{p}$
T	T	T	T
T	F	F	F
F	T	T	T
F	F	T	T

shows that $p \to q$ and $\bar{q} \to \bar{p}$ are logically equivalent. ∎

EXERCISES

In Exercises 1–8, restate each proposition in the form (A.2.2) of a conditional proposition.

1H.   All Cubs are great baseball players.

2.   For the real number x, $|x| < 2$ provided that $0 < x < 2$.

3H.   When better cars are built, Buick will build them.

4.   A sufficient condition for the function f to be integrable is that f be continuous.

5H.   The Cubs will win the World Series when they get a left-handed relief pitcher.

6.   A necessary condition for the graph G to have a Hamiltonian circuit is that G be connected.

7H.   The audience will go to sleep if the chairperson gives the lecture.

8.   The program is readable only if it is well structured.

9H.   Write the converse of each proposition in Exercises 1–8.

10.   Write the contrapositive of each proposition in Exercises 1–8.

Assuming that p and r are false and that q and s are true, find the truth value of each proposition in Exercises 11–18.

11H. $p \to q$ **12.** $\bar{p} \to \bar{q}$ **13H.** $\overline{p \to q}$

14. $(p \to q) \land (q \to r)$

15H. $(p \to q) \to r$

16. $p \to (q \to r)$

17H. $(s \to (p \land \bar{r})) \land ((p \to (r \lor q)) \land s)$

18. $((p \land \bar{q}) \to (q \land r)) \to (s \lor \bar{q})$

Assume that a, b, and c are real numbers. In Exercises 19–24, represent the given statement symbolically by letting

$$p: a < b, \qquad q: b < c, \qquad r: a < c.$$

19H. If $a < b$, then $b \geq c$.

20.   If ($a < b$ and $b < c$), then $a < c$.

21H. If ($a \geq b$ and $b < c$), then $a \geq c$.

22.   If it is not true that ($a < c$ and $b < c$), then $a \geq c$.

23H. $a < b$ if and only if ($b < c$ and $a < c$).

24.   If it is not true that ($a < b$ and (either $a < b$ or $b < c$)), then (if $a \geq b$, then $a < c$).

In Exercises 25–30, formulate the symbolic expression in words using

$$p: \text{Today is Monday.}$$
$$q: \text{It is raining.}$$
$$r: \text{It is hot.}$$

25H. $p \to q$ **26.** $\bar{q} \to (r \land q)$

27H. $\bar{p} \to (q \lor r)$ **28.** $p \lor q \leftrightarrow r$

29H. $(p \wedge (q \vee r)) \rightarrow (r \vee (q \vee p))$

30. $(p \vee (\bar{p} \wedge \overline{(q \vee r)})) \rightarrow (p \vee (r \vee q))$

In Exercises 31–34, write each conditional proposition symbolically. Write the converse and contrapositive of each statement symbolically and in words. Also, find the truth value of each conditional proposition, its converse, and its contrapositive.

31H. If $4 < 6$, then $9 > 12$.

32. If $4 > 6$, then $9 > 12$.

33H. $|1| < 3$ if $-3 < 1 < 3$.

34. $|4| < 3$ if $-3 < 4 < 3$.

For each pair of propositions P and Q in Exercises 35–44, pick all correct answers from among

(a) $P \equiv Q$ (b) $P \Rightarrow Q$ (c) $Q \Rightarrow P$ (d) None

35H. $P = p, Q = p \vee q$

36. $P = p \wedge q, Q = \bar{p} \vee \bar{q}$

37H. $P = p \rightarrow q, Q = \bar{p} \vee q$

38. $P = p \wedge (\bar{q} \vee r), Q = p \vee (q \wedge \bar{r})$

39H. $P = p \wedge (q \vee r), Q = (p \vee q) \wedge (p \vee r)$

40. $P = p \rightarrow q, Q = \bar{q} \rightarrow \bar{p}$

41H. $P = p \rightarrow q, Q = p \leftrightarrow q$

42. $P = (p \rightarrow q) \wedge (q \rightarrow r), Q = p \rightarrow r$

43H. $P = (p \rightarrow q) \rightarrow r, Q = p \rightarrow (q \rightarrow r)$

44. $P = (s \rightarrow (p \wedge \bar{r})) \wedge ((p \rightarrow (r \vee q)) \wedge s), Q = p \vee t$

45H. Define the truth table for *impl* by

p	q	p *impl* q
T	T	T
T	F	F
F	T	F
F	F	T

Show that

$$p \text{ } impl \text{ } q \equiv q \text{ } impl \text{ } p.$$

46.   Define the truth table for *imp2* by

p	q	p *imp2* q
T	T	T
T	F	F
F	T	T
F	F	F

(a) Show that

$$(p \text{ } imp2 \text{ } q) \wedge (q \text{ } imp2 \text{ } p) \not\equiv p \leftrightarrow q. \tag{A.2.9}$$

(b) Show that (A.2.9) remains true if we alter *imp2* so that if p is false and q is true, then p *imp2* q is false.

47H.   Verify the second DeMorgan law $\overline{p \wedge q} \equiv \bar{p} \vee \bar{q}$.

48.   Show that $(p \rightarrow q) \equiv (\bar{p} \vee q)$.

A.3

Proofs and Arguments

A **mathematical system** consists of **axioms**, **definitions**, and **undefined terms**. The axioms are assumed true. Definitions are used to create new concepts in terms of existing ones. Some terms are not explicitly defined but rather are implicitly defined by the axioms.

> **EXAMPLE A.3.1.**   Euclidean geometry furnishes an example of a mathematical system. Among the axioms are
>
> 1. Given two distinct points, there is exactly one line that contains them.
> 2. Given a line and a point not on the line, there is exactly one line parallel to the line through the point.
>
> The terms *point* and *line* are undefined terms which are implicitly defined by the axioms that describe their properties.
>   Among the definitions are
>
> 1. Two triangles are *congruent* if their vertices can be paired so that the corresponding sides and corresponding angles are equal.
> 2. Two angles are *supplementary* if the sum of their measures is 180°.

> **EXAMPLE A.3.2.**   The real numbers furnish another example of a mathematical system. Among the axioms are
>
> 1. For all real numbers x and y, $xy = yx$.
> 2. There is a subset **P** of real numbers satisfying

(a) If x and y are in **P**, then $x + y$ and xy are in **P**.
(b) If x is a real number, then exactly one of the following statements is true:

$$x \text{ is in } \mathbf{P}, \qquad x = 0, \qquad -x \text{ is in } \mathbf{P}.$$

Multiplication is implicitly defined by axiom 1 (and others) which describe the properties multiplication is assumed to obey.

Among the definitions are

1. The elements in **P** (of axiom 2) are called *positive real numbers*.
2. The *absolute value* $|x|$ of a real number x is defined to be x if x is positive or 0, and $-x$ otherwise.

Within a mathematical system we can derive theorems. A **theorem** is a result that can be deduced from the axioms, definitions, and previously derived theorems. The argument establishing the truth of the theorem is called a **proof**. Special kinds of theorems are referred to as lemmas and corollaries. A **lemma** is a theorem that is usually not too interesting in its own right but which is useful in proving some other theorem. A **corollary** is a theorem that follows quickly from some other theorem.

EXAMPLE A.3.3.   Examples of theorems in Euclidean geometry are

1. If two sides of a triangle are equal, then the angles opposite them are equal.
2. If the diagonals of a quadrilateral bisect each other, then the quadrilateral is a parallelogram.

EXAMPLE A.3.4.   An example of a corollary in Euclidean geometry is

1. If a triangle is equilateral, then it is equiangular.

This corollary follows immediately from the first theorem of Example A.3.3. (Do you see why?)

EXAMPLE A.3.5.   Examples of theorems about real numbers are

1. $x \cdot 0 = 0$ for every real number x.
2. For all real numbers x, y, and z, if $x < y$ and $y < z$, then $x < z$.

EXAMPLE A.3.6.   An example of a lemma about real numbers is

1. If n is a positive integer, then either $n - 1$ is a positive integer or $n - 1 = 0$.

Surely this result is not that interesting in its own right, but it can be used to prove other results.

Theorems are often of the form

$$\text{If } p, \text{ then } q. \tag{A.3.1}$$

A proof of (A.3.1) is an argument which shows that (A.3.1) is true. If p is false, then, by Definition A.2.4, (A.3.1) is true; thus we need only consider the case that p is true. A **direct proof** assumes that p is true and then, using p as well as other axioms, definitions, and previously derived theorems, shows directly that q is true.

EXAMPLE A.3.7.   We will give a direct proof of the following proposition (which is often used in calculus) about real numbers d, d_1, d_2, and x:

If $d = \min \{d_1, d_2\}$ and $x < d$, then $x < d_1$ and $x < d_2$.

Proof.   From the definition of min, it follows that $d \leq d_1$ and $d \leq d_2$. From $x < d$ and $d \leq d_1$, we may derive $x < d_1$ from a previous theorem (almost theorem 2 of Example A.3.5). From $x < d$ and $d \leq d_2$, we may derive $x < d_2$ from the same previous theorem. Therefore, $x < d_1$ and $x < d_2$.   ∎

A second technique of proof is **proof by contradiction**. A proof by contradiction establishes (A.3.1) by proving the logically equivalent proposition

$$(p \wedge \bar{q}) \rightarrow (r \wedge \bar{r}) \tag{A.3.2}$$

whose conclusion is a contradiction (see Example A.2.15). Thus a proof by contradiction of (A.3.1) proceeds by assuming not only that p is true, but also that q is false and then, using p and $\bar{q}$ as well as other axioms, definitions, and previously derived theorems, deduces a contradiction. A proof by contradiction is sometimes called an **indirect proof** since to establish (A.3.1) using proof by contradiction, one follows an indirect route: prove (A.3.2), then conclude that (A.3.1) is true.

The only difference between the assumptions in a direct proof and a proof by contradiction is the negated conclusion. In a direct proof, the negated conclusion is not assumed, whereas in a proof by contradiction the negated conclusion is assumed.

EXAMPLE A.3.8.   We will give a proof by contradiction of the following proposition about real numbers x and y:

If $x + y \geq 2$, then either $x \geq 1$ or $y \geq 1$.

Proof.   Assume the hypothesis and suppose that the conclusion is false. Then $x < 1$ and $y < 1$. (Remember that negating *or*s results in *and*s. See Example A.2.13.) Using a previous theorem, we may add these inequalities to obtain

$$x + y < 1 + 1 = 2.$$

At this point, we have derived the contradiction $p \wedge \bar{p}$, where

$$p\text{: } x + y \geq 2.$$

Thus we conclude that the proposition is true.   ∎

Suppose that we give a proof by contradiction of (A.3.1) in which, as in Example A.3.8, we deduce $\bar{p}$. In effect, we have proved

$$\bar{q} \rightarrow \bar{p}. \tag{A.3.3}$$

This special case of proof by contradiction is called **proof by contrapositive**. Some authors define an indirect proof to be a proof by contrapositive. Proof by contrapositive is justified because it is a special case of proof by contradiction; however, proof by contrapositive can be directly justified by noting the equivalence of (A.3.1) and (A.3.3) (see Theorem A.2.19).

In constructing a proof, we must be sure that the arguments used are **valid**. In the remainder of this section, we provide an introduction to the concept of a valid argument and explore this concept in some detail.

Consider the following sequence of propositions.

> The bug is either in module 17 or in module 81.
> The bug is a numerical error.
> Module 81 has no numerical error. (A.3.4)

Assuming that these statements are true, it is reasonable to conclude:

> The bug is in module 17. (A.3.5)

This process of drawing a conclusion from a sequence of propositions is called **deductive reasoning**. The given propositions, such as (A.3.4), are called **hypotheses** or **premises** and the proposition that follows from the hypotheses, such as (A.3.5), is called the **conclusion**. A (**deductive**) **argument** consists of hypotheses together with a conclusion. Many proofs in mathematics and computer science are deductive arguments.

Any argument has the form

$$\text{If } p_1 \text{ and } p_2 \text{ and } \cdots \text{ and } p_n, \text{ then } q. \tag{A.3.6}$$

Argument (A.3.6) is said to be valid if the conclusion follows from the hypothesis; that is, if p_1 and p_2 and $\cdots$ and p_n are all true, then q is also true. In the terminology of Section A.2, an argument is valid if p_1 and p_2 and $\cdots$ and p_n logically imply q. This discussion motivates the following definition.

DEFINITION A.3.9. An *argument* is a sequence of propositions written

$$p_1$$
$$p_2$$
$$\cdot$$
$$\cdot$$
$$\cdot$$
$$\underline{p_n}$$
$$\therefore q$$

or

$$p_1, p_2, \ldots, p_n \, / \, \therefore q. \tag{A.3.7}$$

The propositions $p_1, p_2, \ldots, p_n$ are called the *hypotheses* (or *premises*) and the proposition q is called the *conclusion*. The argument is *valid* provided that

$$p_1 \wedge p_2 \wedge \cdots \wedge p_n \Rightarrow q \qquad (A.3.8)$$

holds; otherwise, the argument is *invalid* (or a *fallacy*).

Note that (A.3.8) is true if and only if whenever $p_1 \wedge p_2 \wedge \cdots \wedge p_n$ is true, then q is also true. But $p_1 \wedge p_2 \wedge \cdots \wedge p_n$ is true precisely when p_1 is true and p_2 is true and $\cdots$ and p_n is true. Thus the argument (A.3.7) is valid provided that whenever p_1 and p_2 and $\cdots$ and p_n are all true, q is also true.

Another characterization of a valid argument is that the argument (A.3.7) is valid precisely when

$$p_1 \wedge p_2 \wedge \cdots \wedge p_n \rightarrow q$$

is a tautology.

In a valid argument, we sometimes say that the conclusion follows from the hypotheses. Notice that we are not saying that the conclusion is true; we are only saying that if you grant the hypotheses, you must also grant the conclusion. An argument is valid because of its form and not because of its content.

EXAMPLE A.3.10.   Determine whether the argument

$$p \rightarrow q$$
$$\underline{p}$$
$$\therefore q$$

is valid.

[First solution.] We construct a truth table for all the propositions involved:

p	q	$p \rightarrow q$	p	q
T	T	T	T	T
T	F	F	T	F
F	T	T	F	T
F	F	T	F	F

We observe that whenever the hypotheses $p \rightarrow q$ and p are true, the conclusion q is also true; therefore, the argument is valid.

[Second solution.] We can avoid writing the truth table by directly verifying that whenever the hypotheses are true the conclusion is also true.

Suppose that $p \rightarrow q$ and p are true. Then q must be true for otherwise $p \rightarrow q$ would be false. Therefore, the argument is valid.

EXAMPLE A.3.11. Represent the argument

If 2 = 3, then I ate my hat.
I ate my hat.
──────────────────
∴ 2 = 3

symbolically and determine whether the argument is valid.
If we let

$$p: 2 = 3, \qquad q: \text{I ate my hat.}$$

the argument may be written

$$p \to q$$
$$q$$
──────
$$\therefore p$$

The argument is valid provided that if $p \to q$ and q are true, then p is also true. Suppose that $p \to q$ and q are true. This is possible if p is false and q is true. In this case p is not true, thus the argument is invalid.

We can also determine the validity of the argument in Example A.3.11 by examining the truth table of Example A.3.10. In the third row of the table, the hypotheses are true and the conclusion is false; thus the argument is invalid.

EXERCISES

1H. Give an example (different from those of Example A.3.1) of an axiom in Euclidean geometry.

2. Give an example (different from those of Example A.3.2) of an axiom in the system of real numbers.

3H. Give an example (different from those of Example A.3.1) of a definition in Euclidean geometry.

4. Give an example (different from those of Example A.3.2) of a definition in the system of real numbers.

5H. Give an example (different from those of Example A.3.3) of a theorem in Euclidean geometry.

6. Give an example (different from those of Example A.3.5) of a theorem in the system of real numbers.

★7H. Show that if x is a real number, then $x \cdot 0 = 0$ by giving a direct proof. Assume that the following are previous theorems: If a, b, and c are real numbers, then $b + 0 = b$ and $a(b + c) = ab + ac$. If $a + b = a + c$, then $b = c$.

★8. Show that if x and y are real numbers, then $|x + y| \leq |x| + |y|$ by giving a direct proof.

9H. Show that if $xy = 0$, then either $x = 0$ or $y = 0$ by giving a proof by contradiction. Assume that if a, b, and c are real numbers with $ab = ac$ and $a \neq 0$, then $b = c$.

10. Show that if 100 balls are placed into nine boxes, some box contains 12 or more balls by giving a proof by contradiction.

Formulate the arguments of Exercises 11–15 symbolically and determine whether each is valid. Let

p: I study hard. q: I get A's. r: I get rich.

11H. If I study hard, then I get A's.
I study hard.
∴ I get A's.

12. If I study hard, then I get A's.
If I don't get rich, then I don't get 'A's.
∴ I get rich.

13H. I study hard if and only if I get rich.
I get rich.
∴ I study hard.

14. If I study hard or I get rich, then I get A's.
I get A's.
∴ If I don't study hard, then I get rich.

15H. If I study hard, then I get A's or I get rich.
I don't get A's and I don't get rich.
∴ I don't study hard.

In Exercises 16–20, write the given argument in words and determine whether each argument is valid. Let

p: 64K is better than no memory at all.
q: We will buy more memory.
r: We will buy a new computer.

16. $p \rightarrow r$
$p \rightarrow q$
$\therefore p \rightarrow (r \wedge q)$

17H. $p \rightarrow (r \vee q)$
$r \rightarrow \bar{q}$
$\therefore p \rightarrow r$

18. $p \rightarrow r$
$r \rightarrow q$
$\therefore q$

19H. $\bar{r} \rightarrow \bar{p}$
r
$\therefore p$

20. $p \rightarrow r$
$r \rightarrow q$
p
$\therefore q$

Determine whether each argument in Exercises 21–25 is valid.

21H. $p \rightarrow q$
$\overline{p}$
$\overline{}$
$\therefore \overline{q}$

22. $p \rightarrow q$
q
$\overline{}$
$\therefore \overline{p}$

23H. $p \wedge \overline{p}$
$\overline{\phantom{p \wedge \overline{p}}}$
$\therefore q$

24. $p \rightarrow (q \rightarrow r)$
$q \rightarrow (p \rightarrow r)$
$\overline{}$
$\therefore (p \vee q) \rightarrow r$

25H. $(p \rightarrow q) \wedge (r \rightarrow s)$
$p \vee r$
$\overline{}$
$\therefore q \vee s$

26. Show that

$$p_1, p_2, \ldots, p_n / \therefore c$$

is a valid argument if $p_1 \wedge p_2 \Rightarrow p$ holds and the argument

$$p, p_3, \ldots, p_n / \therefore c$$

is valid.

27H. Suppose that

$$p_1, p_2, \ldots, p_n / \therefore c$$

is a valid argument and that $p_1 \equiv p$. Show that the argument

$$p, p_2, \ldots, p_n / \therefore c$$

is also valid.

28. Comment on the following argument.

Cassette tape storage is better than nothing.
Nothing is better than a hard disk drive.
$\overline{}$
$\therefore$ Cassette tape storage is better than a hard disk drive.

A.4

Categorical Propositions

A proposition such as

p: Some computer scientists are musicians. (A.4.1)

relates members of one class to members of a second class. In (A.4.1), the first class C_1 is the class of all computer scientists and the second class C_2 is the class of all musicians. Proposition (A.4.1) states that some members of C_1 are members of C_2.

DEFINITION A.4.1.   Let C_1 and C_2 be two classes. A *categorical proposition* is a proposition of one of the forms (a)–(d).

(a) All C_1 is C_2.
(b) No C_1 is C_2.
(c) Some C_1 is C_2.
(d) Some C_1 is not C_2.

EXAMPLE A.4.2.   If we let C_1 denote the class of computer scientists and C_2 denote the class of musicians, proposition (A.4.1) is of type (c) of Definition A.4.1.

EXAMPLE A.4.3.   Classify each categorical proposition as of type (a)–(d) according to Definition A.4.1 and describe the classes C_1 and C_2.

(i)   All trees are connected graphs.
(ii)   Some graphs are trees.
(iii) All squares are rectangles.
(iv)   No two distinct parallel lines intersect.
(v)   Some context-free languages are not regular languages.
(vi)   All prime numbers greater than 2 are odd.

(i)   This is a type (a) categorical proposition with

$$C_1 = \text{class of all trees.}$$
$$C_2 = \text{class of all connected graphs.}$$

(ii) This is a type (c) categorical proposition with

$$C_1 = \text{class of all graphs.}$$
$$C_2 = \text{class of all trees.}$$

(iii) This is a type (a) categorical proposition with

$$C_1 = \text{class of all squares.}$$
$$C_2 = \text{class of all rectangles.}$$

(iv) This is a type (b) categorical proposition with

$$C_1 = \text{class of all distinct pairs of parallel lines.}$$
$$C_2 = \text{class of all distinct pairs of intersecting lines.}$$

(v) This is a type (d) categorical proposition with

$$C_1 = \text{class of all context-free languages.}$$
$$C_2 = \text{class of all regular languages.}$$

(vi) This is a type (a) categorical proposition with

$$C_1 = \text{class of all prime numbers greater than 2.}$$
$$C_2 = \text{class of all odd numbers.}$$

The descriptor *all* in Definition A.4.1 type (a) and the descriptor *no* in Definition A.4.1 type (b) are called **universal quantifiers** since either a type (a) or type (b) categorical proposition asserts something about all elements in the *universe*. A type (a) categorical proposition states that for all elements x in the universe, if x is in class C_1, then x is also in class C_2. A type (b) categorical proposition states that for all elements x in the universe, if x is in class C_1, then x is not in class C_2. The descriptor *some* in Definition A.4.1 type (c) and (d) categorical propositions is called an **existential quantifier,** since a type (c) or (d) categorical proposition asserts the *existence* of a particular element. For example, a type (d) categorical proposition asserts that there exists an element in C_1 which is not in C_2.

There are ways to state propositions which are logically equivalent to the categorical propositions of Definition A.4.1. It is important to recognize that these alternative forms are really categorical propositions in disguise. We regard any proposition equivalent to one of (a)–(d) of Definition A.4.1 as a categorical proposition.

EXAMPLE A.4.4. Equivalent forms of

$$\text{All } C_1 \text{ is } C_2.$$

are

(a) If x is a member of C_1, then x is a member of C_2.
(b) For all x, if x is a member of C_1, then x is a member of C_2.

EXAMPLE A.4.5. Equivalent forms of

$$\text{No } C_1 \text{ is } C_2.$$

are

(a) If x is a member of C_1, then x is not a member of C_2.
(b) For all x, if x is a member of C_1, then x is not a member of C_2.

The word *some* in Definition A.4.1(c) and (d) is interpreted as meaning at least one; thus an equivalent formulation of proposition (A.4.1) is

At least one computer scientist is a musician.

Examples A.4.6 and A.4.7 give general equivalent formulations of Definition A.4.1(c) and (d).

EXAMPLE A.4.6. Equivalent forms of

$$\text{Some } C_1 \text{ is } C_2.$$

are

(a) Some C_2 is C_1.
(b) There is at least one member of C_1 which is also a member of C_2.
(c) There exists x such that x is a member of both C_1 and C_2.

EXAMPLE A.4.7. Equivalent forms of

$$\text{Some } C_1 \text{ is not } C_2.$$

are

(a) Some non-C_2 is C_1.
(b) There is at least one member of C_1 that is not a member of C_2.
(c) There exists x such that x is a member of C_1 and x is not a member of C_2.

EXAMPLE A.4.8. By defining classes C_1 and C_2, state each proposition as a categorical proposition in the form of Definition A.4.1.

(a) There exists x such that $x^2 \geq 4$ and $x < 2$.
(b) For all x, if $x \geq 2$, then $x^2 \geq 4$.
(c) If x is an irrational number, then x is not representable as a repeating decimal.
(d) There exists an isosceles triangle that is not equilateral.

(a) Some C_1 is C_2.

$$C_1 = \text{class of numbers } x \text{ with } x^2 \geq 4.$$
$$C_2 = \text{class of numbers } x \text{ with } x < 2.$$

(b) All C_1 is C_2.

$$C_1 = \text{class of numbers } x \text{ with } x \geq 2.$$
$$C_2 = \text{class of numbers } x \text{ with } x^2 \geq 4.$$

(c) No C_1 is C_2.

$$C_1 = \text{class of irrational numbers.}$$
$$C_2 = \text{class of numbers representable as repeating decimals.}$$

(d) Some C_1 is not C_2.

$$C_1 = \text{class of isosceles triangles.}$$
$$C_2 = \text{class of equilateral triangles.}$$

Categorical propositions can be pictorially represented using **Venn diagrams**. A Venn diagram represents classes as overlapping circles. In Figure A.4.1, we see a Venn diagram for two classes C_1 and C_2. Members of the class C_1, if any, are considered to be within the circle labeled C_1 and members of the class C_2, if any, are considered to be within the circle labeled C_2. Four regions are represented in Figure A.4.1: a is a member of neither class C_1 nor C_2; b is a member of C_1 but not of C_2; c belongs to both C_1 and C_2; and d belongs to C_2 but not C_1.

Consider a Venn diagram for a type (a) categorical proposition

$$\text{All } C_1 \text{ is } C_2.$$

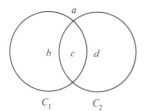

FIGURE A.4.1

If every member of class C_1 is in class C_2, there can be no elements like b in Figure A.4.1. We shade this region to indicate that it is empty. The Venn diagram for a type (a) categorical proposition is shown in Figure A.4.2. Similarly, we obtain the Venn diagram shown in Figure A.4.2 for a type (b) categorical proposition. The Venn diagram for a type (c) categorical proposition

$$\text{Some } C_1 \text{ is } C_2.$$

shows an "x" in the region representing C_1 and C_2 (see Figure A.4.2). The presence of the element x shows that at least one member of C_1 is in C_2. Similarly, we obtain the Venn diagram shown in Figure A.4.2 for a type (d) categorical proposition.

Venn diagrams are useful for testing the validity of arguments involving categorical propositions, as our next examples illustrate.

EXAMPLE A.4.9. Use a Venn diagram to determine whether the following argument is valid.

$$\text{All } C_1 \text{ is } C_2.$$
$$\underline{\text{All } C_2 \text{ is } C_3.}$$
$$\therefore \text{All } C_1 \text{ is } C_3.$$

We assume that the hypotheses are true. To construct the Venn diagram, we first draw three overlapping circles to represent the classes C_1, C_2, and C_3 (see Figure A.4.3). Next, we shade the appropriate empty region (shown with vertical lines) to represent the proposition "All C_1 is C_2." Finally, we shade the appropriate empty region (shown with horizontal lines) to represent the proposition "All C_2 is C_3." We now see that all members of C_1 (if any) must be in C_3; that is, the conclusion "All C_1 is C_3" is true. The argument is valid.

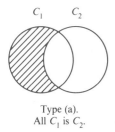

Type (a).
All C_1 is C_2.

Type (b).
No C_1 is C_2.

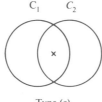

Type (c).
Some C_1 is C_2.

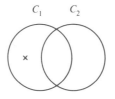

Type (d).
Some C_1 is not C_2.

FIGURE A.4.2

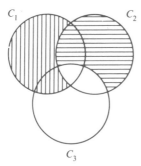

FIGURE A.4.3

EXAMPLE A.4.10. Write the following argument using the forms of Definition A.4.1 for the categorical propositions. Use a Venn diagram to determine whether the argument is valid.

$$\text{All professors are humans.}$$
$$\underline{\text{All good teachers are humans.}}$$
$$\therefore \text{All professors are good teachers.}$$

If we let

$$C_1 = \text{class of professors.}$$
$$C_2 = \text{class of humans.}$$
$$C_3 = \text{class of good teachers.}$$

we may rewrite the argument as

$$\text{All } C_1 \text{ is } C_2.$$
$$\underline{\text{All } C_3 \text{ is } C_2.}$$
$$\therefore \text{All } C_1 \text{ is } C_3.$$

The Venn diagram representing the hypotheses is shown in Figure A.4.4. Does the conclusion follow? Consider the region with a check in it. Since this region is not shaded, it is not necessarily empty; that is, there may be an element in C_1 but not in C_3. Therefore, the conclusion does not necessarily follow. The argument is invalid.

We next consider negations of categorical propositions. Consider the negation of

$$\text{All } C_1 \text{ is } C_2. \tag{A.4.2}$$

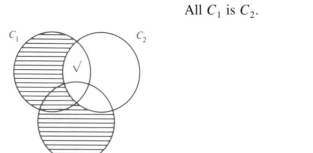

FIGURE A.4.4

If this proposition is false, then the shaded region in Figure A.4.2(a) must have at least one member; that is, the proposition

$$\text{Some } C_1 \text{ is not } C_2. \tag{A.4.3}$$

is true. Similarly, if (A.4.2) is true, then (A.4.3) is false; thus (A.4.3) is the negation of (A.4.2). By similar reasoning (see Exercise 2), we can show that the negation of

$$\text{No } C_1 \text{ is } C_2.$$

is

$$\text{Some } C_1 \text{ is } C_2.$$

EXAMPLE A.4.11. The negations of

(a) All squares are rectangles.
(b) No two parallel lines intersect.
(c) Some graphs are trees.
(d) Some context-free languages are not regular languages.

are

(a) Some square is not a rectangle.
(b) Some two parallel lines intersect.
(c) No graph is a tree.
(d) All context-free languages are regular languages.

Notice that the negation of a proposition involving a universal quantifier is a proposition involving an existential quantifier and that the negation of a proposition involving an existential quantifier is a proposition involving a universal quantifier.

The interpretation of certain universal quantifiers can be a bit tricky. In positive statements, "any," "all," "each," and "every" are equivalent.

$$\text{Any } C_1 \text{ is } C_2.$$
$$\text{All } C_1 \text{ is } C_2.$$
$$\text{Each } C_1 \text{ is } C_2.$$
$$\text{Every } C_1 \text{ is } C_2.$$

are considered to have the same meaning. In negative statements, the situation changes.

$$\text{Not all } C_1 \text{ is } C_2.$$
$$\text{Not each } C_1 \text{ is } C_2.$$
$$\text{Not every } C_1 \text{ is } C_2.$$

are all considered to have the same meaning as

$$\text{Some } C_1 \text{ is not } C_2.$$

whereas

$$\text{Not any } C_1 \text{ is } C_2.$$

means

$$\text{No } C_1 \text{ is } C_2.$$

We conclude this section by discussing some methods of proof used to establish theorems that can be stated as categorical propositions. To give a direct proof of a theorem of the form (A.4.2), we must examine an *arbitrary* element x of C_1 and show that x is also in C_2. To say that x is an arbitrary element in C_1 we mean that x could be *any* element in C_1. After showing that an arbitrarily chosen element in C_1 is in C_2, we can conclude that (A.4.2) is true.

EXAMPLE A.4.12. The theorem

A quadratic equation has at most two solutions.

may be rephrased in the form (A.4.2) as

All quadratic equations are equations with at most two solutions. (A.4.4)

Proof. To prove (A.4.4), we consider an arbitrary quadratic equation

$$ax^2 + bx + c = 0.$$

The symbols a, b, and c denote arbitrary numbers. The quadratic formula states that

$$x = \frac{-b \pm \sqrt{b^2 - 4ac}}{2a};$$

thus, there are at most two solutions, namely

$$\frac{-b + \sqrt{b^2 - 4ac}}{2a} \quad \text{and} \quad \frac{-b - \sqrt{b^2 - 4ac}}{2a}. \qquad \blacksquare$$

In some cases we are asked to show that (A.4.2) is *false*. This is the same as showing that the negation (A.4.3) of (A.4.2) is true. To show that (A.4.3) is true, we must exhibit an element in C_1 but not in C_2. Such an element is called a **counterexample** to (A.4.2).

EXAMPLE A.4.13. To show that

For all integers n, $n^2 + n + 41$ is prime. (A.4.5)

is false, we must find a counterexample; that is, an integer n for which $n^2 + n + 41$ is not prime. For $n = 41$,

$$n^2 + n + 41 = 41^2 + 41 + 41 = 41(41 + 1 + 1) = 41 \cdot 43$$

is not prime. Thus $n = 41$ is a counterexample to (A.4.5).

It is an interesting fact that $n^2 + n + 41$ is prime for $n = -40, -39, \ldots,$ $-1, 0, 1, \ldots, 39$. It is not known whether there are any other second-degree polynomials that are prime for more than 80 consecutive values. The remarkable polynomial $n^2 + n + 41$ was discovered by Euler in 1772.

EXAMPLE A.4.14 **Fermat's Last Theorem.** The seventeenth-century number theorist Pierre de Fermat wrote in the margin of a book that if n is an integer greater than 2, then

$$x^n + y^n = z^n \tag{A.4.6}$$

has no positive integer solutions. He also stated that he had a proof, but that the margin was too narrow to contain it. Fermat should have gotten some extra paper. To this day, it is unknown whether Fermat's Last Theorem is actually a theorem or whether this proposition is false. To show it is false, one would need a counterexample; that is, one would have to find positive integers x, y, and z and an integer $n \geq 3$, satisfying (A.4.6). Fermat's proposition has been shown true for $n < 125{,}000$, so if there is a counterexample, the numbers involved would be quite large.

EXAMPLE A.4.15. Euler conjectured that, if m and n are positive integers satisfying $m < n$ and $n > 2$, then

$$z^n = x_1^n + x_2^n + \cdots + x_m^n$$

has no solution in positive integers. In words, no positive nth power is the sum of fewer than n positive nth powers. In 1966, about 200 years after Euler made his conjecture, Leon Lander and Thomas Parkin provided a counterexample:

$$27^5 + 84^5 + 110^5 + 133^5 = 144^5,$$

thus showing that Euler's conjecture is false.

To prove the proposition

$$\text{Some } C_1 \text{ is } C_2. \tag{A.4.7}$$

we must find an element in C_1 and C_2. A proof in which the element is shown to exist is called an **existence proof**.

EXAMPLE A.4.16. To prove that some real numbers are irrational it suffices to exhibit an irrational number such as $\sqrt{2}$ (and prove that $\sqrt{2}$ is irrational).

The truth of categorical propositions can sometimes be established using proof by contradiction. For example, if we want to establish (A.4.2) using proof by contradiction, we assume that (A.4.2) is false and deduce a contradiction. The negation of (A.4.2) is (A.4.3). Assuming (A.4.3), we know that there exists an element in C_1 that is not in C_2. Using this assumption as well as axioms, definitions, and previously derived theorems, we deduce a contradiction. Similarly, to establish (A.4.7) using proof by contradiction, we assume its negation

$$\text{No } C_1 \text{ is } C_2.$$

EXAMPLE A.4.17. By defining classes C_1 and C_2, state the proposition

For all x and y, if x is a rational number and y is an
irrational number, then $x + y$ is an irrational number.

as a categorical proposition in the form of Definition A.4.1. Then, prove this proposition using proof by contradiction. Assume as a previous theorem: If a and b are rational numbers, then $a - b$ is a rational number.

Proof. If we let C_1 be the class of numbers of the form $x + y$, where x is rational and y is irrational, and C_2 be the class of irrational numbers, we can rephrase the theorem as

$$\text{All } C_1 \text{ is } C_2. \tag{A.4.8}$$

Assume that the given statement is false. We are then assuming that the negation of (A.4.8)

$$\text{Some } C_1 \text{ is not } C_2.$$

is true, that is, that there exist numbers x and y with x rational and y irrational such that $z = x + y$ is rational. By the previous theorem, $y = z - x$ is rational. But now y is both rational and irrational—a contradiction. The theorem is proved. ∎

EXERCISES

1. Find examples of categorical propositions in a newspaper.
2. Give an argument to show that the negation of

$$\text{No } C_1 \text{ is } C_2.$$

is

$$\text{Some } C_1 \text{ is } C_2.$$

In Exercises 3–12, classify each categorical proposition as of type (a)–(d) according to Definition A.4.1 and describe the classes C_1 and C_2. Also, give the negation of each proposition.

3H. All isosceles triangles are equilateral triangles.
4. Some similar triangles are not congruent.
5H. Some expert bridge players can also play a mean game of cribbage.
6. Some chips are not functional.
7H. No one over 30 is trustworthy.

8. Some successful persons never attended college.

9H. All violent movies are R-rated movies.

10. All connected graphs are graphs with spanning trees.

11H. Some linear programming algorithms are polynomial-time algorithms.

12. Nobody is despised who can manage a crocodile.

By defining classes C_1 and C_2, state each proposition in Exercises 13–16 as a categorical proposition in the form of Definition A.4.1. Also, give the negation of each proposition.

13H. If G is a planar map, then G can be colored using at most four colors.

14. There is at least one Cubs fan who is rational.

15H. There exists a continuous function that is not differentiable at any point.

16. If software has poor documentation, then it is not worth much.

Use Venn diagrams to determine whether the arguments in Exercises 17–26 are valid.

17H. Some C_1 is C_2.
All C_2 is C_3.
$\therefore$ Some C_3 is C_1.

18. No C_1 is C_2.
Some C_2 is C_3.
$\therefore$ Some C_3 is not C_1.

19H. All C_1 is C_2.
No C_3 is C_1
$\therefore$ No C_3 is C_2.

20. No C_1 is C_2.
Some C_3 is C_1.
$\therefore$ Some C_3 is not C_2.

21H. Some C_1 is not C_2.
All C_2 is C_3.
$\therefore$ Some C_3 is not C_1.

22. Some C_1 is C_2.
All C_3 is C_1.
$\therefore$ Some C_3 is C_2.

23H. No C_1 is C_2.
All C_3 is C_1.
$\therefore$ No C_3 is C_2.

24. All C_1 is C_2.
Some C_1 is not C_3.
$\therefore$ Some C_3 is not C_2.

25H. No C_1 is C_2.
Some C_1 is C_3.
$\therefore$ Some C_3 is not C_2.

26. All C_1 is C_2.
Some C_3 is not C_1.
$\therefore$ Some C_3 is not C_2.

Write each argument in Exercises 27–32 using the forms of Definition A.4.1 for the categorical propositions. Use Venn diagrams to determine whether the arguments are valid.

27H. No Cubs are White Sox.
All White Sox are baseball players.
$\therefore$ Some baseball players are not Cubs.

28. Some voters are not Republicans.
Some people are not voters.
∴ Some people are not Republicans.

29H. All trees are graphs.
Some structures are not graphs.
∴ Some structures are not trees.

30. All dogs are animals.
No dogs are horses.
∴ No horses are animals.

31H. Some integers are not perfect numbers.
All integers are real numbers.
∴ Some real numbers are not perfect numbers.

32. All movies are violent epics.
No violent epics are worth watching.
∴ Nothing worth watching is a movie.

Provide counterexamples to the propositions in Exercises 33–36.

33H. Every equation that has exactly two distinct solutions is a quadratic equation.

34. For every real number $x > 0$, $(x^2 + 1)/x < 100$.

35H. Every personal computer is a machine costing less than \$3000.

36. For any prime p, $n^2 + n + p$ is prime for $n = 0, 1, \ldots, p - 1$.

Give existence proofs of the propositions in Exercises 37–40.

37H. There is a solution in positive integers to

$$x^2 + y^2 = z^2.$$

38. There is a polynomial $p(n)$ with $p(1) = p(2) = p(3) = 0$ and $p(4) = 162$.

39H. If a and b are real numbers satisfying $0 < a < 1$ and $0 < b$, there is a positive integer n with $a^n < b$.

★40. If a and b are real numbers with $a < b$, there exists a rational number r with $a < r < b$.

41. Verify the counterexample of Example A.4.15.

Appendix B: Matrices

It is a common practice to organize data into rows and columns. In mathematics, such an array of data is called a **matrix**. In this appendix we summarize some definitions and elementary properties of matrices.

Matrices

We begin with the definition of matrix.

DEFINITION B.1.1. A *matrix*

$$
A = \begin{pmatrix}
a_{11} & a_{12} & \cdots & a_{1n} \\
a_{21} & a_{22} & \cdots & a_{2n} \\
\cdot & \cdot & & \cdot \\
\cdot & \cdot & & \cdot \\
\cdot & \cdot & & \cdot \\
a_{m1} & a_{m2} & \cdots & a_{mn}
\end{pmatrix}
\tag{B.1.1}
$$

is a rectangular array of data.

If A has m rows and n columns, we say that the *size* of A is m by n (written $m \times n$).

We will often abbreviate equation (B.1.1) to $A = (a_{ij})$. In this equation, a_{ij} denotes the element of A appearing in the ith row and jth column.

EXAMPLE B.1.2. The matrix

$$A = \begin{pmatrix} 2 & 1 & 0 \\ -1 & 6 & 14 \end{pmatrix}$$

has two rows and three columns, so its size is 2×3. If we write $A = (a_{ij})$, we would have, for example,

$$a_{11} = 2, \quad a_{21} = -1, \quad a_{13} = 0.$$

EXAMPLE B.1.3. A matrix can be used to represent a relation R from X to Y. We label the rows with the elements of X (in some arbitrary order) and we label the columns with the elements of Y (again, in some arbitrary order). We then set the entry in row x and column y to 1 if xRy and to 0 otherwise. This matrix is called the **matrix of the relation** R (relative to the choices of orderings of X and Y).

EXAMPLE B.1.4. The matrix of the relation R from $\{2, 3, 4\}$ to $\{5, 6, 7, 8\}$ defined by

$$xRy \qquad \text{if } x \text{ divides } y$$

is

$$\begin{array}{c} \\ 2 \\ 3 \\ 4 \end{array} \begin{array}{cccc} 5 & 6 & 7 & 8 \\ \begin{pmatrix} 0 & 1 & 0 & 1 \\ 0 & 1 & 0 & 0 \\ 0 & 0 & 0 & 1 \end{pmatrix} \end{array}.$$

DEFINITION B.1.5. Two matrices A and B are *equal*, written $A = B$, if they are the same size and their corresponding entries are equal.

EXAMPLE B.1.6. Determine w, x, y, and z so that

$$\begin{pmatrix} x + y & y \\ w + z & w - z \end{pmatrix} = \begin{pmatrix} 5 & 2 \\ 4 & 6 \end{pmatrix}.$$

According to Definition B.1.5, since the matrices are the same size, they will be equal provided that

$$x + y = 5 \qquad y = 2$$
$$w + z = 4 \qquad w - z = 6.$$

Solving these equations, we obtain

$$w = 5, \quad x = 3, \quad y = 2, \quad z = -1.$$

We describe next some operations that can be performed on matrices. The **sum** of two matrices is obtained by adding the corresponding entries. The **scalar product** is obtained by multiplying each entry in the matrix by a fixed number.

DEFINITION B.1.7. Let $A = (a_{ij})$ and $B = (b_{ij})$ be two $m \times n$ matrices. The *sum* of A and B is defined as

$$A + B = (a_{ij} + b_{ij}).$$

The *scalar product* of a number c and a matrix $A = (a_{ij})$ is defined as

$$cA = (ca_{ij}).$$

If A and B are matrices, we define $-A = (-1)A$ and $A - B = A + (-B)$.

EXAMPLE B.1.8. If

$$A = \begin{pmatrix} 4 & 2 \\ -1 & 0 \\ 6 & -2 \end{pmatrix}, \quad B = \begin{pmatrix} 1 & -3 \\ 4 & 4 \\ -1 & -3 \end{pmatrix},$$

then

$$A + B = \begin{pmatrix} 5 & -1 \\ 3 & 4 \\ 5 & -5 \end{pmatrix}, \quad 2A = \begin{pmatrix} 8 & 4 \\ -2 & 0 \\ 12 & -4 \end{pmatrix}, \quad -B = \begin{pmatrix} -1 & 3 \\ -4 & -4 \\ 1 & 3 \end{pmatrix}.$$

Multiplication of matrices is another important matrix operation.

DEFINITION B.1.9. Let $A = (a_{ij})$ be an $m \times n$ matrix and let $B = (b_{jk})$ be an $n \times l$ matrix. The *matrix product* is defined as the $m \times l$ matrix

$$AB = (c_{ik}),$$

where

$$c_{ik} = \sum_{j=1}^{n} a_{ij}b_{jk}.$$

To multiply the matrix A by the matrix B, Definition B.1.9 requires that the number of columns of A be equal to the number of rows of B.

EXAMPLE B.1.10. Let

$$A = \begin{pmatrix} 1 & 6 \\ 4 & 2 \\ 3 & 1 \end{pmatrix}, \quad B = \begin{pmatrix} 1 & 2 & -1 \\ 4 & 7 & 0 \end{pmatrix}.$$

The matrix product AB is defined since the number of columns of A is the same as the number of rows of B; both are equal to 2. Entry c_{ik} in the product AB is obtained by using the ith row of A and the kth column of B. For example, the entry c_{31} will be computed using the third row

$$(3 \quad 1)$$

of A and the first column

$$\begin{pmatrix} 1 \\ 4 \end{pmatrix}$$

of B. We then multiply, consecutively, each element in the third row of A by each element in the first column of B and then sum to obtain

$$3 \cdot 1 + 1 \cdot 4 = 7.$$

Since the number of columns of A is the same as the number of rows of B, the elements pair up correctly. Proceeding in this way, we obtain the product

$$AB = \begin{pmatrix} 25 & 44 & -1 \\ 12 & 22 & -4 \\ 7 & 13 & -3 \end{pmatrix}.$$

EXAMPLE B.1.11.   The matrix product

$$\begin{pmatrix} a & b \\ c & d \end{pmatrix} \begin{pmatrix} x \\ y \end{pmatrix}$$

is

$$\begin{pmatrix} ax + by \\ cx + dy \end{pmatrix}.$$

EXAMPLE B.1.12.   Let

$$R_1 = \{(1, a), (2, b), (3, a), (3, b)\}$$

be a relation from $X = \{1, 2, 3\}$ to $Y = \{a, b\}$ and let

$$R_2 = \{(a, x), (a, y), (b, y), (b, z)\}$$

be a relation from Y to $Z = \{x, y, z\}$. The matrices of R_1 and R_2 are, respectively,

$$A_1 = \begin{matrix} 1 \\ 2 \\ 3 \end{matrix} \begin{pmatrix} \overset{a}{1} & \overset{b}{0} \\ 0 & 1 \\ 1 & 1 \end{pmatrix}, \qquad A_2 = \begin{matrix} a \\ b \end{matrix} \begin{pmatrix} \overset{x}{1} & \overset{y}{1} & \overset{z}{0} \\ 0 & 1 & 1 \end{pmatrix}.$$

The product is

$$A_1 A_2 = \begin{pmatrix} 1 & 1 & 0 \\ 0 & 1 & 1 \\ 1 & 2 & 1 \end{pmatrix}.$$

Let us interpret this product.

The *ik*th entry in $A_1 A_2$ is computed as

$$i\ (s\quad t)\begin{pmatrix} \overset{a}{} & \overset{b}{} & \overset{k}{\begin{pmatrix} u \\ v \end{pmatrix}} \end{pmatrix} = su + tv.$$

If this value is nonzero, then either *su* or *tv* is nonzero. Suppose that $su \neq 0$. (The argument is similar if $tv \neq 0$.) This means that $(i, a) \in R_1$ and $(a, k) \in R_2$. This implies that $(i, k) \in R_2 \circ R_1$. We have shown that if the *ik*th entry in $A_1 A_2$ is nonzero, then $(i, k) \in R_2 \circ R_1$. The converse is also true, as we now show.

Assume that $(i, k) \in R_2 \circ R_1$. Then, either

1. $(i, a) \in R_1$ and $(a, k) \in R_2$

or

2. $(i, b) \in R_1$ and $(b, k) \in R_2$.

If case 1 holds, then $s = 1$ and $u = 1$, so $su = 1$ and $su + tv$ is nonzero. Similarly, if case 2 holds, $tv = 1$ and again we have $su + tv$ nonzero.

We have shown that $A_1 A_2$ is "almost" the matrix of the relation $R_2 \circ R_1$. To obtain the matrix of the relation $R_2 \circ R_1$ we need only change all nonzero entries in $A_1 A_2$ to 1. Thus the matrix of the relation $R_2 \circ R_1$ is

$$\begin{array}{c} \\ 1 \\ 2 \\ 3 \end{array}\begin{array}{c} x\quad y\quad z \\ \begin{pmatrix} 1 & 1 & 0 \\ 0 & 1 & 1 \\ 1 & 1 & 1 \end{pmatrix} \end{array}.$$

EXERCISES

1H. Compute the sum

$$\begin{pmatrix} 2 & 4 & 1 \\ 6 & 9 & 3 \\ 1 & -1 & 6 \end{pmatrix} + \begin{pmatrix} a & b & c \\ d & e & f \\ g & h & i \end{pmatrix}.$$

In Exercises 2–8, let

$$A = \begin{pmatrix} 1 & 6 & 9 \\ 0 & 4 & -2 \end{pmatrix}, \qquad B = \begin{pmatrix} 4 & 1 & -2 \\ -7 & 6 & 1 \end{pmatrix}$$

and compute each expression.

2.	$A + B$	**3H.**	$B + A$
4.	$-A$	**5H.**	$3A$
6.	$-2B$	**7H.**	$2B + A$
8.	$B - 6A$		

In Exercises 9–13, compute the products.

9H. $\begin{pmatrix} 1 & 2 & 3 \\ -1 & 2 & 3 \\ 0 & 1 & 4 \end{pmatrix} \begin{pmatrix} 2 & 8 \\ -1 & 1 \\ 6 & 0 \end{pmatrix}$

10. $\begin{pmatrix} 1 & 6 \\ -8 & 2 \\ 4 & 1 \end{pmatrix} \begin{pmatrix} 4 & 1 \\ 7 & -6 \end{pmatrix}$

11H. $A^2 (= AA)$, where

$$A = \begin{pmatrix} 1 & -2 \\ 6 & 2 \end{pmatrix}$$

12.

$$(2 \quad -4 \quad 6 \quad 1 \quad 3) \begin{pmatrix} 1 \\ 3 \\ -2 \\ 6 \\ 4 \end{pmatrix}$$

13H. $\begin{pmatrix} 2 & 4 & 1 \\ 6 & 9 & 3 \\ 1 & -1 & 6 \end{pmatrix} \begin{pmatrix} a & b \\ c & d \\ e & f \end{pmatrix}$

14. **(a)** Give the size of each matrix.

$$A = \begin{pmatrix} 1 & 4 & 6 \\ 0 & 1 & 7 \end{pmatrix}, \quad B = \begin{pmatrix} 1 & 4 & 7 \\ 8 & 2 & 1 \\ 0 & 1 & 6 \end{pmatrix}, \quad C = \begin{pmatrix} 4 & 2 \\ 0 & 0 \\ 2 & 9 \end{pmatrix}$$

(b) Using the matrices of part (a), decide which of the products

$$A^2, AB, BA, AC, CA, AB^2, BC, CB, C^2$$

are defined and then compute these products.

15H. Determine x, y, and z so that the equation

$$\begin{pmatrix} x+y & 3x+y \\ x+z & x+y-2z \end{pmatrix} = \begin{pmatrix} -1 & 1 \\ 9 & -17 \end{pmatrix}$$

holds.

16. Determine w, x, y, and z so that the equation

$$\begin{pmatrix} 2 & 1 & -1 & 7 \\ 6 & 8 & 0 & 3 \end{pmatrix} \begin{pmatrix} x & 2x \\ y & -y+z \\ x+w & w-2y+x \\ z & z \end{pmatrix} = -\begin{pmatrix} 45 & 46 \\ 3 & 87 \end{pmatrix}$$

holds.

In Exercises 17–19, find the matrix of the relation.

17H. $R = \{(1, a), (1, c), (2, b), (3, a)\}$ from $\{1, 2, 3\}$ to $\{a, b, c, d\}$.

18. R on $\{1, 2, 3, 4, 5\}$ defined by

$$x R y \qquad \text{if } x < y.$$

19H. R on $\{1, 2, 3, 4, 5\}$ defined by

$$x R y \qquad \text{if } x \leq y.$$

20. Given the matrix

$$
\begin{array}{c}
 \\
1 \\
2 \\
3
\end{array}
\begin{array}{ccccc}
a & b & c & d & e \\
\begin{pmatrix}
1 & 1 & 0 & 0 & 1 \\
0 & 1 & 0 & 1 & 0 \\
1 & 1 & 0 & 0 & 1
\end{pmatrix}
\end{array}
$$

of a relation R from $X = \{1, 2, 3\}$ to $Y = \{a, b, c, d, e\}$, represent R as a set of ordered pairs.

21H. Let

$$R_1 = \{(a, x), (b, x), (c, y), (d, z), (a, z), (c, z)\}$$

be a relation from $\{a, b, c, d\}$ to $Y = \{w, x, y, z\}$ and let

$$R_2 = \{(w, 1), (x, 1), (y, 3), (z, 3), (x, 2), (x, 3)\}$$

be a relation from Y to $\{1, 2, 3\}$.
(a) Find the matrix of each of the relations R_1 and R_2.
(b) By multiplying the appropriate matrices, as described in Example B.1.12, find the matrix of the relation $R_2 \circ R_1$.

22. Let A be the matrix of a relation R on X. What conditions must A satisfy for R to be reflexive? transitive? symmetric? antisymmetric? a function?

23H. Let A be the matrix of a function f from X to Y.
(a) What conditions must A satisfy for f to be onto Y?
(b) What conditions must A satisfy for f to be one-to-one?

24. Show that the ikth nonzero entry in the matrix $A_1 A_2$ of Example B.1.12 counts the number of elements in the set

$$\{(i, k) \mid (i, m) \in R_1 \quad \text{and} \quad (m, k) \in R_2\}.$$

25H. Define the $n \times n$ matrix $I_n = (a_{ij})$ by

$$
a_{ij} = \begin{cases} 1 & \text{if } i = j \\ 0 & \text{if } i \neq j. \end{cases}
$$

The matrix I_n is called the $n \times n$ **identity matrix**.
   Show that if A is an $n \times n$ matrix (such a matrix is called a **square matrix**), then

$$A I_n = A = I_n A.$$

An $n \times n$ matrix A is said to be **invertible** if there exists an $n \times n$ matrix B satisfying

$$AB = I_n = BA.$$

(The matrix I_n is defined in Exercise 25.)

26. Show that the matrix

$$\begin{pmatrix} 2 & 1 \\ 1 & 1 \end{pmatrix}$$

is invertible.

★27H. Show that the matrix

$$\begin{pmatrix} a & b \\ c & d \end{pmatrix}$$

is invertible if and only if $ad - bc \neq 0$.

28. Suppose that we want to solve the system

$$AX = C$$

where

$$A = \begin{pmatrix} a_{11} & a_{12} \\ a_{21} & a_{22} \end{pmatrix}, \qquad X = \begin{pmatrix} x \\ y \end{pmatrix}, \qquad C = \begin{pmatrix} c_1 \\ c_2 \end{pmatrix}$$

for x and y.

Show that if A is invertible, the system has a solution.

29H. The **transpose** of a matrix $A = (a_{ij})$ is the matrix $A^T = (a'_{ji})$, where $a'_{ji} = a_{ij}$.

EXAMPLE

$$\begin{pmatrix} 1 & 3 \\ 4 & 6 \end{pmatrix}^T = \begin{pmatrix} 1 & 4 \\ 3 & 6 \end{pmatrix}.$$

If A and B are $m \times k$ and $k \times n$ matrices, respectively, show that

$$(AB)^T = B^T A^T.$$

References

AHO, A., J. HOPCROFT, and J. ULLMAN, *The Design and Analysis of Computer Algorithms*, Addison-Wesley, Reading, Mass., 1974.

AINSLIE, T., *Ainslie's Complete Hoyle*, Simon and Schuster, New York, 1975.

APPEL, K., and W. HAKEN, "Every planar map is 4-colorable," *Bull. Am. Math. Soc.*, 82 (1976), 711–712.

ARBIB, M. A., *Theories of Abstract Automata*, Prentice-Hall, Englewood Cliffs, N.J., 1969.

ARBIB, M. A., A. J. KFOURY, and R. N. MOLL, *A Basis for Theoretical Computer Science*, Springer-Verlag, New York, 1981.

BAASE, S., *Computer Algorithms: Introduction to Design and Analysis*, Addison-Wesley, Reading, Mass., 1978.

BABAI, L., and T. KUCERA, "Canonical labelling of graphs in linear average time," Proc. 20th Symposium on the Foundations of Computer Science, 1979, 39–46.

BELL, R. C., *Board and Table Games from Many Civilizations*, revised ed., Dover, New York, 1979.

BELLMAN, R., K. L. COOKE, and J. A. LOCKETT, *Algorithms, Graphs, and Computers*, Academic Press, New York, 1970.

BELLMORE, M., and G. L. NEMHAUSER, "The traveling salesman problem," *Oper. Res.*, 16 (1968), 538–558.

BERGE, C., *The Theory of Graphs and Its Applications*, Wiley, New York, 1962.

BERLEKAMP, E. R., J. H. CONWAY, and R. K. GUY, *Winning Ways*, Vols. 1 and 2, Academic Press, New York, 1982.

BERLINER, H., "Computer backgammon," *Sci. Am.*, June 1980, 64–72.

405

BERZTISS, A. T., *Data Structures: Theory and Practice*, 2nd ed., Academic Press, New York, 1975.

BIRKHOFF, G., and T. C. BARTEE, *Modern Applied Algebra*, McGraw-Hill, New York, 1970.

BOBROW, L. S., and M. A. ARBIB, *Discrete Mathematics*, Saunders, Philadelphia, 1974.

BOGART, K. P., *Introductory Combinatorics*, Pitman, Marshfield, Mass., 1983.

BOOLE, G., *The Laws of Thought*, reprinted by Dover, New York, 1951.

BRUALDI, R. A., *Introductory Combinatorics*, North-Holland, New York, 1977.

BUSACKER, R. G., and T. L. SAATY, *Finite Graphs and Networks: An Introduction with Applications*, McGraw-Hill, New York, 1965.

CARBERRY, M. S., H. M. KHALIL, J. F. LEATHRAM, and L. S. LEVY, *Foundations of Computer Science*, Computer Science Press, Rockville, Md., 1979.

Curriculum Committee on Computer Science, "Curriculum '68—Recommendations for academic programs in computer science," *Commun. ACM*, 11 (1968), 151–197.

Curriculum Committee on Computer Science, "Curriculum '78—Recommendations for the undergraduate program in computer science," *Commun. ACM*, 22 (1979), 147–166.

D'ANGELO, H., *Microcomputer Structures*, McGraw-Hill, New York, 1981.

DE CARTEBLANCHE, F., "Piles of cubes," *Eureka*, 1947.

DEO, N., *Graph Theory with Applications to Engineering and Computer Science*, Prentice-Hall, Englewood Cliffs, N.J., 1974.

DIJKSTRA, E., "Cooperating sequential processes," in *Programming Languages,* F. Genuys, ed., Academic Press, New York, 1968.

DORNHOFF, L. L., and F. E. HOHN, *Applied Modern Algebra*, Macmillan, New York, 1978.

DUDA, R. O., and P. E. HART, *Pattern Classification and Scene Analysis*, Wiley, New York, 1973.

EULER, L., "Leonhard Euler and the Koenigsberg bridges," J. R. NEWMAN, ed., *Sci. Am.*, July 1953, 66–70.

EVEN, S., *Algorithmic Combinatorics*, Macmillan, New York, 1973.

EVEN, S., *Graph Algorithms*, Computer Science Press, Rockville, Md., 1979.

FISHER, J. L., *Application-Oriented Algebra*, Harper & Row, New York, 1977.

FORD, L. R., JR., and D. R. FULKERSON, *Flows in Networks*, Princeton University Press, Princeton, N.J., 1962.

FREEMAN, J., *The Playboy Winner's Guide to Board Games,* Playboy, Chicago, 1979.

FREY, P., "Machine-problem solving—Part 3: The alpha-beta procedure," *Byte*, 5 (November 1980), 244–264.

GARDNER, M., *Mathematical Puzzles and Diversions*, Simon and Schuster, New York, 1959.

GARDNER, M., *Mathematical Circus*, Knopf, New York, 1979.

GERSTING, J. L., *Mathematical Structures for Computer Science*, W. H. Freeman, San Francisco, 1982.

GILL, A., *Applied Algebra for the Computer Sciences*, Prentice-Hall, Englewood Cliffs, N.J., 1976.

GOLDSTINE, H. H., *The Computer from Pascal to von Neumann*, Princeton University Press, Princeton, N.J., 1972.

GOLOMB, S., and L. BAUMERT, "Backtrack programming," *J. ACM*, 12 (1965), 516–524.

GOSE, E. E., "Introduction to biological and mechanical pattern recognition," pp. 203–252 in *Methodologies of Pattern Recognition*, S. Watanabe, ed., Academic Press, New York, 1969.

GRILLO, J. P., and J. D. ROBERTSON, *Data Management Techniques*, Wm. C. Brown, Dubuque, Iowa, 1981.

HAILPERIN, T., "Boole's algebra isn't Boolean algebra," *Math. Mag.*, 54 (1981), 173–184.

HALMOS, P. R., *Lectures on Boolean Algebras*, D. Van Nostrand, New York, 1967.

HALMOS, P. R., *Naive Set Theory*, Springer-Verlag, New York, 1974.

HARARY, F., *Graph Theory*, Addison-Wesley, Reading, Mass., 1969.

HELL, P., "Absolute retracts in graphs," in *Graphs and Combinatorics*, R. A. Bari and F. Harary, eds., Lecture Notes in Mathematics, Vol. 406, Springer-Verlag, New York, 1974.

HILL, F. J., and G. R. PETERSON, *Switching Theory and Logical Design*, 2nd ed., Wiley, New York, 1974.

HILLIER, F. S., and G. J. LIEBERMAN, *Introduction to Operations Research*, Holden-Day, San Francisco, 1974.

HOHN, F. E., *Applied Boolean Algebra*, 2nd ed., Macmillan, New York, 1966.

HOPCROFT, J. E., and J. D. ULLMAN, *Formal Languages and Their Relation to Automata*, Addison-Wesley, Reading, Mass., 1969.

HOPCROFT, J. E., and J. D. ULLMAN, *Introduction to Automata Theory, Languages, and Computation*, Addison-Wesley, Reading, Mass., 1979.

HOROWITZ, E., and S. SAHNI, *Fundamentals of Data Structures*, Computer Science Press, Rockville, Md., 1976.

HOROWITZ, E., and S. SAHNI, *Fundamentals of Computer Algorithms*, Computer Science Press, Rockville, Md., 1978.

HU, T. C., *Integer Programming and Network Flows*, Addison-Wesley, Reading, Mass., 1969.

HU, T. C., *Combinatorial Algorithms*, Addison-Wesley, Reading, Mass., 1982.

JENSEN, K., and N. WIRTH, *PASCAL: User Manual and Report*, 3rd ed., revised by A. B. Mickel and J. F. Miner, Springer-Verlag, New York, 1985.

JOHNSONBAUGH, R., and T. MURATA, "Petri nets and marked graphs—mathematical models of concurrent computation," *Am. Math. Mon.*, 89 (1982), 552–566.

KAIN, R. Y., *Automata Theory: Machines and Languages*, McGraw-Hill, New York, 1972.

KLINE, M., *Mathematical Thought from Ancient to Modern Times*, Oxford University Press, New York, 1972.

KNUTH, D. E., *The Art of Computer Programming*, Vol. 1: *Fundamental Algorithms*, 2nd ed., Addison-Wesley, Reading, Mass., 1973.

KNUTH, D. E., *The Art of Computer Programming*, Vol. 3: *Sorting and Searching*, Addison-Wesley, Reading, Mass., 1973.

KNUTH, D. E., "Computer science and its relation to mathematics," *Am. Math. Mon.*, 81 (1974), 323–343.

KNUTH, D. E., "Algorithms," *Sci. Am.*, April 1977, 63–80.

KNUTH, D. E., *The Art of Computer Programming*, Vol. 2: *Seminumerical Algorithms*, 2nd ed., Addison-Wesley, Reading, Mass., 1981.

KOHAVI, Z., *Switching and Finite Automata Theory*, McGraw-Hill, New York, 1970.

KÖNIG, D., *Theorie der endlichen und unendlichen Graphen*, Leipzig, 1936. (Reprinted in 1950 by Chelsea, New York.)

LEVY, L. S., *Discrete Structures of Computer Science*, Wiley, New York, 1980.

LIPSCHUTZ, S., *Theory and Problems of Set Theory and Related Topics*, Schaum, New York, 1964.

LIPSCHUTZ, S., *Discrete Mathematics*, Schaum, New York, 1976.

LIPSCHUTZ, S., *Essential Computer Mathematics*, Schaum, New York, 1982.

LIU, C. L., *Introduction to Combinatorial Mathematics*, McGraw-Hill, New York, 1968.

LIU, C. L., *Elements of Discrete Mathematics*, 2nd ed., McGraw-Hill, New York, 1985.

MCNAUGHTON, R., *Elementary Computability, Formal Languages, and Automata*, Prentice-Hall, Englewood Cliffs, N.J., 1982.

MENDELSON, E., *Boolean Algebra and Switching Circuits*, Schaum, New York, 1970.

MINSKY, M., *Computation: Finite and Infinite Machines*, Prentice-Hall, Englewood Cliffs, N.J., 1967.

NELSON, R. J., *Introduction to Automata*, Wiley, New York, 1968.

NIEVERGELT, J., J. C. FARRAR, and E. M. REINGOLD, *Computer Approaches to Mathematical Problems*, Prentice-Hall, Englewood Cliffs, N.J., 1974.

NILSSON, N. J., *Problem-Solving Methods in Artificial Intelligence*, McGraw-Hill, New York, 1971.

NIVEN, I., *Mathematics of Choice*, Mathematical Association of America, Washington, D.C., 1965.

ORE, O., *Graphs and Their Uses*, Random House, New York, 1963.

PARRY, R. T., and H. PFEFFER, "The infamous traveling-salesman problem: a practical approach," *Byte*, 6 (July 1981), 252–290.

PEARL, J., "The solution for the branching factor of the alpha-beta pruning algorithm and its optimality," *Commun. ACM*, 25 (1982), 559–564.

PETERSON, J. L., *Petri Net Theory and the Modeling of Systems*, Prentice-Hall, Englewood Cliffs, N.J., 1981.

PETRI, C., "Kommunikation mit Automaten," Ph.D. dissertation, University of Bonn, Bonn, West Germany, 1962 (in German). Translated by C. F. Greene, Jr., "Communication with automata," Supplement to Technical Report RADC-TR-65-377, Vol. 1, Rome Air Development Center, Griffiss Air Force Base, New York, 1966.

PRATHER, R. E., *Discrete Mathematical Structures for Computer Science*, Houghton Mifflin, Boston, 1976.

READ, R. C., and D. G. CORNEIL, "The graph isomorphism disease," *J. Graph Theory*, 1 (1977), 339–363.

Recommendations for a General Mathematical Sciences Program, CUPM, Mathematical Association of America, Washington, D.C., 1981.

Recommendations on the Mathematical Training of Teachers, preliminary draft, CUPM Teacher Training Panel, March 1982.

REINGOLD, E., J. NIEVERGELT, and N. DEO, *Combinatorial Algorithms*, Prentice-Hall, Englewood Cliffs, N.J., 1977.

RIORDAN, J., *An Introduction to Combinatorial Analysis*, Wiley, New York, 1958.

SAHNI, S., *Concepts in Discrete Mathematics*, Camelot, Frindley, Minn., 1981.

SHANNON, C. E., "A symbolic analysis of relay and switching circuits," *Trans. Am. Inst. Electr. Eng.*, 57 (1938), 713–723.

SLAGLE, J. R., *Artificial Intelligence: The Heuristic Programming Approach*, McGraw-Hill, New York, 1971.

STANAT, D. F., and D. F. MCALLISTER, *Discrete Mathematics in Computer Science*, Prentice-Hall, Englewood Cliffs, N.J., 1977.

STANDISH, T. A., *Data Structure Techniques*, Addison-Wesley, Reading, Mass., 1980.

STOLL, R. R., *Set Theory and Logic*, W. H. Freeman, San Francisco, 1963.

STONE, H. S., *Discrete Mathematical Structures and Their Applications*, Science Research Associates, Palo Alto, Calif., 1973.

TREMBLAY, J. P., and R. MANOHAR, *Discrete Mathematical Structures with Applications to Computer Science*, McGraw-Hill, New York, 1975.

TSICHRITZIS, D. C., and F. H. LOCHOVSKY, *Data Base Management Systems*, Academic Press, New York, 1977.

TUCKER, A., *Applied Combinatorics*, 2nd ed, Wiley, New York, 1985.

VILENKIN, N. Y., *Combinatorics*, Academic Press, New York, 1971.

WAND, M., *Induction, Recursion, and Programming*, North-Holland, New York, 1980.

WILLIAMS, G., and R. MEYER, "The Panasonic and Quasar hand-held computers: beginning a new generation of consumer computers," *Byte*, 6 (January 1981), 34–45.

Hints and Solutions to Selected Exercises

Section 1.1

1. $\{1, 2, 3, 4, 5, 7, 10\}$ **3.** $\{7, 10\}$ **5.** $\{2, 3, 5, 6, 8, 9\}$ **7.** $\varnothing$ **9.** $\varnothing$

11. B **13.** $\{6, 8\}$ **15.** $\{2, 3, 4, 5, 6, 7, 8, 9, 10\}$

17. $\{(1, a), (1, b), (1, c), (2, a), (2, b), (2, c)\}$

19. $\{(1, 1), (1, 2), (2, 1), (2, 2)\}$

21. $\{(1, a, \alpha), (1, a, \beta), (2, a, \alpha), (2, a, \beta)\}$

23. $\{(1, 1, 1), (1, 2, 1), (2, 1, 1), (2, 2, 1), (1, 1, 2), (1, 2, 2), (2, 1, 2), (2, 2, 2)\}$

25. $\{\{1\}\}$

27. $\{\{a, b, c\}\}, \{\{a, b\}, \{c\}\}, \{\{a, c\}, \{b\}\}, \{\{b, c\}, \{a\}\}, \{\{a\}, \{b\}, \{c\}\}$

29. True **31.** True **33.** Equal **35.** Equal **37.** Not equal

39. $\varnothing, \{a\}, \{b\}, \{c\}, \{d\}, \{a, b\}, \{a, c\}, \{a, d\}, \{b, c\}, \{b, d\}, \{c, d\}, \{a, b, c\},$
$\{a, b, d\}, \{a, c, d\}, \{b, c, d\}, \{a, b, c, d\}$. All except $\{a, b, c, d\}$ are proper subsets.

41. $2^n - 1$

43. False. $X = \{1, 2\}, Y = \{2, 3\}$.

45. True **47.** True **49.** False. $X = \{1, 2, 3\}, Y = \{2\}, Z = \{3\}$.

51. True **53.** False. Take $U = \{1, 2, 3, 4, 5\}, X = \{2, 3\}, Y = \{3, 4\}$.

55. True **57.** True **59.** True

61. False. Take $X = \{1, 2\}, Y = \{1, 3\}, Z = \{1, 4\}$.

63. Suppose that $\varnothing \subseteq X$ is false. **65.** $B \subseteq A$ **67.** $B \subseteq A$

69. The symmetric difference of A and B consists of those elements in A or B but not both.

71. $|A| + |B|$ counts the elements in A and B but counts the elements in $A \cap B$ twice.

73. The center of C

Section 1.2

1. {(8840, Hammer), (9921, Pliers), (452, Paint), (2207, Carpet)}

3. {(Sally, Math), (Ruth, Physics), (Sam, Econ)}

5.			**7.**	
a	6		1	1
b	2		2	1
a	1		3	1
c	1		4	1
			2	2
			3	2
			4	2
			2	3
			3	3
			4	3
			2	4
			3	4
			4	4

9.

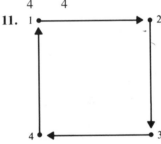

13. {(a, b), (a, c), (b, a), (b, d), (c, c), (c, d)}

15. { }

17. (For Exercise 1) domain = {8840, 9921, 452, 2207}, range = {Hammer, Pliers, Paint, Carpet}

19. {(1, 1), (1, 4), (2, 2), (2, 5), (3, 3), (4, 1), (4, 4), (5, 2), (5, 5)}

21. {1, 2, 3, 4, 5}   **23.** {1, 2, 3, 4, 5}

25. $R = R^{-1}$ = {(1, 1), (1, 2), (1, 3), (1, 4), (1, 5), (2, 1), (2, 2), (2, 3), (2, 4), (3, 1), (3, 2), (3, 3), (4, 1), (4, 2), (5, 1)}

domain R = range R = domain R^{-1} = range R^{-1} = {1, 2, 3, 4, 5}

27. Symmetric   **29.** Antisymmetric

31. Reflexive, antisymmetric, transitive, partial order

33. Symmetric

35. Reflexive, antisymmetric, transitive, partial order

37. $R_1 \circ R_2$ = {(1, 1), (1, 2), (2, 1), (2, 2), (3, 1), (3, 2), (4, 2)}
$R_2 \circ R_1$ = {(1, 1), (1, 2), (3, 4), (4, 1), (4, 2)}

39. {(1, 1), (2, 2), (3, 3), (4, 4), (1, 2), (2, 3)}

41. {(1, 1), (1, 2), (2, 1), (2, 2)}

43. False. Let R = {(1, 2)}, S = {(2, 3)}.

45. False. Let R = {(2, 3), (4, 5)}, S = {(1, 2), (3, 4)}.

47. True   **49.** True   **51.** True

53. False. Let R = {(2, 3), (3, 2)}, S = {(1, 2), (2, 1)}.

55. False. Let R = {(1, 2)}, S = {(2, 1)}.

57. False. Let R = {(2, 3), (1, 1)}, S = {(1, 2), (3, 1)}.

59. It may be the case that for some $x \in X$, there is *no* $y \in X$ such that $(x, y) \in R$. Consider, for example, X = {1, 2, 3}, R = {(1, 1), (2, 2), (1, 2), (2, 1)}, and $x = 3$.

Section 1.3

1. Equivalence relation; $[1] = [3] = \{1, 3\}$, $[2] = \{2\}$, $[4] = \{4\}$, $[5] = \{5\}$

3. Not an equivalence relation (not reflexive)

5. Equivalence relation; all equivalence classes are equal to $\{1, 2, 3, 4, 5\}$

7. Not an equivalence relation (neither transitive nor reflexive)

9. $\{(1, 1), (1, 2), (2, 1), (2, 2), (3, 3), (3, 4), (4, 3), (4, 4)\}$, $[1] = [2] = \{1, 2\}$, $[3] = [4] = \{3, 4\}$

11. $\{(1, 1), (2, 2), (3, 3), (4, 4)\}$, $[i] = \{i\}$ for $i = 1, \ldots, 4$

13. $\{(1, 1), (1, 2), (1, 3), (1, 4), (2, 1), (2, 2), (2, 3), (2, 4), (3, 1), (3, 2), (3, 3), (3, 4), (4, 1), (4, 2), (4, 3), (4, 4)\}$

$$[1] = [2] = [3] = [4] = \{1, 2, 3, 4\}$$

15. We show transitivity only. Suppose $b_1 R b_2$ and $b_2 R b_3$. Then the first four bits of b_1 and b_2 coincide and the first four bits of b_2 and b_3 coincide. Therefore, the first four bits of b_1 and b_3 coincide. Thus $b_1 R b_3$ and R is transitive.

17. 11110000, 11100000, 11010000, 11000000, 10110000, 10100000, 10010000, 10000000, 01110000, 01100000, 01010000, 01000000, 00110000, 00100000, 00010000, 00000000

19. Since R is reflexive, $(x, x) \in R$ for every $x \in X$. Thus domain $R = $ range $R = X$.

21. $R = \{(x, x) \mid x \in X\}$

23. Five, corresponding to the partitions $\{\{1\}, \{2\}, \{3\}\}$, $\{\{1\}, \{2, 3\}\}$, $\{\{1, 2\}, \{3\}\}$, $\{\{1, 3\}, \{2\}\}$, $\{\{1, 2, 3\}\}$

25. (a) We show transitivity only. Suppose that $(a, b)R(c, d)$ and $(c, d)R(e, f)$. Then $ad = bc$ and $cf = de$. Multiplying the first equation by e and the second equation by a, we obtain $acf = ade = bce$. Canceling c, we obtain $af = be$. Therefore, $(a, b)R(e, f)$ and R is transitive.

(b) $(1, 1), (1, 2), (1, 3), (1, 4), (1, 5), (1, 6), (1, 7), (1, 8), (1, 9), (1, 10), (2, 1), (2, 3), (2, 5), (2, 7), (2, 9), (3, 1), (3, 2), (3, 4), (3, 5), (3, 7), (3, 8), (3, 10), (4, 1), (4, 3), (4, 5), (4, 7), (4, 9), (5, 1), (5, 2), (5, 3), (5, 4), (5, 6), (5, 7), (5, 8), (5, 9), (6, 1), (6, 5), (6, 7), (7, 1), (7, 2), (7, 3), (7, 4), (7, 5), (7, 6), (7, 8), (7, 9), (7, 10), (8, 1), (8, 3), (8, 5), (8, 7), (8, 9), (9, 1), (9, 2), (9, 4), (9, 5), (9, 7), (9, 8), (9, 10), (10, 1), (10, 3), (10, 7), (10, 9)$

(c) $(a, b)R(c, d)$ if and only if the fractions a/c and b/d are equal.

27. (a) We show symmetry only. Let $(x, y) \in R_1 \cap R_2$. Then $(x, y) \in R_1$ and $(x, y) \in R_2$. Since R_1 and R_2 are symmetric, $(y, x) \in R_1$ and $(y, x) \in R_2$. Thus $(y, x) \in R_1 \cap R_2$ and, therefore, $R_1 \cap R_2$ is symmetric.

(b) A is an equivalence class of $R_1 \cap R_2$ if and only if there are equivalence classes A_1 of R_1 and A_2 of R_2 such that $A = A_1 \cap A_2$.

29. (b) Cylinder

31. $\rho(R_1) = \{(1, 1), (2, 2), (3, 3), (4, 4), (1, 2), (3, 4), (4, 2)\}$

$\sigma(R_1) = \{(1, 1), (2, 1), (1, 2), (3, 4), (4, 3), (4, 2), (2, 4)\}$

$\tau(R_1) = \{(1, 1), (1, 2), (3, 4), (4, 2), (3, 2)\}$

$\tau(\sigma(\rho(R_1))) = \{(x, y) \mid x, y \in \{1, 2, 3, 4\}\}$

33. Let $(x, y) \in R \cup R^{-1}$. If $(x, y) \in R$, then $(y, x) \in R^{-1}$, so $(y, x) \in R \cup R^{-1}$. If $(x, y) \in R^{-1}$, then $(y, x) \in R$, so $(y, x) \in R \cup R^{-1}$. In any case, if $(x, y) \in R \cup R^{-1}$, $(y, x) \in R \cup R^{-1}$, so $R \cup R^{-1}$ is symmetric.

35. Since

$$R \subseteq \rho(R), \qquad R \subseteq \sigma(R), \qquad R \subseteq \tau(R), \qquad (*)$$

it follows that $R \subseteq \tau(\sigma(\rho(R)))$.

By $(*)$, $\rho(R) \subseteq \tau(\sigma(\rho(R)))$ and by Exercise 32, $\rho(R)$ is reflexive. Therefore, $\tau(\sigma(\rho(R)))$ is reflexive.

By Exercise 33, $\sigma(\rho(R))$ is symmetric. We show that if R' is any symmetric relation, $\tau(R')$ is symmetric. We can then conclude that $\tau(\sigma(\rho(R)))$ is symmetric.

Let R' be a symmetric relation. Let $(x, y) \in \tau(R')$. Then there exist $x = x_0, \ldots, x_n = y \in X$ such that $(x_{i-1}, x_i) \in R'$ for $i = 1, \ldots, n$. Since R' is symmetric, $(x_i, x_{i-1}) \in R'$ for $i = 1, \ldots, n$. Thus $(y, x) \in \tau(R')$ and $\tau(R')$ is symmetric.

By Exercise 34, $\tau(\sigma(\rho(R)))$ is transitive; hence $\tau(\sigma(\rho(R)))$ is an equivalence relation containing R.

37. Suppose that R is transitive. If $(x, y) \in \tau(R) = \cup\{R^n\}$; then there exist $x = x_0, \ldots, x_n = y \in X$ such that $(x_{i-1}, x_i) \in R$ for $i = 1, \ldots, n$. Since R is transitive, it follows that $(x, y) \in R$. Thus $R \supseteq \tau(R)$. Since we always have $R \subseteq \tau(R)$, it follows that $R = \tau(R)$.

Suppose that $\tau(R) = R$. By Exercise 34, $\tau(R)$ is transitive. Therefore, R is transitive.

39. False. Let $R_1 = \{(1, 1), (1, 2)\}$, $R_2 = \{(1, 1), (2, 1)\}$.
41. False. Let $R_1 = \{(1, 2), (2, 3)\}$, $R_2 = \{(1, 3), (3, 4)\}$.
43. True

Section 1.4
1. It is a function from X to Y; domain $= X$, range $= \{a, b, c\}$; it is neither one-to-one nor onto.
3. It is a function from X to Y; domain $= X$, range $= Y$; it is both one-to-one and onto. The inverse function is

$$\{(c, 1), (d, 2), (a, 3), (b, 4)\}.$$

For the inverse function, domain $= Y$, range $= X$.
5. It is a function from X to Y; domain $= X$, range $= \{b\}$; it is neither one-to-one nor onto.
7. Let f be the function from $X = \{a, b\}$ to $Y = \{y\}$ given by $f = \{(a, y), (b, y)\}$.
9. $f \circ g = \{(1, x), (2, z), (3, x)\}$
11. 4; one-to-one functions: $\{(1, a), (2, b)\}$ and $\{(1, b), (2, a)\}$. In this case, the onto and one-to-one functions are the same.
13. $f = \{(0, 0), (1, 4), (2, 3), (3, 2), (4, 1)\}$; f is one-to-one and onto.
15. The greatest common divisor of m and n must be 1.

In the solutions to Exercises 17 and 19, $a : b$ means store item a in cell b.
17. 714 : 0, 631 : 2, 26 : 9, 373 : 16, 775 : 10, 906 : 5, 509 : 1, 2032 : 11, 42 : 8, 4 : 4, 136 : 3, 1028 : 12
19. 714 : 0, 631 : 6, 26 : 5, 373 : 1, 775 : 8, 906 : 13, 509 : 2, 2032 : 7, 42 : 4, 4 : 3, 136 : 9, 1028 : 10
21. No. If the data item is present, it will be found before an empty cell is encountered.
23. False. Take $g = \{(a, x), (b, x)\}$ and $f = \{(x, 1)\}$.
25. True   **27.** True
29. False. Let $X = \{x, y\}$, $Y = \{a, b\}$, $Z = \{1\}$. Define g from X to Y by $g = \{(x, a), (y, a)\}$ and f from Y to Z by $f = \{(a, 1), (b, 1)\}$.
31. Suppose that f is not one-to-one. Then, for some x and y, $f(x) = f(y)$, but $x \neq y$. Let $A = \{x\}$, $B = \{y\}$.

Suppose that f is one-to-one. Let $y \in f(A \cap B)$. Then $y = f(x)$ for some $x \in A \cap B$. Thus $y \in f(A) \cap f(B)$. Let $y \in f(A) \cap f(B)$. Then $y = f(a) = f(b)$, for some $a \in A$, $b \in B$. Since f is one-to-one, $a = b$. Therefore, $y \in f(A \cap B)$.
33. If $x \in X$, then $x \in f^{-1}(f(\{x\}))$. Thus $\cup \{S \mid S \in \mathcal{S}\} = X$. Suppose that

$$a \in f^{-1}(\{y\}) \cap f^{-1}(\{z\})$$

for some $y, z \in Y$. Then $f(a) = y$ and $f(a) = z$. Thus $y = z$. Therefore, $\mathscr{S}$ is a partition of X. The equivalence relation that generates this partition is given in Exercise 32.

35. Suppose that $[x] = [y]$. Then xRy. Therefore, $g(x) = g(y)$.

37. Suppose that f is onto Y. Let g be a function from Y onto Z. We must show that $g \circ f$ is onto Z. Let $z \in Z$. Since g is onto Z, there exists $y \in Y$, with $g(y) = z$. Since f is onto Y, there exists $x \in X$, with $f(x) = y$. Now $g \circ f(x) = z$. Therefore, $g \circ f$ is onto.

Suppose that whenever g is a function from Y onto Z, $g \circ f$ is onto Z. Suppose that f is not onto Y. Then there exists $y_0 \in Y$ such that for no $x \in X$ do we have $f(x) = y_0$. Let $Z = \{0, 1\}$. Define g from Y to Z by $g(y_0) = 0$ and $g(y) = 1$, if $y \neq y_0$. Then g is onto Z, but $g \circ f$ is not onto Z.

39. If $x \in X - Y$, then

$$C_{X \cup Y}(x) = 1 = 1 + 1 - 1 \cdot 1 = C_X(x) + C_Y(x) - C_X(x)C_Y(x).$$

Similarly, if $x \in Y - X$, the equation holds. If $x \in X \cap Y$, then

$$C_{X \cup Y}(x) = 1 = 1 + 1 - 1 \cdot 1 = C_X(x) + C_Y(x) - C_X(x)C_Y(x).$$

If $x \notin X \cup Y$, then

$$C_{X \cup Y}(x) = 0 = 0 + 0 - 0 \cdot 0 = C_X(x) + C_Y(x) - C_X(x)C_Y(x).$$

Thus the equation holds for all $x \in U$.

41. If $x \in X - Y$, then

$$C_{X-Y}(x) = 1 = 1 \cdot [1 - 0] = C_X(x)[1 - C_Y(x)].$$

If $x \notin X - Y$, then either $x \notin X$ or $x \in Y$. In case $x \notin X$,

$$C_{X-Y}(x) = 0 = 0 \cdot [1 - C_Y(x)] = C_X(x)[1 - C_Y(x)].$$

In case $x \in Y$,

$$C_{X-Y}(x) = 0 = C_X(x)[1 - 1] = C_X(x)[1 - C_Y(x)].$$

Thus the equation holds for all $x \in U$.

43. By Exercise 39,

$$C_{X \cup Y}(x) = C_X(x) + C_Y(x) \qquad \text{for all } x \in U$$

if and only if

$$C_X(x)C_Y(x) = 0 \qquad \text{for all } x \in U.$$

By Exercise 38, this last equation holds if and only if

$$C_{X \cap Y}(x) = 0 \qquad \text{for all } x \in U.$$

By definition, this last equation holds if and only if

$$x \notin X \cap Y \qquad \text{for all } x \in U,$$

that is, if and only if $X \cap Y = \varnothing$.

45. f is onto by definition. Suppose that $f(X) = f(Y)$. Then $C_X(x) = C_Y(x)$ for all $x \in U$. Suppose that $x \in X$. Then $C_X(x) = 1$. Thus $C_Y(x) = 1$. Therefore, $x \in Y$. This argument shows that $X \subseteq Y$. Similarly, $Y \subseteq X$. Therefore, $X = Y$ and f is one-to-one.

47. A set is equivalent to itself by the identity function.

If X is equivalent to Y, there is a one-to-one, onto function f from X to Y. Now f^{-1} is a one-to-one, onto function from Y to X.

If X is equivalent to Y, there is a one-to-one, onto function f from X to Y. If Y is equivalent to Z, there is a one-to-one, onto function g from Y to Z. Now $g \circ f$ is a one-to-one, onto function from X to Z.

50. Assume that X is equivalent to $\mathcal{P}(X)$. Then there is a one-to-one, onto function f from X to $\mathcal{P}(X)$. Let

$$Y = \{x \in X \mid x \notin f(x)\}.$$

Then $f(y) = Y$ for some $y \in X$. Consider the possibilities $y \in Y$ and $y \notin Y$.

51. Suppose that there is a one-to-one function f from X to Y. Let R be the range of f and choose $a \in X$. If $y \in R$, let $g(y) = f^{-1}(y)$. If $y \in Y - R$, let $g(y) = a$. Then g is a function from Y onto X.

Suppose that there is a function f from Y onto X. For each $x \in X$, select one y with $f(y) = x$. Define $g(x) = y$. Then g is a one-to-one function from X to Y.

53. λ, b, a, c, ba, ab, bc, bab, abc, $babc$

55. f is not a binary operator since the range of f is not contained in X.

57. f is not a binary operator since $f(x, 0)$ is not defined.

59. $g(x) = -x$

Section 1.5

1. Change LARGE to SMALL in Algorithm 1.5.1. In step 2, change $>$ to $<$. Alter comments appropriately.

3. 1. Set $J := 0$, $I := 1$.

    2. If $S(I) = $ KEY, set $J := I$ and stop. (J is the first occurrence of the value KEY.)

    3. Set $I := I + 1$. If $I > N$, then stop (KEY is not found); otherwise, go to step 2.

5. 1. Set $I := 1$, $M := \dfrac{N}{2}$.

    2. If $I > M$, stop (the sequence is reversed); otherwise, set TEMP $:= S(I)$, $S(I) := S(N - I + 1)$, $S(N - I + 1) := $ TEMP.

    3. Set $I := I + 1$ and go to step 2.

7. 1. Set $I := N + 1$.

    2. Use the algorithm of Exercise 6 to determine whether I is prime. If I is prime, stop; otherwise, set $I := I + 1$ and go to step 2.

9. Set $I = 1$ and $S = 0$. Since $I \leq N$ is true, we set $S = S + I = 1$ and $I = 2$. Since $I \leq N$ is true, we set $S = S + I = 3$ and $I = 3$. Since $I \leq N$ is true, we set $S = S + I = 6$ and $I = 4$. This time $I \leq N$ is false, so we output S, whose value is 6.

11. Assume that values for M and N are given.

    1. Set $R := $ remainder when M is divided by N.

    2. If $R = 0$, stop (N contains the output); otherwise, set $M := N$ and $N := R$ and go to step 1.

13. The greatest common divisor of M and N

14. (For Exercise 11) We divide M by N to get a remainder R. By definition, $0 \leq R < N$. If $R \neq 0$, we set $M = N$ and $N = R$; thus the new value of N is less than the previous value of N. It follows that the new value of R is less than the previous value of R. Eventually, $R = 0$ and the algorithm terminates.

17. Finiteness, input, output, generality

18. Assume that $S(N)S(N - 1) \cdots S(1)$ and $T(N)T(N - 1) \cdots T(1)$ are the decimal representations of the two numbers to be added.

    1. Set $I := 1$ and $C := 0$.

    2. If $I > N$, set $U(N + 1) := C$ and stop. (The sum is in U.)

    3. Let XY be the decimal representation of the sum $C + S(I) + T(I)$.

    4. Set $U(I) := Y$, $C := X$, and $I := I + 1$. Go to step 2.

Section 1.6
1. $O(n)$ **3.** $O(n^3)$ **5.** $O(n \lg n)$ **7.** $O(n^2)$ **9.** $O(n \lg n)$ **11.** $O(n)$
13. $O(n^2)$ **15.** $O(n^{5/2})$ **17.** $O(N^2)$ **19.** $O(N^2)$ **21.** $O(\lg N)$
23. Use the change-of-base formula for logarithms.
25. We are given

$$|f(n)| \le C_1|h(n)| \qquad \text{and} \qquad |g(n)| \le C_2|h(n)|,$$

for all but finitely many positive integers n. Let $C = \max \{C_1, C_2\}$. Then

$$|f(n)| + |g(n)| \le |f(n)| + |g(n)| \le C_1|h(n)| + C_2|h(n)| \le C|h(n)| + C|h(n)| = 2C|h(n)|$$

for all but finitely many positive integers n.
  The second relation is proved similarly.
27. $2^n = 2(2^{n-1}) = 2(2 \cdot 2 \cdot \cdots \cdot 2) \le 2(2 \cdot 3 \cdot \cdots \cdot n) = 2(n!)$
29. $n! = n(n - 1) \cdot \cdots \cdot 2 \cdot 1 \le n \cdot n \cdot \cdots \cdot n \cdot n = n^n$, so

$$\lg (n!) \le n \lg n.$$

31. $n^2 \le 3n^2 - 1 \le 3n^2$
33. Since $(2n - 1)(n - 1) = 2 + 1/(n - 1)$, $(2n - 1)/(n - 1) \le 3$. Therefore,

$$12n = 2(6n) \le 2(6n + 1) = \frac{(2n - 2)(6n + 1)}{n - 1}$$

$$\le \frac{(2n - 1)(6n + 1)}{n - 1} \le 3(6n + 1) \le 3(6n + 6n) = 36n.$$

35. No

Section 1.7
1. BASIS. $1 = 1^2$
  INDUCTIVE STEP.   Assume true for n.

$$1 + \cdots + (2n - 1) + (2n + 1) = n^2 + 2n + 1 = (n + 1)^2$$

4. BASIS. $1^2 = 1 \cdot 2 \cdot \frac{3}{6}$
  INDUCTIVE STEP.   Assume true for n.

$$1^2 + \cdots + n^2 + (n + 1)^2 = \frac{n(n + 1)(2n + 1)}{6} + (n + 1)^2$$

$$= \frac{(n + 1)(n + 2)(2n + 3)}{6}$$

7. BASIS. $1/(1 \cdot 3) = \frac{1}{3}$
  INDUCTIVE STEP.   Assume true for n.

$$\frac{1}{1 \cdot 3} + \cdots + \frac{1}{(2n - 1)(2n + 1)} + \frac{1}{(2n + 1)(2n + 3)}$$

$$= \frac{n}{2n + 1} + \frac{1}{(2n + 1)(2n + 3)}$$

$$= \frac{n + 1}{2n + 3}$$

10. BASIS. $\cos x = \dfrac{\cos [(x/2) \cdot 2] \sin (x/2)}{\sin (x/2)}$

INDUCTIVE STEP. Assume true for n. Then

$$\cos x + \cdots + \cos nx + \cos (n + 1)x$$
$$= \frac{\cos [(x/2)(n + 1)] \sin (nx/2)}{\sin (x/2)} + \cos (n + 1)x. \quad (*)$$

We must show that the right-hand side of $(*)$ is equal to

$$\frac{\cos [(x/2)(n + 2)] \sin [(n + 1)x/2]}{\sin (x/2)}.$$

This is the same as showing that [after multiplying by the term $\sin (x/2)$]

$$\cos \left[\frac{x}{2} (n + 1)\right] \sin \frac{nx}{2} + \cos (n + 1)x \sin \frac{x}{2} = \cos \left[\frac{x}{2} (n + 2)\right] \sin \left[\frac{(n + 1)x}{2}\right].$$

If we let $\alpha = (x/2)(n + 1)$ and $\beta = x/2$, we must show that

$$\cos \alpha \sin (\alpha - \beta) + \cos 2\alpha \sin \beta = \cos (\alpha + \beta) \sin \alpha.$$

This last equation can be verified by reducing each side to terms involving α and β.

13. BASIS. $\frac{1}{2} \leq \frac{1}{2}$

INDUCTIVE STEP. Assume true for n.

$$\frac{1 \cdot 3 \cdot 5 \cdots (2n - 1)(2n + 1)}{2 \cdot 4 \cdot 6 \cdots (2n)(2n + 2)} \geq \frac{1}{2n} \cdot \frac{2n + 1}{2n + 2} = \frac{2n + 1}{2n} \cdot \frac{1}{2n + 2} \geq \frac{1}{2n + 2}$$

16. BASIS. $(n = 4) \ 2^4 = 16 \geq 16 = 4^2$

INDUCTIVE STEP. Assume true for n.

$$(n + 1)^2 = n^2 + 2n + 1 \leq 2^n + 2n + 1$$
$$\leq 2^n + 2^n \quad \text{by Exercise 15}$$
$$= 2^{n+1}.$$

17. At the Inductive Step, apply the case n to $\{a_1, \ldots, a_{2n}\}$ and to $\{a_{2n+1}, \ldots, a_{2n+1}\}$. Multiply the inequalities and then apply the case $n = 1$.

19. BASIS. $7^1 - 1 = 6$ is divisible by 6.

INDUCTIVE STEP. Suppose that 6 divides $7^n - 1$. Now

$$7^{n+1} - 1 = 7 \cdot 7^n - 1 = 7^n - 1 + 6 \cdot 7^n.$$

Since 6 divides both $7^n - 1$ and $6 \cdot 7^n$, it divides their sum, which is $7^{n+1} - 1$.

21. BASIS. $6 \cdot 7^1 - 2 \cdot 3^1 = 36$ is divisible by 4.

INDUCTIVE STEP. Suppose that 4 divides $6 \cdot 7^n - 2 \cdot 3^n$. Now
$$6 \cdot 7^{n+1} - 2 \cdot 3^{n+1} = 7(6 \cdot 7^n) - 3(2 \cdot 3^n)$$
$$= 6 \cdot 7^n - 2 \cdot 3^n + 6(6 \cdot 7^n) - 2(2 \cdot 3^n)$$
$$= 6 \cdot 7^n - 2 \cdot 3^n + 36 \cdot 7^n - 4 \cdot 3^n.$$

Since 4 divides $6 \cdot 7^n - 2 \cdot 3^n$, $36 \cdot 7^n$, and $-4 \cdot 3^n$, it divides their sum, which is $6 \cdot 7^{n+1} - 2 \cdot 3^{n+1}$.

23. $n/(n + 1)$

27. At the Inductive Step when the $(n + 1)$st line is added, because of the assumptions, the line will intersect each of the other n lines. Now, imagine traveling along the $(n + 1)$st line. Each time we pass through one of the original regions, it is divided into two regions.

29. Verify directly the cases 24–28. Assume that the statement is true for postage i satisfying $24 \leq i < n$. We must show that we can achieve n cents postage using only 5-cent and 7-cent stamps. We may assume that $n > 28$. Then $n > n - 5 > 23$. By assumption, we can achieve $n - 5$ cents postage using 5-cent and 7-cent stamps. Add a 5-cent stamp to obtain n cents postage.

31. If

$$\frac{p}{q} = \frac{1}{n_1} + \frac{1}{n_2} + \cdots + \frac{1}{n_k}$$

where $n_1 < n_2 < \cdots < n_k$, then another representation is

$$\frac{p}{q} = \frac{1}{n_1} + \frac{1}{n_2} + \cdots + \frac{1}{n_{k-1}} + \frac{1}{n_k + 1} + \frac{1}{n_k(n_k + 1)}.$$

33. $\frac{3}{8} = \frac{1}{3} + \frac{1}{24}$, $\frac{5}{7} = \frac{1}{2} + \frac{1}{5} + \frac{1}{70}$, $\frac{13}{19} = \frac{1}{2} + \frac{1}{6} + \frac{1}{57}$

35. (a) $S_1 = 0 \neq 2$; $2 + \cdots + 2n + 2(n + 1) = S_n + 2n + 2 = (n + 2)(n - 1) + 2n + 2 = (n + 3)n = S_{n+1}$.

(b) We must have $S'_n = S'_{n-1} + 2n$; thus

$$S'_n = S'_{n-1} + 2n = [S'_{n-2} + 2(n - 1)] + 2n$$

$$= S'_{n-2} + 2n + 2(n - 1)$$

$$= S'_{n-3} + 2n + 2(n - 1) + 2(n - 2) = \cdots$$

$$= S'_1 + 2[n + (n - 1) + \cdots + 2]$$

$$= C' + 2\left[\frac{n(n + 1)}{2} - 1\right] = n^2 + n + C.$$

37. In the argument given, the constant is dependent on n.

38. Let the nth statement be: "If $n \in X$, then X contains a least element."

39. Suppose, by way of contradiction, that some statement $S(n)$ is false. Let X be the set of positive integers n for which $S(n)$ is false. Now apply the Well-Ordering Theorem.

40. The strong form of induction clearly implies the form of induction where the Inductive Step is: "If $S(n)$ is true, then $S(n + 1)$ is true." For the converse, use Exercises 38 and 39.

Section 2.1

1. $8 \cdot 4 \cdot 5$ **3.** 3 **5.** $6 + 2$ **7.** 10 **9.** 5^2 **11.** $10 \cdot 5$ **13.** $50 \cdot 49$
15. 2^4 **17.** $3 \cdot 2^6$ **19.** $(\frac{1}{2})(8 \cdot 7)$ **21.** 2^4 **23.** $4 \cdot 3 \cdot 2$
25. $6 \cdot 5 \cdot 4 - 3 \cdot 2 \cdot 4$ **27.** $3 \cdot 2 \cdot 4 + 4 \cdot 3 \cdot 2$ **29.** 5^3 **31.** 5^2
33. 4^3 **35.** $5^3 - 4^3$ **37.** $200 - 5 + 1$ **39.** $(\frac{1}{2})196$ **41.** $200 - 72$
43. One one-digit number contains 7. The distinct two-digit numbers that contain 7 are 17, 27, . . . , 97 and 70, 71, . . . , 76, 78, 79. There are 18 of these. The distinct three-digit numbers that contain 7 are 107 and $1xy$, where xy is one of the two-digit numbers listed above. The answer is $1 + 18 + 19$.
45. $1 + 1 \cdot 9 \cdot 9 - 2$ **47.** $2 + 3 + \cdots + 9 = 44$ **49.** 10! **51.** $5!5!$
53. First, arrange the computer science and mathematics books. This can be done in 8! ways. Then place the art books in two of the nine in-between or end positions. This can be done in $9 \cdot 8$ ways. The answer is $(8!)9 \cdot 8$.
55. The first letter can be selected in 26 ways. The second alphanumeric can be

selected or not selected in 37 ways. The last symbol can be selected or not selected in five ways. Thus the total number of possibilities is $(26 \cdot 37 - 5)5$. (We must subtract the five disallowed possibilities.)

57. $(26 + 26 \cdot 36 + 26 \cdot 36^2 + \cdots + 26 \cdot 36^7)(1 + 26 + 26 \cdot 36 + 26 \cdot 36^2)(4)$

59. 2^{10}

61. A subset X has n elements or less if and only if $\overline{X}$ has more than n elements. Thus exactly half of the subsets have n elements or less. Therefore, the number of subsets is $(\frac{1}{2})2^{2n+1} = 2^{2n}$.

Section 2.2

1. $P(4, 3) = 4 \cdot 3 \cdot 2$ **3.** $C(4, 3) = 4!/(3!1!) = 4$

5.

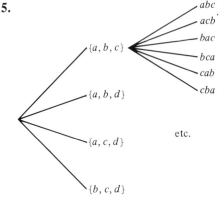

etc.

7. $C(11, 3)$ **9.** $C(6, 3)C(7, 4)$ **11.** $6C(7, 3) + C(7, 4)$

13. $C(13, 4) - C(11, 2)$ **15.** $C(8, 3)$

17. $2^8 - [1 + 8 + C(7, 2) + C(6, 3) + C(5, 4)]$

19. $4C(13, 5)$ **21.** $4C(13, 2)13^3$ **23.** $4 \cdot 9$ **25.** $13 \cdot 12 \cdot C(4, 3)C(4, 2)$

27. $C(52, 13)$ **29.** $C(4, 2)[C(26, 13) - 2]$

31. $C(13, 5)C(13, 4)C(13, 3)C(13, 1)$ **33.** $C(4, 3)C(13, 4)^3C(13, 1)$

35. The Second Counting Principle must be applied to a family of pairwise disjoint sets. Here the sets involved are *not* pairwise disjoint. For example, if X is the set of hands containing clubs, diamonds, and spades and Y is the set of hands containing clubs, diamonds, and hearts, $X \cap Y$ contains hands that contain clubs and diamonds.

37. $2 \cdot 4!$ **39.** $5! - (2 \cdot 4! - 3!)$ **41.** $C(10, 3)$ **43.** 2^9

45. The solution counts *ordered* hands.   **47.** $C(46, 4)$ **49.** $C(50, 4) - C(46, 4)$

51. $5!/2$ **53.** $5!P(6, 5)$ **55.** $4!5!$ **57.** $7!P(8, 5)$

59. Each route can be described by a string of four R's (right) and four U's (up). For example, the route shown can be described by $RRURUUUR$. Count the number of such strings.

61. Look at the formula for $C(n, k)$.

Section 2.3

1. $5!$ **3.** $12!/[4!2!]$ **5.** $C(10 + 3 - 1, 10)$

7. $C(4 + 3 - 1, 4)$ **9.** $C(8 + 2 - 1, 8)$

11. Four, since the possibilities are $(0, 0)$, $(2, 1)$, $(4, 2)$, and $(6, 3)$, where the pair (r, g) designates r red and g green balls.

13. $C(12 + 3 - 1, 12)$ **15.** $C(13 + 2 - 1, 13)$

17. There are $C(14 + 3 - 1, 14)$ solutions satisfying $0 \leq x_1, 1 \leq x_2, 0 \leq x_3$. Of these, $C(8 + 3 - 1, 8)$ have $x_1 \geq 6$; $C(6 + 3 - 1, 6)$ have $x_2 \geq 9$; and there is one with $x_1 \geq 6$ and $x_2 \geq 9$. Thus there are $C(8 + 3 - 1, 8) + C(6 + 3 - 1, 6) - 1$ solutions with $x_1 \geq 6$ or $x_2 \geq 9$. Therefore, there are $C(14 + 3 - 1, 14) - [C(8 + 3 - 1, 8) + C(6 + 3 - 1, 6) - 1]$ of the desired type.

19. We must count the number of solutions of

$$x_1 + x_2 + x_3 + x_4 + x_5 + x_6 = 15$$

satisfying $0 \leq x_i \leq 9$, $i = 1, \ldots, 6$. There are $C(15 + 6 - 1, 15)$ solutions with $x_i \geq 0$, $i = 1, \ldots, 6$. There are $C(5 + 6 - 1, 5)$ with $x_1 \geq 10$. There are $6C(5 + 6 - 1, 5)$ with some $x_i \geq 10$. (Note that there is no double counting, since we cannot have $x_i \geq 10$ and $x_j \geq 10$, $i \neq j$.) Thus the solution is $C(15 + 6 - 1, 15) - 6C(5 + 6 - 1, 5)$.

21. $52!/(13!)^4$ **23.** $C(7 + 2 - 1, 2)$ **25.** $C(5 + 3 - 1, 5)$ **27.** $C(20, 5)^2$
29. $[C(20, 5) - C(14, 5)][C(14, 5) + 6C(14, 4)]$
31. $C(15 + 6 - 1, 15)$ **33.** $C(12, 10)$
35. Consider the number of orderings of kn objects where there are n identical objects of each of k types.
37. Apply the result of Example 2.3.9 to the inner $K - 1$ nested loops of that example. Next, write out the number of iterations for $I_1 = 1$; then $I_1 = 2$; and so on. By Example 2.3.9, this sum is equal to $C(K + N - 1, K)$.
39. See *Math. Mag.*, May 1975.

Section 2.4

1. $x^4 + 4x^3y + 6x^2y^2 + 4xy^3 + y^4$ **3.** $C(11, 7)x^4y^7$
5. $C(10, 2)C(8, 3) = 10!/(2!3!5!)$ **7.** $C(5, 2)$
9. $C(7, 3) + C(5, 2)$, since

$$(a + \sqrt{ax} + x)^2(a + x)^5 = [(a + x) + \sqrt{ax}]^2(a + x)^5$$

$$= (a + x)^7 + 2\sqrt{ax}(a + x)^6 + ax(a + x)^5.$$

11. $C(12 + 4 - 1, 12)$ **13.** 1 8 28 56 70 56 28 8 1
15. Take $a = -1$ and $b = 1$ in the Binomial Theorem.

17. $C(n, k - 1) + C(n, k) = \dfrac{n!}{(k - 1)!(n - k + 1)!} + \dfrac{n!}{k!(n - k)!}$

$$= \dfrac{(n!)k}{k!(n - k + 1)!} + \dfrac{(n!)(n - k + 1)}{k!(n - k + 1)!}$$

$$= \dfrac{(n!)(n + 1)}{k!(n - k + 1)!} = C(n + 1, k).$$

19. The number of solutions in nonnegative integers of

$$x_1 + x_2 + \cdots + x_{k+2} = n - k$$

is $C(k + 2 + n - k - 1, n - k) = C(n + 1, k + 1)$. The number of solutions is also the number of solutions $C(k + 1 + n - k - 1, n - k) = C(n, k)$ with $x_{k+2} = 0$ plus the number of solutions $C(k + 1 + n - k - 1 - 1, n - k - 1) = C(n - 1, k)$ with $x_{k+2} = 1$ plus $\cdots$ plus the number of solutions $C(k + 1 + 0 - 1, 0) = C(k, k)$ with $x_{k+2} = n - k$. The result now follows.
21. Use $k^2 = 2C(k, 2) + C(k, 1)$.
23. Use Exercise 15 and equation (2.4.3).
24. Imitate the combinatorial proof of the Binomial Theorem.
27. Think of $C(n, k)^2$ as $C(n, k)C(n, n - k)$. Let X and Y be disjoint sets each having

n elements. Now, $C(2n, n)$ is the number of ways of picking n-element subsets of $X \cup Y$. Picking an n-element subset of $X \cup Y$ is the same as picking a k-element subset of X and an $(n - k)$-element subset of Y.

28. Use induction on m.

29. Use Exercise 28 to write

$$n^m = \sum_{i=0}^{m} b_i C(n, i).$$

Now apply Theorem 2.4.4.

Section 2.5

1. (a) $A_n = (1.14)A_{n-1}, A_0 = 2000$ **(b)** $A_n = (1.14)^n 2000$

3. $P_n = nP_{n-1}, P_1 = 1$

5. Suppose that we have n dollars. If we buy orange juice the first day, we have $n - 1$ dollars left, which may be spent in C_{n-1} ways. Similarly, if the first day we buy milk or beer, there are C_{n-2} ways to spend the remaining dollars. Since these cases are disjoint, $C_n = C_{n-1} + 2C_{n-2}$.

7. n lines divide the plane into C_n regions. Because of the assumptions, when the $(n + 1)$st line L is added, it will intersect the other n lines. If we imagine traveling along L, each time we pass through one of the original regions, we divide it into two regions. Since we pass through $n + 1$ regions, $C_{n+1} = C_n + n + 1$.

9. $S_n = \frac{2}{3}[1 - (-\frac{1}{2})^{n-1}]$

11. Use induction.

     BASIS. $f_0 f_2 - 1 = 1 \cdot 2 - 1 = 1 = f_1^2$.

     INDUCTIVE STEP

$$f_n f_{n+2} + (-1)^{n+1} = f_n(f_{n+1} + f_n) + (-1)^{n+1}$$
$$= f_n f_{n+1} + f_n^2 + (-1)^{n+1}$$
$$= f_n f_{n+1} + f_{n-1} f_{n+1} + (-1)^n + (-1)^{n+1}$$
$$= f_{n+1}(f_n + f_{n-1}) = f_{n+1}^2.$$

13. Show that $C_n = C_{n-1} + C_{n-2}, C_1 = 2, C_2 = 3$.

15. A string α of length n with $C(\alpha) \leq 2$ either begins with 1 (there are S_{n-1} of these), 01 (there are S_{n-2} of these), or 001 (there are S_{n-3} of these). Thus $S_n = S_{n-1} + S_{n-2} + S_{n-3}$.

16. Let α be an n-bit string not containing 010. We count the number of such strings having i leading 0's. For $i = 0$, there are S_{n-1} such strings. For $i = 1$, the string begins 011, so there are S_{n-3} such strings. Similarly, for $i = 2$, there are S_{n-4} such strings; ... for $i = n - 3$, there are S_1 such strings. For $i = n - 2, n - 1$, or n, there is one such string. Thus

$$S_n = S_{n-1} + S_{n-3} + S_{n-4} + \cdots + S_1 + 3.$$

19. A function f from $X = \{1, \ldots, n\}$ into X will be denoted $(i_1, i_2, \ldots, i_n)$, which means that $f(k) = i_k$. The problem then is to count the number of ways to select $i_1, \ldots, i_n$ so that if i occurs, so do $1, 2, \ldots, i - 1$.

     We shall count the number of such functions having exactly j 1's. Such functions can be constructed in two steps: Pick the positions for the j 1's; then place the other numbers. There are $C(n, j)$ ways to place the 1's. The remaining numbers must be selected so that if i appears, so do $1, \ldots, i - 1$. There are F_{n-j} ways to select the remaining numbers, since the remaining numbers must be selected from $\{2, \ldots, n\}$. Thus there are $C(n, j)F_{n-j}$ functions of the desired type having exactly j 1's. Therefore, the total number of functions from X into X having the property that if i is in the range of f, then so are $1, \ldots, i - 1$, is

$$\sum_{j=1}^{n} C(n, j)F_{n-j} = \sum_{j=1}^{n} C(n, n - j)F_{n-j} = \sum_{j=0}^{n-1} C(n, j)F_{j}.$$

21. Use Theorem 2.4.4 to write $S(k, n) = \sum_{i=1}^{n} S(k - 1, i)$.

24. The input to the algorithm is the board with n disks on one peg denoted X and a designated peg $Y \neq X$. The algorithm moves the disks from X to Y.

    1. If $n = 1$, move the disk from X to Y and return.

    2. Let Z be the peg that is not X or Y.

    3. Fix the bottom disk. Call this algorithm for the case of $n - 1$ disks on peg X and designated peg Z.

    4. ($n - 1$ disks are now on peg Z.) Move the disk on peg X to peg Y.

    5. Call this algorithm for the case of $n - 1$ disks on peg Z and designated peg Y. Return.

25. 7, 9, 29

27. BASIS. $A(2, 0) = A(1, 1) = 3$ by Exercise 26.

    INDUCTIVE STEP. Assume the case n.

$$A(2, n + 1) = A(1, A(2, n))$$
$$= A(1, 3 + 2n) \qquad \text{by assumption}$$
$$= 3 + 2n + 2 \qquad \text{by Exercise 26}$$
$$= 2n + 5$$

28. $A(3, n) = 2^{n+3} - 3$

31. BASIS. $AO(2, 2, 1) = 2(1) = 2$

    INDUCTIVE STEP. Assume the case x.

$$AO(x + 1, 2, 1) = AO(x, 2, AO(x + 1, 2, 0)) = AO(x, 2, 1) = 2.$$

33. The cases $x = 0, 1, 2$ are Exercise 30; thus the given statement must be proved for $x \geq 3$, $y \geq 0$. The proof is by induction on x.

    BASIS. The case $x = 3$ is Exercise 30.

    INDUCTIVE STEP. Assume the given statement. We must now show that

$$A(x + 1, y) = AO(x + 1, 2, y + 3) - 3 \qquad \text{for all } y \geq 0.$$

Use induction on y to establish this equation.

36. Use the series

$$e^x = 1 + \frac{x}{1!} + \frac{x^2}{2!} + \cdots$$

with $x = -1$ together with the following estimate for the alternating series

$$a_1 - a_2 + a_3 - \cdots,$$

where $\{a_n\}$ is decreasing to zero: If

$$S = a_1 - a_2 + a_3 - \cdots,$$

then

$$\left|S - (a_1 - a_2 + \cdots + (-1)^{n+1}a_n)\right| \le a_{n+1}.$$

37. Use $S_n = C(n, k)D_{n-k}$ and (2.5.9).

Section 2.6

1. Yes; order 1 **3.** No **5.** Yes; order 3 **7.** No **9.** Yes; order 2

11. $a_n = -3a_{n-1} = (-3)^2 a_{n-2} = \cdots = (-3)^n a_0 = 2(-3)^n$

13. $a_n = a_{n-1} + n = a_{n-2} + (n - 1) + n = \cdots$

$$= a_0 + 1 + 2 + \cdots + n = 1 + 2 + \cdots + n = \frac{n(n + 1)}{2}$$

15. The roots of $2x^2 - 7x + 3$ are $x = \frac{1}{2}, 3$. Thus the general solution is

$$a_n = c_1(\tfrac{1}{2})^n + c_2 3^n.$$

To satisfy the initial conditions, we must have

$$1 = a_0 = c_1 + c_2$$

$$1 = a_1 = \frac{c_1}{2} + 3c_2.$$

Solving this system, we obtain $c_1 = \frac{4}{5}, c_2 = \frac{1}{5}$. Thus the solution is

$$a_n = \frac{2^{2-n} + 3^n}{5}.$$

17. $a_n = a_{n-1} + 1 + 2^{n-1}$

$$= (a_{n-2} + 1 + 2^{n-2}) + 1 + 2^{n-1}$$

$$= a_{n-2} + 2 + 2^{n-1} + 2^{n-2} = \cdots$$

$$= a_0 + n + 2^{n-1} + 2^{n-2} + \cdots + 1 = n + 2^n - 1$$

19. $a_n = \dfrac{3 - 5\sqrt{3}i}{6}\left(\dfrac{-1 + \sqrt{3}i}{2}\right)^n + \dfrac{3 + 5\sqrt{3}i}{6}\left(\dfrac{-1 - \sqrt{3}i}{2}\right)^n$

21. Solving the recurrence relation $C_n = C_{n-1} + 2C_{n-2}$, with initial conditions $C_1 = 1, C_2 = 3$, gives $C_n = \frac{1}{3}[(-1)^n + 2^{n+1}]$.

23. $S_n = \frac{2}{3}[1 - (-\frac{1}{2})^{n-1}]$

25. Let $b_n = \sqrt{a_n}$; then solve $b_n = b_{n-1} + 2b_{n-2}$. This yields $a_n = \frac{1}{9}[2^{n+1} + (-1)^n]^2$.

27. Set $b_n = a_n/n!$ to obtain $b_n = -2b_{n-1} + 3b_{n-2}$. Solving gives $a_n = n!b_n = (n!/4)[5 - (-3)^n]$.

29. The argument is identical to that given in Theorem 2.6.6.

31. Let $a_n = r^n$. Then

$$c_1 a_{n-1} + \cdots + c_k a_{n-k} = c_1 r^{n-1} + \cdots + c_k r^{n-k}$$

$$= r^n = a_n.$$

Section 2.7

1. 13 **3.** -14 **5.** 16

7. Let $n = \lfloor x - 1 \rfloor$. Then $n \le x - 1 < n + 1$. Therefore, $n + 1 \le x < n + 2$. Therefore, $\lfloor x \rfloor = n + 1$. Thus $\lfloor x \rfloor - 1 = n = \lfloor x - 1 \rfloor$. The second equation is proved similarly.

9. Let $a = \lg n$ and $k = \lfloor \lg n \rfloor$. Show that $k < a$. Then $2^k < 2^a = n$. Thus $2^k \le n - 1$. Therefore, $k \le \lg (n - 1) < \lg n$. The conclusion follows.

15. 1

17. (a) If $I + J$ is even, M is unchanged. If $I + J$ is odd, we simply divide the array in half by letting the upper part of the array contain one less than the lower part of the array rather than the other way around, which corresponds to $M = \lfloor (I + J)/2 \rfloor$.
(b) $\lfloor N/2 \rfloor$

19. At step 2, $M = 2$. At step 3, we call Algorithm 2.7.6 with input $S(1)$, $S(2)$. By Exercise 18, two comparisons are required. We next call Algorithm 2.7.6 with input $S(3)$ which, by Exercise 18, requires zero comparisons. At steps 4 and 5, two more comparisons are required. Therefore, $a_3 = 4$.

21. If $N = 2^k$, then (2.7.9) becomes

$$a_{2^{k-1}} + a_{2^{k-1}} + 2 = a_{2^k}.$$

Letting $b_k = a_{2^k}$, we obtain

$$b_k = 2b_{k-1} + 2 = 2[2b_{k-2} + 2] + 2 = 2^2 b_{k-2} + 2^2 + 2$$

$$= \cdots = 2^k b_0 + 2^k + 2^{k-1} + \cdots + 2$$

$$= 2^k + 2^{k-1} + \cdots + 2 = \frac{2^{k+1} - 2}{2 - 1}.$$

Thus $a_N = a_{2^k} = b_k = 2^{k+1} - 2 = 2N - 2$.

23. $a_1 = 0$ since, if $N = 1$, we return at step 1a. $a_2 = 1$ since, if $N = 2$, there is one comparison at step 1c before we return.

25. Consider the cases N odd and N even.

28. We will use the following fact, which can be verified by considering the cases x even and x odd:

$$\left\lceil \frac{3x}{2} - 2 \right\rceil + \left\lceil \frac{3(x + 1)}{2} - 2 \right\rceil = 3x - 2 \qquad \text{for } x = 1, 2, \ldots .$$

Let a_N denote the number of comparisons required by the algorithm in the worst case. We will show that $a_N \le \lceil (3N/2) - 2 \rceil$ by induction. The cases $N = 1$ and 2 are left to the reader. (The case $N = 2$ is the *Basis Step*.)

INDUCTIVE STEP. Assume that $a_k \le \lceil (3k/2) - 2 \rceil$ for $2 \le k < N$. We must show the inequality holds for $k = N$.

In case N is odd, the algorithm partitions the array into subclasses of size $(N - 1)/2$ and $(N + 1)/2$. Now

$$a_N = a_{(N-1)/2} + a_{(N+1)/2} + 2 \le \left\lceil \frac{3}{2} \frac{N - 1}{2} - 2 \right\rceil$$

$$+ \left\lceil \frac{3}{2} \frac{N + 1}{2} - 2 \right\rceil + 2 = 3 \frac{N - 1}{2} - 2 + 2$$

$$= \frac{3N}{2} - \frac{3}{2} = \left\lceil \frac{3N}{2} - 2 \right\rceil.$$

The case N even is treated similarly.

29. $a_N = a_{\lfloor (1+N)/2 \rfloor} + a_{\lfloor N/2 \rfloor} + 3$

31. Let $C_k = a_{2^k}$. Then $C_k = 2C_{k-1} + 3$. If $N = 2^k$,

$$a_N = C_k = 2C_{k-1} + 3 = 2(2C_{k-2} + 3) + 3 = \cdots$$
$$= 2^k C_0 + 3(2^{k-1} + 2^{k-2} + \cdots + 1) = 2^k 0 + 3(2^k - 1)$$
$$= 3(N - 1).$$

33. Use the method of Exercise 31.

35. We will show that $a_N \le a_{N+1}$, $N = 1, 2, \ldots$. We have the recurrence relation

$$a_N = a_{\lfloor (1+N)/2 \rfloor} + a_{\lfloor N/2 \rfloor} + b_{\lfloor (1+N)/2 \rfloor, \lfloor N/2 \rfloor}.$$

BASIS. $a_2 = 2a_1 + b_{1,1} \ge 2a_1 \ge a_1$.

INDUCTIVE STEP. Assume that the statement holds for $k < N$. In case N is even, we have $a_N = 2a_{N/2} + b_{N/2, N/2}$; so $a_{N+1} = a_{(N+2)/2} + a_{N/2} + b_{(N+2)/2, N/2} \ge a_{N/2} + a_{N/2} + b_{N/2, N/2} = a_N$. The case N odd is similar.

36. Use Exercise 33 to show that if N is a power of 2, $a_N = N \lg N$. Now let N be arbitrary. Choose k so that $2^k < m \le 2^{k+1}$. By Exercise 35,

$$a_m \le a_{2^{k+1}}.$$

Now

$$a_{2^{k+1}} = 2^{k+1}(k + 1) \le 2^{k+1}(k + k) = 4(2^k k) \le 4 m \lg m.$$

Section 3.1

1. (a) No; 6   **(b)** Yes   **3.** No. There are vertices of odd degree.

5. The degrees of $v_1, \ldots, v_9$ are 2, 2, 2, 4, 6, 2, 4, 4, and 2, respectively. It is possible to traverse the graph. One circuit is

$$(v_1, v_2, v_5, v_6, v_9, v_8, v_7, v_4, v_7, v_5, v_8, v_4, v_5, v_3, v_1).$$

7. No. Since $\delta(C) = 3$, the path would have to either begin or end at C. But a similar remark can be made for A, B, and D. This is impossible.

9.

K_3 K_4 K_5

11. When n is odd. In K_n every vertex has degree $n - 1$; thus every vertex has even degree precisely when n is odd.

13. mn

15. $(a, e, d, c, b, h, i, j, k, l, m, n, o, t, s, r, q, p, g, f, a)$

17. $(a, b, c, d, e, f, n, p, m, l, k, j, o, i, h, g, a)$

19. We would have to eliminate one edge at f, three edges at c, one edge at b, one edge at i, three edges at j, and three edges at m, leaving 15 edges. Since there are 16 vertices, a Hamiltonian circuit would have 16 edges.

21. $(a, b, c, j, i, m, k, d, e, f, l, g, h, a)$

23. There is no Hamiltonian circuit. We would have to eliminate two edges at c, three edges at e, and one edge at f, leaving six edges. Since there are seven vertices, a Hamiltonian circuit would have seven edges.

25.

27. If n is even and $m > 1$ or if m is even and $n > 1$, there is a Hamiltonian circuit. The sketch shows the solution in case n is even.

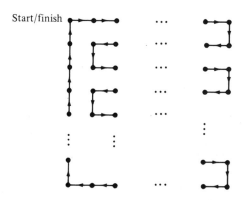

If $n = 1$ or if $m = 1$, there is no circuit and, in particular, there is no Hamiltonian circuit. Suppose that n and m are both odd and that the graph has a Hamiltonian circuit. Since there are nm vertices, this circuit has nm edges; therefore, the Hamiltonian circuit contains an odd number of edges. However, we note that in a Hamiltonian circuit, there must be as many "up" edges as "down" edges and as many "left" edges as "right" edges. Thus a Hamiltonian circuit must have an even number of edges. This contradiction shows that if n and m are both odd, the graph does not have a Hamiltonian circuit.

29. When $m = n$ and $n > 1$

33. Let C be a Hamiltonian circuit in G. Consider a traversal of C. When we traverse an edge from a vertex $v_1 \in V_1$ to a vertex $v_2 \in V_2$, this uniquely associates one vertex v_2 with v_1. Since C traverses all vertices, $|V_1| = |V_2|$.

35. Two classes

37. Every member of a class C is similar to every other member in C.

41. There are six choices for the top and having chosen the top, there are four choices for the front, for a total of $6 \cdot 4$ choices.

43. One edge can be chosen in $C(2 + 4 - 1, 2) = 10$ ways. The three edges labeled 1 can be chosen in $C(3 + 10 - 1, 3) = 220$ ways. Thus the total number of graphs is 220^4.

45. (a)

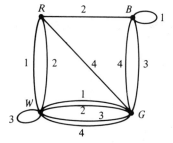

(b)

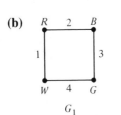
G_1

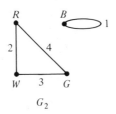
G_2

(c) The other subgraphs are

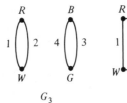
G_3

G_4

(d) Since G_1 meets G_3 and G_4; G_2 meets G_3 and G_4; and G_3 meets G_4; only G_1 and G_2 are disjoint.

47.

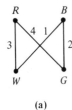
(a)

(b)

57. 5

59. The solution is unique.

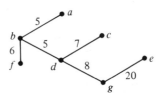

Section 3.2

1.
$$\begin{array}{c c c c c c} & a & b & c & d & e \\ a & 0 & 1 & 1 & 1 & 1 \\ b & 1 & 0 & 1 & 0 & 0 \\ c & 1 & 1 & 0 & 1 & 1 \\ d & 1 & 0 & 1 & 0 & 1 \\ e & 1 & 0 & 1 & 1 & 0 \end{array}$$

3.
$$\begin{array}{c c c c c c} & a & b & c & d & e \\ a & 0 & 1 & 0 & 0 & 0 \\ b & 1 & 0 & 0 & 0 & 0 \\ c & 0 & 0 & 0 & 1 & 1 \\ d & 0 & 0 & 1 & 0 & 1 \\ e & 0 & 0 & 1 & 1 & 0 \end{array}$$

5. If $V_1 = \{a, b\}$ and $V_2 = \{c, d, e\}$ are the vertex sets, we obtain

$$\begin{array}{c c c c c c} & a & b & c & d & e \\ a & 0 & 0 & 1 & 1 & 1 \\ b & 0 & 0 & 1 & 1 & 1 \\ c & 1 & 1 & 0 & 0 & 0 \\ d & 1 & 1 & 0 & 0 & 0 \\ e & 1 & 1 & 0 & 0 & 0 \end{array}$$

7.

	x_1	x_2	x_3	x_4	x_5	x_6	x_7	x_8	x_9	x_{10}
a	1	0	0	0	0	0	0	0	0	0
b	1	1	0	1	0	0	0	0	0	0
c	0	1	1	0	0	0	0	1	0	0
d	0	0	0	0	1	1	1	0	0	0
e	0	0	0	1	0	0	1	1	0	1
f	0	0	0	0	0	0	0	0	1	1
g	0	0	0	0	0	1	0	0	1	0

9.

a	1	1	1	1	0	0	0	0	0	0
b	1	0	0	0	1	1	1	0	0	0
c	0	1	0	0	1	0	0	1	1	0
d	0	0	1	0	0	1	0	1	0	1
e	0	0	0	1	0	0	1	0	1	1

11.

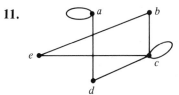

13.

15. For K_5

$$\begin{pmatrix} 4 & 3 & 3 & 3 & 3 \\ 3 & 4 & 3 & 3 & 3 \\ 3 & 3 & 4 & 3 & 3 \\ 3 & 3 & 3 & 4 & 3 \\ 3 & 3 & 3 & 3 & 4 \end{pmatrix}$$

17. The graph consists of two subgraphs with no vertices in common.

19. Let entry ij be the *number* of edges incident on vertices i and j.

21.

23. No edge is incident on that vertex.

26. Use the fact that

$$\begin{pmatrix} d_{n+1} & a_{n+1} & \cdots & a_{n+1} \\ & & & \\ \vdots & & & \vdots \\ & & & \\ a_{n+1} & a_{n+1} & \cdots & d_{n+1} \end{pmatrix} = A^{n+1} = \begin{pmatrix} d_n & a_n & \cdots & a_n \\ a_n & d_n & \cdots & a_n \\ \vdots & & & \vdots \\ & & & \\ a_n & a_n & \cdots & d_n \end{pmatrix} \begin{pmatrix} 0 & 1 & 1 & 1 & 1 \\ 1 & 0 & 1 & 1 & 1 \\ \vdots & & & & \vdots \\ & & & & \\ 1 & 1 & 1 & 1 & 0 \end{pmatrix}.$$

27. First, obtain a_n from the recurrence relation $a_{n+1} = 3a_n + 4a_{n-1}$. Obtain d_n from the relation $d_{n+1} = 4a_n$.

Section 3.3
1. **(a)** Yes, **(b)** No, **(c)** Yes, **(d)** Yes   **3.** **(a)** No, **(b)** No, **(c)** No, **(d)** No
5. **(a)** No, **(b)** No, **(c)** No, **(d)** No   **7.** **(a)** Yes, **(b)** Yes, **(c)** No, **(d)** No
9. There is no such graph, since there are always an even number of vertices of odd degree.
11.    **13.**   **15.**

17. (a, a), (b, c, g, b), (b, c, d, f, g, b), (b, c, d, e, f, g, b), (c, g, f, d, c), (c, g, f, e, d, c), (d, f, e, d)
19.

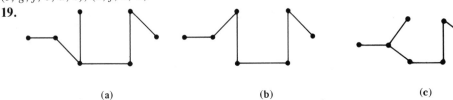

(a)   (b)   (c)

The second is a path and a simple path. Neither is a circuit or a simple circuit.
21. $(a, b, c, b, d, e, h, f, i, j, k, i, h, g, f, e, g, d, c, a)$
23. d and e are the only vertices of odd degree.
25. $(d, a, b, d, e, b, c, e, h, g, d, f, g, j, k, i, e)$
27. $(b, c, d, g, b, a, f, g, i, f)$; (c, h, e)
29. True. In the path, for all repeated a,

$$(\ldots, a, \ldots, b, a, \ldots)$$

eliminate $a, \ldots, b$.
31. False. Consider the circuit (a, b, c, a) for the graph

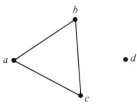

33. •————————•

35. A path that is not a simple path repeats a vertex, say a:

$$(v, \ldots, a, \ldots, a, \ldots, w).$$

Now $(a, \ldots, a)$ is a circuit.
37. The union of all connected subgraphs containing G' is a component.
39. Since this graph is connected, the given matrix is the only component.
43. No. Suppose that it is possible. Consider a graph where the vertices are the persons and an edge connects two vertices (people) if the people get along. Now there are an odd number of vertices of odd degree.
45. Let G be a simple, disconnected graph with n vertices having the maximum

number of edges. Show that G has two components. If one component has i vertices, show that the components are K_i and K_{n-i}. Use Exercise 10, Section 3.1, to find a formula for the number of edges in G as a function of i. Show that the maximum occurs when $i = 1$.

47.

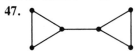

53. Use Exercises 50 and 52.

Section 3.4

1. $7; (a, b, c, f)$ **3.** $10; (a, b, c, d, z)$ **5.** $10; (h, f, c, d)$

7. Change step 2 of Algorithm 3.4.1 to

If $T = \varnothing$, stop. $(L(x)$ gives the length of a shortest path from a to x for every x.)

9. Assume that the vertices of G are $1, 2, \ldots, n$. If there is no edge between x and y, set $w(x, y) = \infty$.

1. Set DIST$(J, K) := w(J, K)$ for all $J, K = 1, \ldots, n$.
2. Set $I := 1$.
3. If $I = n + 1$, stop. (DIST(J, K) is the shortest distance between vertices J and K.)
4. Set $J := 1$.
5. If $J = n + 1$, then set $I := I + 1$ and go to step 3.
6. Set $K := 1$.
7. If $K = n + 1$, then set $J := J + 1$ and go to step 5.
8. If DIST $(J, I) +$ DIST$(I, K) <$ DIST(J, K), then set DIST$(J, K) :=$ DIST(J, I) $+$ DIST(I, K). Set $K := K + 1$ and go to step 7.

11. See the hints for Exercise 9.

13. False

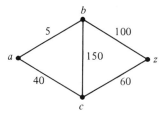

14. In this case Algorithm 3.4.1 finds the length of the longest and shortest path between two vertices, since the path is unique.

16. $g(n) = n^2$

18. See the hints to Exercises 7 and 14.

20. See the hint to Exercise 9.

Section 3.5

1. The graphs are not isomorphic, since they do not have the same number of vertices.

3. The graphs are not isomorphic, since G_1 has a simple circuit of length 3 and G_2 does not.

5. The graphs are isomorphic.

$$f(a) = c', \quad f(b) = d', \quad f(c) = a', \quad f(d) = e', \quad f(e) = b'$$

7. The graphs are not isomorphic, since G_1 has a vertex of degree 2, but G_2 does not.

9. The graphs are not isomorphic, since G_1 has two simple circuits of length 3, but G_2 has only one simple circuit of length 3 (see also Exercise 14).

In Exercises 11–19, we use the notation of Definition 3.5.1.

11. If $(v_0, v_1, \ldots, v_n)$ is a simple circuit of length n in G_1, then $(f(v_0), f(v_1), \ldots, f(v_n))$ is a simple circuit of length n in G_2. [The vertices $f(v_i)$, $i = 1, \ldots, n - 1$, are distinct, since f is one-to-one.]

13. Suppose that G_1 is connected. We must show that G_2 is connected. Let V and W be distinct vertices in G_2. Then there exist vertices v and w in G_1 with $f(v) = V$ and $f(w) = W$. Since G_1 is connected, there exists a path $(v_0, v_1, \ldots, v_n)$ in G_1 with $v_0 = v$ and $v_n = w$. Now $(f(v_0), f(v_1), \ldots, f(v_n))$ is a path in G_2 from V to W. Therefore, G_2 is connected.

15. Let (v, w) be an edge in G_1 with $\delta(v) = i$ and $\delta(w) = j$. Example 3.5.6 shows that $\delta(f(v)) = i$ and $\delta(f(w)) = j$. Now the edge $(f(v), f(w))$ has the desired property in G_2.

17. The property is an invariant. If $(v_0, v_1, \ldots, v_n)$ is an Euler circuit in G_1, then, since g is onto, $(f(v_0), f(v_1), \ldots, f(v_n))$ is an Euler circuit in G_2.

19. The property is an invariant. If V_1 and V_2 partition the vertex set of G_1, then $\{f(v) \mid v \in V_1\}$ and $\{f(v) \mid v \in V_2\}$ partition the vertex set of G_2.

21.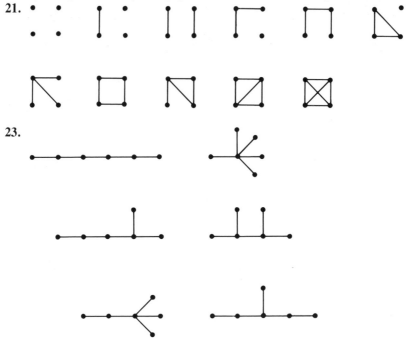

23.

25. Suppose that G is not connected. Let C be a component of G and let V_1 be the set of vertices in G belonging to C. Let V_2 be the set of vertices in G not in V_1. In $\overline{G}$, for every $v_1 \in V_1$, $v_2 \in V_2$, there is an edge e incident on v_1 and v_2. Thus, in G there is a path from v to w if $v \in V_1$ and $w \in V_2$. Suppose that $v, w \in V_1$. Choose $x \in V_2$. Then (v, x), (x, w) is a path from v to w. Similarly, if $v, w \in V_2$, there is a path from v to w. Thus $\overline{G}$ is connected.

27. Suppose that G_1 and G_2 are isomorphic. We use the notation of Definition 3.5.1. We construct an isomorphism for $\overline{G}_1$ and $\overline{G}_2$. The function f is unchanged. Let (v, w) be an edge in $\overline{G}_1$. Set $g((v, w)) = (f(v), f(w))$.

It can be verified that the functions f and g provide an isomorphism of $\overline{G}_1$ and $\overline{G}_2$.

If $\overline{G}_1$ and $\overline{G}_2$ are isomorphic, by the above, $\overline{\overline{G}}_1 = G_1$ and $\overline{\overline{G}}_2 = G_2$ are isomorphic.

29. Define $g((v, w)) = (f(v), f(w))$.
31. $f(a') = a''$, $f(b') = b''$, $f(c') = c''$, $f(d') = d''$, $f(e') = c''$, $f(f') = b''$
33. $f(a) = a'$, $f(b) = b'$, $f(c) = c'$, $f(d) = a'$
35. See [Hell].

Section 3.6

1.

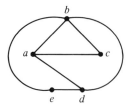

3.

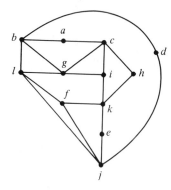

5. Remove (g, e) and (a, c) to obtain a graph homeomorphic to

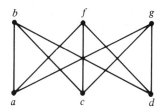

7. Planar

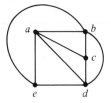

9. $2e = 2 + 2 + 2 + 3 + 3 + 3 + 4 + 4 + 5$, so $e = 14$. $f = e - v + 2 = 14 - 9 + 2 = 7$

11. Let G be a graph having four or fewer vertices. By Exercise 10, the planarity of G is not affected by deleting loops or parallel edges; so we can assume that G has neither loops nor parallel edges. Now G is a subgraph of K_4 and, since K_4 is planar, so is G.

13. Since every circuit has at least three edges, each face is bounded by at least three edges. Thus the number of edges that bound faces is at least $3f$. In a planar graph, each edge belongs to at most two bounding circuits. Therefore, $2e \geq 3f = 3(e - v + 2)$. Thus $3v - 6 \geq e$.

15. If K_5 is planar, $e \leq 3v - 6$ becomes

$$10 \leq 3 \cdot 5 - 6 = 9.$$

17.

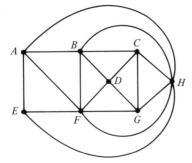

19. Color D, say, red. Now B and C must be different colors and different from red.

23. A, blue; B, green; C, red; D, yellow; E, green; F, red; G, yellow; H, green; I, yellow; J, green; K, blue; L, red.

25. Suppose that G' can be colored with n colors. If we eliminate edges from G' to obtain G, G is colored with n colors.

27. It contains

29. Suppose that G has a vertex v of degree 4. Then we find the configuration shown.

Consider the map G' obtained from G by removing vertex v and the four edges incident on v. By assumption, G' can be colored with four colors. Show that if v_1, v_2, v_3, and v_4 use three or fewer colors, we get an immediate contradiction.

Suppose that v_1, v_2, v_3, and v_4 require four colors and that v_i is colored C_i. Consider the subgraph G_1' of G' consisting of all simple paths starting at v_1 whose vertices are alternately colored C_1 and C_3. If G_1' does not include v_3, we may change each C_1 to C_3 and each C_3 to C_1 in G_1' and produce a coloring of G' with four colors. If this is done, we can then color G with four colors.

Suppose that G_1' includes v_3. Consider the subgraph G_2' of G' consisting of all simple paths starting at v_2 whose vertices are alternately colored C_2 and C_4. Show that G_2' cannot include v_4. We may change each C_2 to C_4 and each C_4 to C_2 in G_2' and produce a coloring of G' with four colors. If this is done, we can then color G with four colors.

Deduce that G cannot have a vertex of degree 4.

30. Assume that G does not have a vertex of degree 5. Show that $2e \geq 6v$. Now use Exercise 13 to deduce a contradiction.

31. Use the methods of Exercises 28–30.

Section 4.1

1. T_1 and T_2 are isomorphic as graphs.

3.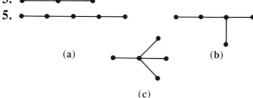

5.

(a)   (b)

(c)

7.

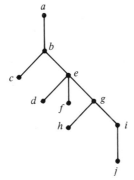

Root   Root

9. ROOT, 0; a, 1; b, 1; c, 1; d, 1; e, 2; f, 3; g, 3; h, 4; i, 2; j, 3.

11. Height $= 5$

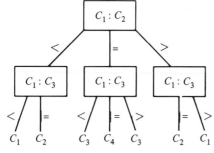

15. The first weighing either compares one coin against another or two coins against the other two. If the first weighing is two coins against the other two, the root is incident on only two edges. Now level 2 has at most six vertices. Since there are eight possibilities, this is a contradiction. Suppose that the first weighing is one coin against another. Consider what happens if equality occurs.

17.

$$C_1 : C_2$$

$<$ $=$ $>$

$$C_1 : C_3 \qquad C_1 : C_3 \qquad C_1 : C_3$$

$<$ $=$ $<$ $=$ $>$ $=$ $>$

C_1 C_2 C_3 C_4 C_3 C_2 C_1

19. If we could solve the 12 coins problem in two weighings, we could solve the 8 coins problem in two weighings.

21. PEN   **23.** DEAL   **25.** 010000001111

27.

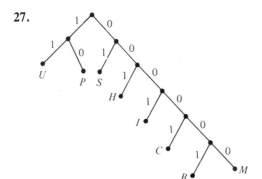

29. The tree shown in the solution of Exercise 27 is optimal.

33. Since $K_{3,3}$ and K_5 contain circuits, a tree cannot contain a subgraph homeomorphic to either; thus a tree is planar.

35. Consider the tree rooted. Color the vertices on even levels one color and the vertices on odd levels a second color.

37. e, g

39. Assume that the centers are not adjacent and deduce a contradiction.

Section 4.2

1.

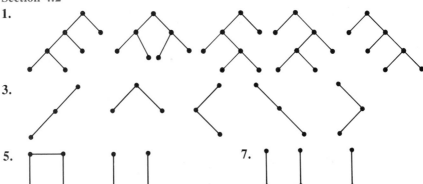

3.

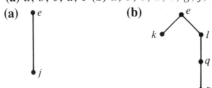

5. □ ⊔ **7.** ⊔ |

9. No such graph. Since the sum of the degrees is 10, there would be five edges. This contradicts Theorem 4.2.1.

11.

13. No such graph. If there were such a graph, it would contradict Theorem 4.2.8.

15. (a) b; d (b) a; c **17.** (a) $h, i; j$ (b) $j; k, l$ **19.** (a) $e, g; i$ (b) None; g, i

21. (a) a, b, c, d, e (b) $a, b, c, d, e, g, j, l, q$

23. (a)   (b)

25. 4

27. Use the notation of Algorithm 4.2.12.

　　1. Get the next word WORD. If no more, stop.

2. If the binary tree is empty, create ROOT and place WORD in ROOT. Go to step 1.

3. Set $PT := \text{ROOT}$.

4. If WORD $<$ VALUE (PT) and LEFT$(PT) = \lambda$, create a left child of PT and insert WORD there. Go to step 1.

5. If WORD $>$ VALUE(PT) and RIGHT$(PT) = \lambda$, create a right child of PT and insert WORD there. Go to step 1.

6. If WORD $<$ VALUE(PT), set $PT := \text{LEFT}(PT)$ and go to step 4.

7. If WORD $>$ VALUE(PT), set $PT := \text{RIGHT}(PT)$ and go to step 4.

29. False

31. $n - m$

33. Let v be a vertex of degree at least 2 in a tree G and let $P = (v_0, \ldots, v_n)$ be a path of maximum length passing through v. Since G is a tree, P is not a circuit and, since v has degree at least 2, $v \neq v_0$ and $v \neq v_n$. If removing v and all edges incident on v leaves a connected graph, then there is a path, distinct from P from v_0 to v_n. Since G is a tree, this is impossible. Therefore, v is an articulation point. For the converse, use Exercise 49, Section 3.3.

35. 1. Let T be a root with two children.

2. Set $n := n - 1$. If $n = 1$, stop.

3. Choose a terminal vertex v. Give v two children and go to step 2.

37. $t - 1$ **39.** Yes   **41.** Yes

43.

Section 4.3

1.

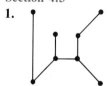

3.

5.

7.

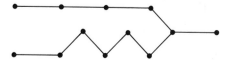

9.

11.

(a)

(b)

13. False. Consider K_4.

15. If G is the graph,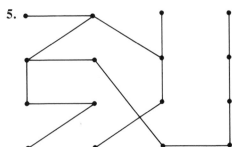
every vertex ordering produces the same spanning tree (namely, G itself).

17. First, show that the graph T constructed is a tree. Now use induction on the level of T to show that T contains all the vertices of G.

19. If the edge is not contained in a circuit of G.

29.

	e_1	e_2	e_6	e_5	e_3	e_4	e_7	e_8
(abca)	1	0	0	0	1	1	0	0
(acda)	0	1	0	0	1	0	0	1
(bcdb)	0	0	1	0	0	1	0	1
(bcdeb)	0	0	0	1	0	1	1	1

31.

	e_3	e_4	e_2	e_9	e_1	e_5	e_6	e_7	e_8
(abdea)	1	0	0	0	1	1	0	1	0
(abdea)'	0	1	0	0	1	1	0	1	0
(bcdb)	0	0	1	0	0	1	1	0	0
(defd)	0	0	0	1	0	0	0	1	1

33. The input is a graph with vertices ordered $v_1, \ldots, v_n$. The steps are as in Algorithm 4.3.7 except for the following modifications

3. If $w = v_1$, go to step 5.

5. If the number of vertices in T is n, return CONNECTED; otherwise, return NOT CONNECTED. Stop.

Section 4.4

1.

3.

5.

7. The kth time at step 3, there are k vertices in T and $n - k$ vertices not in T. This means that we examine $k(n - k)$ edges in the worst case. Thus the number of edges examined is

$$\sum_{k=1}^{n-1} k(n - k) = \sum_{k=1}^{n-1} kn - \sum_{k=1}^{n-1} k^2 = n\left[\frac{(n - 1)n}{2}\right] - \frac{(n - 1)n(2n - 1)}{6} = O(n^3).$$

9. The mth time at step 3, we visit $n - m - 1$ edges in the worst case. Thus the number of edges examined is

$$\sum_{m=1}^{n-2} (n - m - 1) = 1 + 2 + \cdots + n - 2 = \frac{(n - 2)(n - 1)}{2} = O(n^2).$$

11. Consider K_4. If an algorithm never visits edge e, then e is not in the minimal spanning tree T found by the algorithm. Let $w(e) = 1$ and $w(x) = 2$ for all edges $x \neq e$. Then the weight of T is 6. However, a minimal spanning tree must include e and thus has weight 5. Thus T is not a minimal spanning tree. This is a contradiction.

13. False   **15.** False. Consider

with vertex ordering a, b, c, d.

17. The proof is similar to the proof of Theorem 4.4.5. Let G_i be the graph produced at the ith iteration. Use induction to show that G_i contains a minimal spanning tree.

21. (For Exercise 2)   The edges of the minimal spanning tree are chosen in the order (b, c), (g, h), (e, h), (a, b), (c, f), (b, e), (d, g), (f, i).

27. 16 cents postage requires only two stamps—two 8-cent stamps. The algorithm generates one 10-cent stamp and six 1-cent stamps.

Section 4.5

	preorder	inorder	postorder
1.	ABDCE	BDAEC	DBECA
3.	ABHIKLMJCDEFG	ILKMHJBADFEGC	LMKIJHBFGEDCA
5.	ABCDEFG	DCBAEFG	DCBGFEA

7.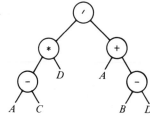

prefix: $/* - A C D + A - B D$
postfix: $A C - D * A B D - + /$

9.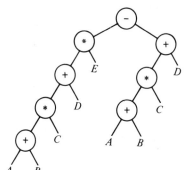

prefix: $- * + * + A B C D E + * + A B C D$
postfix: $A B + C * D + E * A B + C * D + -$

11.

$$\begin{aligned}
\text{prefix:} &\quad -+ABC \\
\text{usual infix:} &\quad A+B-C \\
\text{parened infix:} &\quad ((A+B)-C)
\end{aligned}$$

13.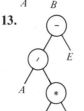

$$\begin{aligned}
\text{prefix:} &\quad -/A*B+CDE \\
\text{usual infix:} &\quad A/(B*(C+D))-E \\
\text{parened infix:} &\quad ((A/(B*(C+D)))-E)
\end{aligned}$$

15.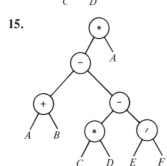

$$\begin{aligned}
\text{prefix:} &\quad *-+AB-*CD/EFA \\
\text{usual infix:} &\quad (A+B-(C*D-E/F))*A \\
\text{parened infix:} &\quad (((A+B)-((C*D)-(E/F)))*A)
\end{aligned}$$

17. 0   **19.** 0   **21.** -6

23. The tree is

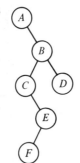

The argument that this is the only tree begins as follows. Because of the preorder listing, A is the root. If A had a left child, the inorder listing would not begin with A. Since A has no left child, the preorder listing tells us that the right child of A is B.

25.

27. Use induction on the number of levels.

29. Call the following algorithm with PT set to the root of the tree.

   1. If PT is empty, return.

   2. Interchange the left and right children of PT.

   3. Call this algorithm with input the subtree whose root is the left child of PT.

   4. Call this algorithm with input the subtree whose root is the right child of PT and return.

31. Define an *initial segment* of a string to be the first $i \geq 1$ characters for some i.

Define $r(x) = 1$, for $x = A, B, \ldots, Z$; and $r(x) = -1$, for $x = +, -, *, /$. If $x_1 \cdots x_n$ is a string over $A, \ldots, Z, +, -, *, /$, define

$$r(x_1 \cdots x_n) = r(x_1) + \cdots + r(x_n).$$

Then a string s is a postfix string if and only if $r(s) = 1$ and $r(s') \geq 1$, for all initial segments s' of s.

33. It is assumed the algorithm is called with PT set to the root of the tree.

1. If PT is empty, then return the null string.

2. If PT has no children, return the contents of PT.

3. Call this algorithm, which returns LEFT, with input the subtree whose root is the left child of PT.

4. Call this algorithm, which returns RIGHT, with input the subtree whose root is the right child of PT.

5. Return

$$\text{'(' } \| \text{ LEFT } \| \text{ contents of } PT \| \text{ RIGHT } \| \text{ ')'}.$$

($\|$ is the concatenation operator.)

35. Call the algorithm of Exercise 34 and then call the algorithm of Exercise 33.

Section 4.6

1.

$\overline{1}$	$\overline{1}$	1
$\overline{9}$	9	3
$\overline{7}$	3	7
$\overline{3}$	7	9
Merge	Merge	
one element	two element	
arrays	arrays	

5. Suppose that the arrays are $A(1), \ldots, A(n)$ and $B(1), \ldots, B(n)$.
(a) $A(1) < B(1) < A(2) < B(2) < \cdots$
(b) $A(n) < B(1)$

7. The array 5 3 1 6 4 2 requires the maximum number of comparisons, which is 11.

9. We use induction on N. For $N = 1$, $a_1 = 0 \leq 0 = b_1$. Assume that $a_k \leq b_k$ for $k < N$. Then

$$a_N \leq a_{\lfloor N/2 \rfloor} + a_{\lfloor (N+1)/2 \rfloor} + N$$

$$\leq b_{\lfloor N/2 \rfloor} + b_{\lfloor (N+1)/2 \rfloor} + N = b_N.$$

11. Algorithm B is superior if $2 \leq n \leq 15$. (For $n = 16$, the run times coincide.)

13. Part of the tree is

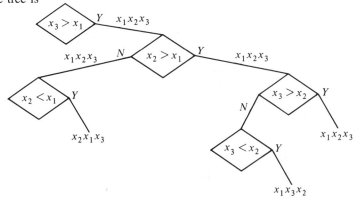

15. Sequences sorted in increasing or decreasing order that require $O(n^2)$ comparisons.

17.

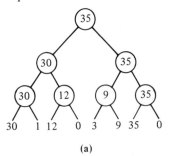 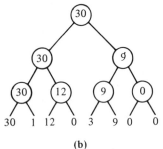

(a)   (b)

19. Suppose that we have an algorithm which finds the largest value among $x_1, \dots,$ x_n. Let $x_1, \dots, x_n$ be the vertices of a graph. An edge exists between x_i and x_j if the algorithm compares x_i and x_j. The graph must be connected. The least number of edges necessary to connect n vertices is $n - 1$.

23. Suppose that there is an algorithm which is $O(g(n))$. Let $f(n)$ be the number of comparisons needed in the worst case. Then, for some C and all but finitely many n, $f(n) \leq Cg(n)$. Also, for all but finitely many n, $g(n)/(n \lg n) < 1/(4C)$. Now use Theorem 4.6.3 to deduce a contradiction.

25. $a_n \leq a_{\lfloor n/2 \rfloor} + a_{\lfloor (n+1)/2 \rfloor} + 2 + \lg (\lfloor n/2 \rfloor \lfloor (n + 1)/2 \rfloor)$. If n is even, $\lfloor n/2 \rfloor \lfloor (n + 1)/2 \rfloor = n^4/4$. If n is odd, $\lfloor n/2 \rfloor \lfloor (n + 1)/2 \rfloor = [(n - 1)/2] [(n + 1)/2] = (n^2 - 1)/4 \leq n^2/4$. Thus we can write

$$a_n \leq a_{\lfloor n/2 \rfloor} + a_{\lfloor (n+1)/2 \rfloor} + 2 + \lg \frac{n^2}{4}$$

$$= a_{\lfloor n/2 \rfloor} + a_{\lfloor (n+1)/2 \rfloor} + 2 \lg n.$$

27. The argument is similar to that for Exercise 9.
29. Choose k with $2^k < n \leq 2^{k+1}$. Then

$$a_n \leq b_n \leq b_{2^{k+1}} \leq 4 \cdot 2^{k+1} - 2(k + 1) - 4 \leq 8 \cdot 2^k \leq 8n.$$

30. Kruskal's algorithm implicitly sorts the edges by weight, so must require at least $O(e \lg e)$ comparisons.

Section 4.7

1.

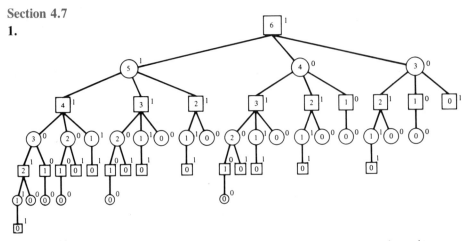

The first player always wins. The winning strategy is to first take one token; then, whatever the second player does, leave one token.

3. The tree is the same as Figure 4.7.1. The terminal vertices are assigned values as in Figure 4.7.2 with 0 and 1 interchanged. After applying the minimax procedure, the root receives the value 1; thus the first player will always win. The optimal strategy is to first leave 2,2. If the second player leaves only one pile, take it; otherwise, leave 1,1.
5. The tree is the same as in Exercise 1. The terminal vertices are assigned values as in the hint for Exercise 1 with 0 and 1 interchanged. After applying the minimax procedure, the root receives the value 1; thus the first player will always win. The optimal strategy is to take 2. No matter how many player 2 chooses, player 1 can take the rest.
7. The strategy for winning play is: Play nim' exactly like nim unless the move would leave an odd number of singleton piles and no other pile. In this case, leave an even number of piles.
9.

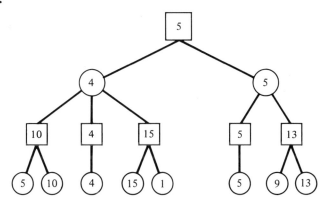

11. The value of the root is 3.
13. (For Exercise 10)

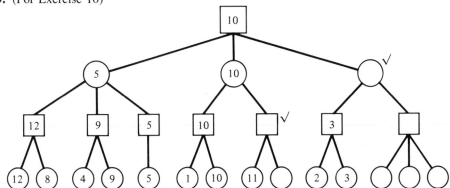

15. $4 - 2 = 2$.

17.

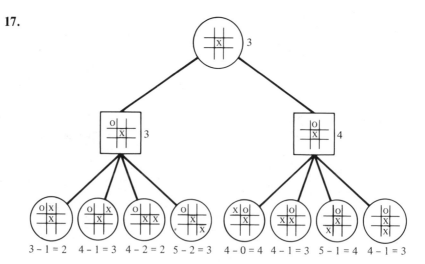

$$3-1=2 \quad 4-1=3 \quad 4-2=2 \quad 5-2=3 \quad 4-0=4 \quad 4-1=3 \quad 5-1=4 \quad 4-1=3$$

O will move to a corner.

19. The input is the level N and a tree with root PT.

1. If $N = 0$ or if PT has no children, set the contents of PT to $E(PT)$ and return.

2. Let $C_1, \ldots, C_k$ be the children of PT. Call this algorithm with input C_1, $N - 1$; $C_2, N - 1$; $\ldots$; $C_k, N - 1$. Suppose that the contents of C_i is E_i.

3. If PT is a box vertex, set the contents of PT to max $\{E_1, \ldots, E_k\}$; otherwise, set the contents of PT to min $\{E_1, \ldots, E_k\}$.

4. Return.

21. We first obtain the values 6, 6, 7 for the children of the root. Thus we order the children of the root with the rightmost child first and use the alpha-beta procedure to obtain

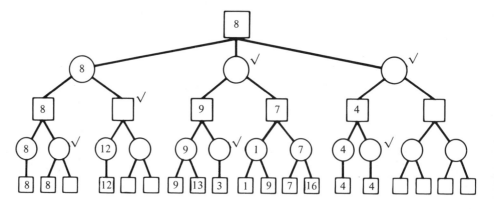

Section 5.1

1. (b, c) is 6,3; (a, d) is 4, 2; (c, e) is 6, 1; (c, z) is 5, 2. The value of the flow is 5.

3. (a, b) is 5, 3; (a, d) is 1, 1; (d, c) is 6, 0; (d, f) is 2, 1; (c, e) is 4, 1; (g, z) is 4, 2; (c, z) is 5, 2. The value of the flow is 6.

5.

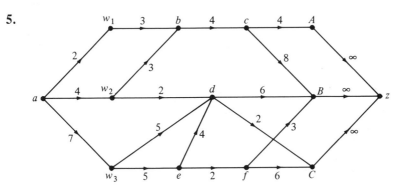

7. Replace vertex d in the network shown in the hint to Exercise 5 by

Change the capacities of (A, z), (B, z), and (C, z) to 4, 3, and 4, respectively.
9. Replace each undirected edge by two directed edges

each having weight equal to the weight of the undirected edge.

Section 5.2
1. 1   **3.** 1
5.

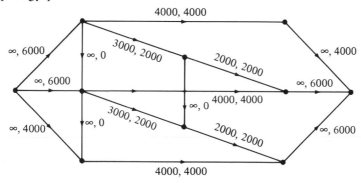

7.

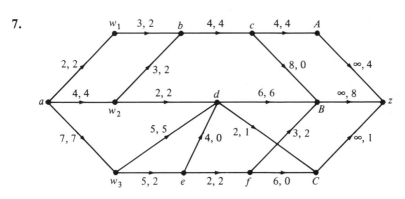

9.

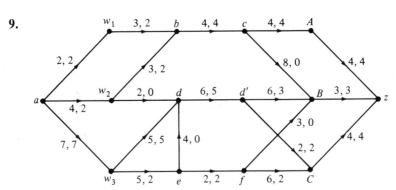

11.

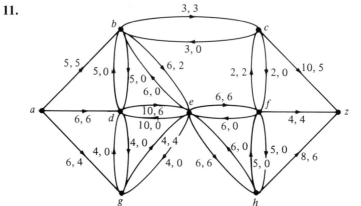

13.

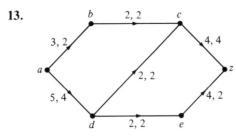

15.

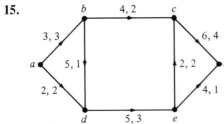

17.

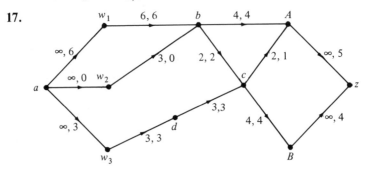

21. Suppose that the sum of the capacities of the edges incident on a is U. Each iteration of Algorithm 4.2.5 increases the flow by 1. Since the flow cannot exceed U, eventually the algorithm must terminate.

Section 5.3
1. 8; minimal   **3.** 15; not minimal
5. $P = \{a, w_1, w_2, w_3, b, d\}$ **7.** $P = \{a, d\}$
9. $P = \{a, b\}$ **11.** $P = \{a\}$
13. $P = \{a, w_1, w_2, w_3, b, c, d, d', e, f, A, B, C\}$
15. $P = \{a, b, d, e, g\}$
17.

$$a \longrightarrow b \longrightarrow z$$

with $C_{ab} = 1$, $C_{bz} = 2$, $m_{ab} = 1$, $m_{bz} = 2$.
19. Use Exercise 18 and imitate the proofs of Theorems 5.3.7 and 5.3.9.
21. The argument is similar to that of Exercise 19.
23. False. Consider the flow

$$a \xrightarrow{1,\,1} b \xrightarrow{2,\,1} z$$

and the cut $P = \{a, b\}$.

Section 5.4
1. $P = \{a, A, B, D, J_2, J_5\}$
3. **(a)(b)**

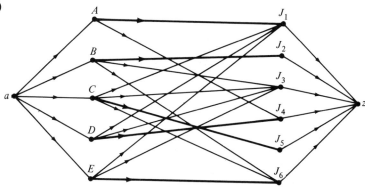

(c) Yes
6. **(a)(d)**

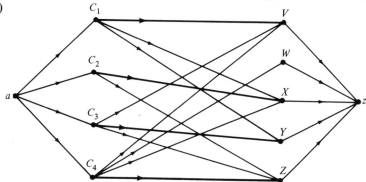

(b) The maximum number of committees that can be represented.
(c) All committees are represented.
(e) Yes

7. Let

$$M^* = \min \{\delta(v) \mid v \in V\}.$$

Let $S \subseteq V$ and suppose that $|S| = k$ and $|R(S)| = j$. Each vertex in S is incident on at least M^* edges; thus S sends at least kM^* edges to $R(S)$. Each vertex in $R(S)$ is incident on at most M edges; thus $R(S)$ receives at most jM edges. It follows that $kM^* \leq jM$. Since $M \leq M^*$, $jM \leq jM^*$. Therefore, $kM^* \leq jM^*$ and $|S| = k \leq j = |R(S)|$. By Theorem 5.4.7, G has a complete matching.

9. Let $V = \{v_1, \ldots, v_m\}$ and $W = \{w_1, \ldots, w_n\}$ be the disjoint vertex sets. Order the vertices

$$v_1, \ldots, v_m, w_1, \ldots, w_n.$$

11. Each row has exactly one label and each column has at most one label.

15. See [Liu, 1968].

Section 5.5

1.

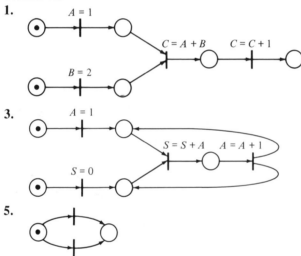

3.

5.

7. Let M_1 denote the marking that results from M by firing t_1. The only transition enabled in M_1 is t_2. Let M_2 denote the marking that results from M_1 by firing t_2. The only transition enabled in M_2 is t_3. If t_3 is fired, we obtain the marking M. It follows that M is live and bounded.

9. (a) t_1

(b)

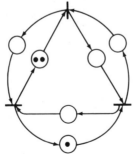

(c) No   **(d)** No   **(e)** Yes

(f) The marking above is the only marking reachable from M.

(g) Take any marking except M and that shown above.

11. (a) t_1

(b)

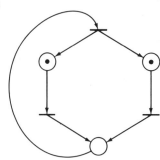

(c) Yes   (d) No   (e) No

(f) Designate the input place to t_i as p_i. The markings reachable from M are of two types:

1. p_3 has at least one token.
2. p_3 has no tokens and p_1 and p_2 each have at least one token.

(g)

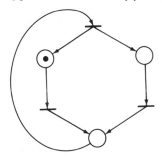

13. Figure 5.5.5

15. Use two places for each philosopher—one representing meditation and the other representing eating.

17. 8 and 9

19.

21. Suppose that M is live. Let C be a simple directed circuit and let v be a vertex in C. Since M is live, eventually v can be fired. Now C contains at least one token. But by Exercise 20, the token count on C does not change.

Suppose that M places at least one token in each simple directed circuit. Let v be a vertex. If there are no token-free edges coming into v, v is enabled and M is live.

If there are token-free edges coming into v, consider the set V of vertices, other than v itself, on which these token-free edges are incident. If every member of V is enabled, after firing each, v will be enabled.

If some member of V is not enabled, for each element of V that is not enabled, we consider the incoming edges which are token-free. We continue in this way selecting vertices and edges to generate a subgraph. This subgraph is circuit-free, since there are no token-free simple directed circuits. Therefore, this subgraph has at least one vertex v_0 with no incoming edges in the subgraph. Thus, in G, there are no token-free incoming edges to v_0. We fire v_0 to obtain a new marking M'.

We repeat the procedure above with the marking M'. The new subgraph produced has fewer vertices than the original subgraph. Eventually, we will enable v. Therefore, M is live.

23.

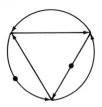

Section 6.1

1. $\overline{x_1 \wedge x_2}$

x_1	x_2	$\overline{x_1 \wedge x_2}$
1	1	0
1	0	1
0	1	1
0	0	1

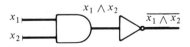

3. $(x_1 \wedge x_2) \vee \overline{x}_3$

x_1	x_2	x_3	$(x_1 \wedge x_2) \vee \overline{x}_3$
1	1	1	1
1	1	0	1
1	0	1	0
1	0	0	1
0	1	1	0
0	1	0	1
0	0	1	0
0	0	0	1

5. $((\overline{x}_1 \vee x_2) \wedge (\overline{x}_3 \vee x_4)) \wedge (\overline{x}_2 \vee x_4)$

7. See Exercise 9.

9. Suppose that $x = 1$ and $y = 0$. Then the input to the AND gate is 1, 0. Thus the output of the AND gate is 0. Since this is then NOTed, $y = 1$. Contradiction. Similarly, if $x = 1$ and $y = 1$, we obtain a contradiction.

11. 1   **13.** 1

15. (For Exercise 10) x_1 and x_2 are Boolean expressions by (6.1.2); $x_1 \wedge x_2$ is a Boolean expression by (6.1.3d); $\overline{x_1 \wedge x_2}$ is a Boolean expression by (6.1.3b).

17. Yes   **19.** No

21. The circuit for Exercise 10 is Exercise 1. The solution to Exercise 1 gives the logic table.

23.

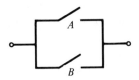

25. $(A \wedge B) \vee (C \wedge \overline{A})$

A	B	C	$(A \wedge B) \vee (C \wedge \overline{A})$
1	1	1	1
1	1	0	1
1	0	1	0
1	0	0	0
0	1	1	1
0	1	0	0
0	0	1	1
0	0	0	0

27. $(A \wedge (B \vee C)) \vee \overline{D}$

29. $A \wedge ((B \vee \overline{D}) \vee (\overline{C} \wedge (A \vee D \vee \overline{C}))) \wedge B$

31.

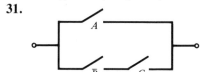

A	B	C	$A \vee (\overline{B} \wedge C)$
1	1	1	1
1	1	0	1
1	0	1	1
1	0	0	1
0	1	1	0
0	1	0	0
0	0	1	1
0	0	0	0

33.

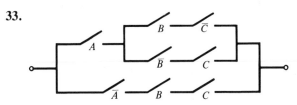

35. No. Consider $\overline{A \wedge B}$.

Section 6.2

1.

x_1	x_2	$\overline{x_1 \wedge x_2}$	$\overline{x}_1 \vee \overline{x}_2$
1	1	0	0
1	0	1	1
0	1	1	1
0	0	1	1

3.

x_1	x_2	x_3	$\overline{(x_1 \vee \bar{x}_2)} \vee (\bar{x}_1 \wedge x_3)$	$\bar{x}_1 \wedge (x_2 \vee x_3)$
1	1	1	0	0
1	1	0	0	0
1	0	1	0	0
1	0	0	0	0
0	1	1	1	1
0	1	0	1	1
0	0	1	1	1
0	0	0	0	0

5. The Boolean expression for the second circuit can be transformed to the Boolean expression for the first circuit

$$(x_1 \vee x_3) \wedge (x_2 \vee x_3) \wedge (x_2 \vee x_4) \wedge (x_1 \vee x_4)$$
$$= (x_3 \vee (x_1 \wedge x_2)) \wedge (x_4 \vee (x_2 \wedge x_1))$$
$$= (x_1 \wedge x_2) \vee (x_3 \wedge x_4)$$

7.

x_1	x_2	$x_1 \vee (x_1 \wedge x_2)$
1	1	1
1	0	1
0	1	0
0	0	0

9.

x_1	x_2	x_3	$x_1 \wedge \overline{(x_2 \wedge x_3)}$	$(x_1 \wedge \bar{x}_2) \vee (x_1 \wedge \bar{x}_3)$
1	1	1	0	0
1	1	0	1	1
1	0	1	1	1
1	0	0	1	1
0	1	1	0	0
0	1	0	0	0
0	0	1	0	0
0	0	0	0	0

11.

x	$\bar{\bar{x}}$
1	1
0	0

13. False. Take $x_1 = x_2 = x_3 = 1$.

15. False. Take $x_1 = x_3 = 1$ and $x_2 = x_4 = 0$.

19. (For Exercise 28, Section 6.1)

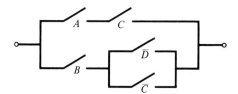

21.

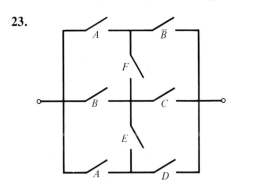

23.

Section 6.3

3. If $(S, +, \cdot, ', 1, 8)$ is a Boolean algebra, then $x + x' = 1$. In this case, we must have

$$\text{lcm } (x, 8/x) = 8 \qquad \text{for } x = 1, 2, 4, 8.$$

However, for $x = 4$, lcm $(4, 8/4) = $ lcm $(4, 2) = 4$. Therefore, this system is not a Boolean algebra.

5. In this solution, will denote the 0 (respectively, 1) of a Boolean algebra by m (respectively, M).

Every Boolean algebra has at least two elements, since m and M are distinct. Thus $n \geq 2$.

Suppose that $n > 2$ and that S_n is a Boolean algebra. Then $n = n \cdot M = $ min $\{n, M\}$. Thus $M = n$. For any $x \in S_n$, we have $M = x + x' = $ max $\{x, x'\}$. It follows that if $x \neq M$, then $x' = M$. Therefore, if $x \neq M$, $x = (x')' = M'$. This says that M' is not unique. Contradiction.

If $n = 2$, we may take $m = 1$, $M = 2$, $1' = 2$, and $2' = 1$, and show that S_n is a Boolean algebra.

7. If $X \cup Y = U$ and $X \cap Y = \emptyset$, then $Y = \overline{X}$.

9. $(x'y')' = x + y$

11. $x + y' = 1$ if and only if $x + y = x$.

13. $x = 1$ if and only if $y = (x + y')(x' + y)$ for all y.

15. (For Exercise 12)

$$0 = x + y = (x + x) + y = x + (x + y) = x + 0 = x.$$

Similarly, $y = 0$.

17. If $xy = 0$ and $x + y = 1$, then $y = x'$. The dual of Theorem 6.3.4 is Theorem 6.3.4.

19. [For part (c)] $x(x + y) = (x + 0)(x + y) = x + 0y = x + y0 = x + 0 = x$.

21. First, show that if $ba = ca$ and $ba' = ca'$, then $b = c$. Now take $a = x$, $b = x + (y + z)$, and $c = (x + y) + z$ and use this result.

23. To show that $0 \in A$, set $x = y = 1$. Set $x = 1$ to show that if $y \in A$, then $\bar{y} \in A$. Replace y by $\bar{y}$ to show that if $x, y \in A$, then $xy \in A$. Now show that if $x, y \in A$, then $x + y \in A$. Notice that parts (a)–(e) of Definition 6.3.1 automatically hold in A, since they hold in S.

Section 6.4
In these hints, $a \wedge b$ is written ab.

1. $xy \vee \bar{x}y \vee \bar{x}\bar{y}$ **3.** $xyz \vee xy\bar{z} \vee x\bar{y}\bar{z} \vee \bar{x}\bar{y}z \vee \bar{x}\bar{y}\bar{z}$

5. $xyz \vee xy\bar{z} \vee x\bar{y}z \vee \bar{x}y\bar{z} \vee \bar{x}\bar{y}z \vee \bar{x}\bar{y}\bar{z}$ **7.** $xyz \vee xy\bar{z} \vee \bar{x}\bar{y}\bar{z}$

9. $wxyz \vee wx\bar{y}z \vee w\bar{x}\bar{y}\bar{z} \vee \bar{w}xyz \vee \bar{w}\bar{x}y$ **11.** $xy \vee x\bar{y}$

13. $xyz \vee x\bar{y}z \vee xy\bar{z} \vee x\bar{y}\bar{z} \vee \bar{x}y\bar{z}$ **15.** $\bar{x}\bar{y}z \vee \bar{x}y\bar{z} \vee \bar{x}yz$

17. $xyz \vee \bar{x}y\bar{z} \vee x\bar{y}\bar{z} \vee \bar{x}y\bar{z}$

19. $wxyz \vee w\bar{x}yz \vee wxy\bar{z} \vee wx\bar{y}\bar{z} \vee wx\bar{y}z \vee \bar{w}xyz \vee \bar{w}x\bar{y}z \vee \bar{w}\bar{x}yz \vee \bar{w}\bar{x}y\bar{z} \vee w\bar{x}y\bar{z}$

21. 2^{2^n} **25.** (For Exercise 3) $(\bar{x} \vee y \vee \bar{z})(x \vee \bar{y} \vee \bar{z})(x \vee \bar{y} \vee z)$

27. If $f(x_1, \ldots, x_n) = m_1 \vee \cdots \vee m_k$, then $\overline{f(x_1, \ldots, x_n)} = \bar{m}_1\bar{m}_2 \cdots \bar{m}_k$. Since each $m_i = y_1y_2 \cdots y_n$, where each y_j is either x_j or $\bar{x}_j$, $\bar{m}_i = \bar{y}_1 \vee \bar{y}_2 \vee \cdots \vee \bar{y}_n$. Thus $\overline{f(x_1, \ldots, x_n)}$ is expressed as the conjunction of maxterms.

29. If $j > k$, then some term m'_t does not occur in the expansion $m_1 \vee \cdots \vee m_k$. Choose $x_i \in Z_2$ so that $m'_t = 1$. Show that $m_i = 0$ for $i = 1, \ldots, n$. Conclude that $j \leq k$. Similarly, $j \geq k$. Therefore, $j = k$.

Give a similar argument to show that each m_i is equal to some m'_t.

Section 6.5

1. AND can be expressed in terms of OR and NOT: $xy = \overline{\bar{x} \vee \bar{y}}$.

3. A combinatorial circuit consisting only of OR gates would output 1 when all inputs are 1.

5. A combinatorial circuit consisting only of OR and AND gates would output 1 when all inputs are 1.

7. $xy = (x \uparrow y) \uparrow (x \uparrow y)$ **9.** $y_1 = x_1x_2 \vee \overline{(x_2 \vee x_3)}; y_2 = \overline{x_2 \vee x_3}$

12. (For Exercise 3)   The dnf may be simplified to $xy \vee x\bar{z} \vee \bar{x}\bar{y}$ and then rewritten as $x(y \vee \bar{z}) \vee \bar{x}\bar{y} = \overline{(x\bar{y}z)} \vee \bar{x}\bar{y} = \overline{\overline{x\bar{y}z}\overline{\bar{x}\bar{y}}}$, which gives the circuit

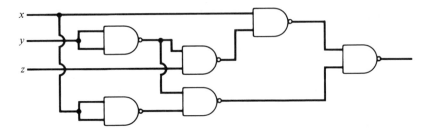

15.

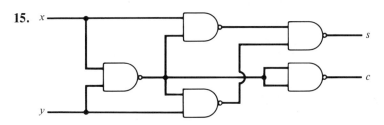

17. $xy = (x \downarrow x) \downarrow (y \downarrow y)$; $x \vee y = (x \downarrow y) \downarrow (x \downarrow y)$; $\bar{x} = x \downarrow x$; $x \uparrow y = [(x \downarrow x) \downarrow (y \downarrow y)] \downarrow [(x \downarrow x) \downarrow (y \downarrow y)]$

19.

x_1	x_2	$x_1 \downarrow x_2$
1	1	0
1	0	0
0	1	0
0	0	1

21. (For Exercise 3)   Write $xy \vee x\bar{z} \vee \bar{x}\bar{y} = x(y \vee \bar{z}) \vee \bar{x}\bar{y} = \overline{\bar{x} \vee \overline{y \vee \bar{z}}} \vee \overline{x \vee y}$, which gives

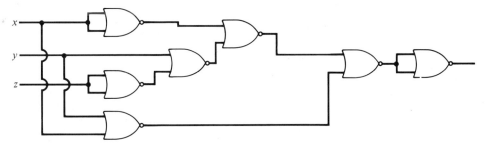

23.

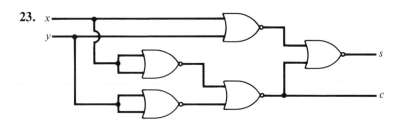

25. The logic table is

x	y	z	Output
1	1	1	1
1	1	0	1
1	0	1	1
1	0	0	0
0	1	1	1
0	1	0	0
0	0	1	0
0	0	0	0

which gives $xyz \vee xy\bar{z} \vee x\bar{y}z \vee \bar{x}yz = xy \vee xz \vee yz$. We obtain the circuit

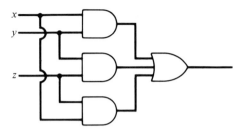

27. The logic table is

b	FLAGIN	c	FLAGOUT
1	1	0	1
1	0	1	1
0	1	1	1
0	0	0	0

Thus $c = b \oplus \text{FLAGIN}$ and $\text{FLAGOUT} = b \vee \text{FLAGIN}$. We obtain the circuit

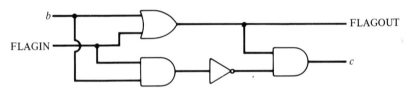

29. 00101

31.

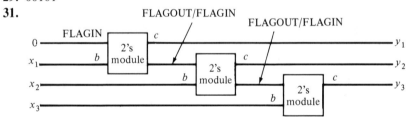

Section 7.1

1.

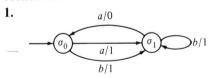

3.

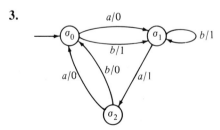

5.

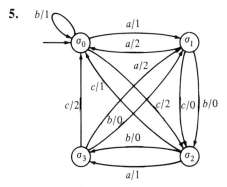

7. $\mathscr{I} = \{a, b\}$; $\mathbb{O} = \{0, 1\}$; $\mathscr{S} = \{A, B, C\}$; initial state $= A$

$\mathscr{S}$ \ $\mathscr{I}$	a	b	a	b
A	A	B	0	1
B	A	C	0	1
C	C	A	1	0

9. $\mathscr{I} = \{a, b\}$; $\mathbb{O} = \{0, 1\}$; $\mathscr{S} = \{\sigma_0, \sigma_1, \sigma_2, \sigma_3\}$; initial state $= \sigma_0$

$\mathscr{S}$ \ $\mathscr{I}$	a	b	a	b
σ_0	σ_1	σ_2	0	0
σ_1	σ_0	σ_2	1	0
σ_2	σ_3	σ_0	0	1
σ_3	σ_1	σ_3	0	0

11. 1110   **13.** 0101100   **15.** 121121   **17.** 001110001
19. 010000000001
21.

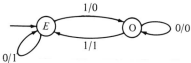

23.

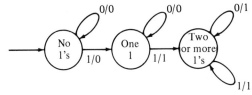

25.

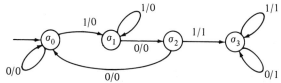

27. When γ is input, the machine outputs x_n, x_{n-1}, . . . until $x_i = 1$. Thereafter, it outputs $\bar{x}_i$. However, according to Algorithm 6.5.16, this is the 2's complement of α.
29. $a = 1, b = 0, c = 0, y = 1$ **31.** $a = 0, b = 1, c = 1, y = 0$

34. Suppose that such a finite-state machine M exists and has m states. Let X and Y each be 1 followed by $m + 2$ 0's. Then the sequence

$$\underbrace{00, 00, \ldots, 00,}_{m + 2} 11, \underbrace{00, 00, \ldots, 00}_{m + 3}$$

is input to M. The product of X and Y is 1 followed by $2m + 4$ 0's. The output is

$$\underbrace{0, 0, \ldots, 0, 1, 0.}_{2m + 4}$$

After 11 is input, a sequence of $m + 1$ 00's is input and the output is 0 each time. Since there are only m states, we must return to a state that we previously visited. That is, the path in the transition diagram contains a circuit. Since the input is constant (00), we must remain on this circuit. Therefore, we continue outputting 0's and we never output 1.

Section 7.2

1. All incoming edges to σ_0 output 1 and all incoming edges to σ_1 output 0; hence the finite-state machine is a finite-state automaton.

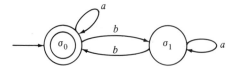

3.

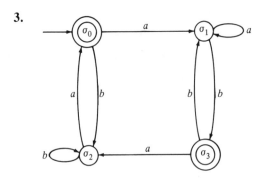

5.

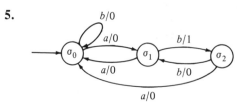

7.

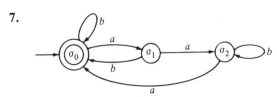

9.

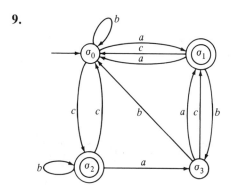

11. 1, 7, 9   **13.** Yes   **15.** Yes   **17.** Yes

19. If a string α, which ends bb is input, no matter which state we are in prior to bb, we will end at state σ_2, as can be seen by checking the three possibilities. Since σ_2 is accepting, α is accepted.

Suppose that α is accepted by Figure 7.2.5. We end in state σ_2. Thus the last character in α is b. There is at least one character before b. If the last two characters are ab and we are in state σ just before the a, by checking the three possibilities for σ, we can show that we will not end in σ_2. Therefore, the last two characters are bb.

21.

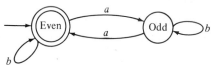

23.

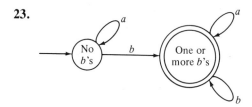

25.

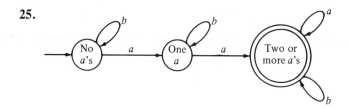

27.

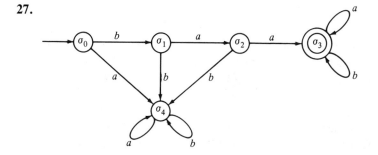

29.

31.

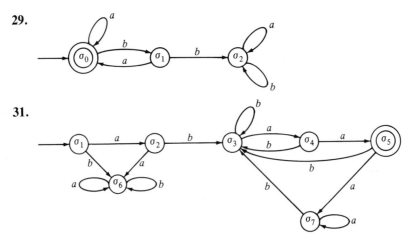

33. If a string consisting only of b's is input to either finite-state automaton, it is accepted. If a string contains an a, in either finite-state automaton we move to a nonaccepting state. In neither finite-state automaton is there an edge from a nonaccepting state to an accepting state. Thus once an a is encountered, both finite-state automata reject the string. Therefore, the set of strings accepted by each finite-state automaton is the same—namely, the set of strings over $\{a, b\}$ that do not contain an a.

35. Make each accepting state nonaccepting and each nonaccepting state accepting.

37. Suppose that $\alpha = x_1 \cdots x_n \in L_1$. Then there exist states $\sigma_{10}, \ldots, \sigma_{1n}$ satisfying

$$\sigma_{10} = \sigma_1^*;$$

$$f_1(\sigma_{1,i-1}, x_i) = \sigma_{1,i} \qquad \text{for } i = 1, \ldots, n;$$

$$\sigma_{1n} \in \mathcal{A}_1.$$

Define

$$\sigma_{20} = \sigma_2^*;$$

$$\sigma_{2,i} = f_2(\sigma_{2,i-1}, x_i) \qquad \text{for } i = 1, \ldots, n;$$

$$\sigma_i = (\sigma_{1i}, \sigma_{2i}) \qquad \text{for } i = 0, \ldots, n.$$

Now

$$\sigma_0 = \sigma^*;$$

$$f(\sigma_{i-1}, x_i) = (\sigma_{1i}, \sigma_{2i}) \qquad \text{for } i = 1, \ldots, n;$$

$$\sigma_n = (\sigma_{1n}, \sigma_{2n}) \in \mathcal{A}.$$

Thus $L_1 \subseteq \text{Ac}(A)$. Similarly, $L_2 \subseteq \text{Ac}(A)$. Therefore, $L_1 \cup L_2 \subseteq \text{Ac}(A)$.

A similar kind of argument may be used to show that $\text{Ac}(A) \subseteq L_1 \cup L_2$.

39. Use Exercises 36 and 37.   **41.** Use Exercises 36 and 37.

Section 7.3

1. Regular, context-free, context-sensitive   **3.** Context-sensitive

5. Regular, context-free, context-sensitive

7. $\sigma* \Rightarrow b\sigma* \Rightarrow bb\sigma* \Rightarrow bbaA \Rightarrow bbabA \Rightarrow bbabbA \Rightarrow bbabba\sigma* \Rightarrow bbabbab$

9. $\sigma* \Rightarrow AAB \Rightarrow AaaB \Rightarrow ABaaB \Rightarrow ABaab$
$\Rightarrow ABBaab \Rightarrow aaBBaab \Rightarrow aabBaab \Rightarrow aabbaab$

11. $<S> \Rightarrow a<A> \Rightarrow ab<B> \Rightarrow aba<S> \Rightarrow abaa<A>$
$\Rightarrow abaab<B> \Rightarrow abaabb<A> \Rightarrow abaabba<S>$
$\Rightarrow abaabbab<S> \Rightarrow abaabbabb<S> \Rightarrow abaabbabba$

13. The productions $\sigma* \to b\sigma*$, $A \to bA$, and $\sigma* \to b$ generate any number of b's. If these are omitted, the only derivations possible are

$$\sigma* \Rightarrow aA \Rightarrow aa\sigma* \Rightarrow \cdots \Rightarrow (aa)^n\sigma* \Rightarrow (aa)^naA \Rightarrow (aa)^{n+1}.$$

15. $S \to aA, A \to aA, A \to bA, A \to a, A \to b, S \to a$

17. $S \to aS, S \to bA, A \to bA, A \to aB, B \to aB, B \to bB, B \to a, B \to b, A \to a$

19. $<\text{digit}> ::= 0|1|2|3|4|5|6|7|8|9$
$<\text{nonzero digit}> ::= 1|2|3|4|5|6|7|8|9$
$<\text{integer}> ::= <\text{signed integer}> \mid <\text{unsigned integer}>$
$<\text{signed integer}> ::= +<\text{unsigned integer}> \mid -<\text{unsigned integer}>$
$<\text{unsigned integer}> ::= <\text{digit}> \mid <\text{nonzero digit}><\text{digit string}>$
$<\text{digit string}> ::= <\text{digit}> \mid <\text{digit}><\text{digit string}>$

21. $<\text{exp number}> ::= <\text{integer}>E<\text{integer}> \mid <\text{float number}> \mid$
$\qquad\qquad\qquad <\text{float number}>E<\text{integer}>$

23. $S \to aS, S \to bS, S \to a, S \to b$

25. Replace each production

$$A \to x_1 \cdots x_n B,$$

where $n > 1$, $x_i \in T$, and $B \in N$, with the productions

$$A \to x_1 A_1$$

$$A_1 \to x_2 A_2$$

$$\cdot$$

$$\cdot$$

$$\cdot$$

$$A_{n-1} \to x_n B,$$

where $A_1, \ldots, A_{n-1}$ are additional nonterminal symbols.

27. $S \to AC, C \to cC, C \to c, A \to aAb, A \to ab$

Section 7.4

1.

3.

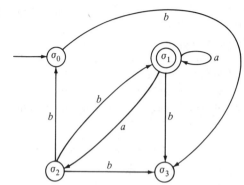

5.

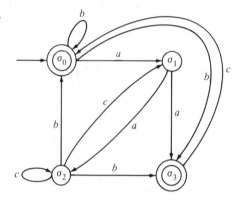

7. $\mathcal{I} = \{a, b\}$, $\mathcal{S} = \{A, B, C\}$, $\mathcal{A} = \{A, C\}$, initial state $= A$

$\mathcal{S}$ \ $\mathcal{I}$	a	b
A	$\{A, C\}$	$\{B\}$
B	$\{C\}$	$\{B, C\}$
C	$\varnothing$	$\varnothing$

9. $\mathcal{I} = \{a, b\}$, $\mathcal{S} = \{\sigma_0, \sigma_1, \sigma_2, \sigma_3\}$, $\mathcal{A} = \{\sigma_3\}$, initial state $= \sigma_0$

$\mathcal{S}$ \ $\mathcal{I}$	a	b
σ_0	$\{\sigma_0\}$	$\{\sigma_0, \sigma_1\}$
σ_1	$\{\sigma_2\}$	$\varnothing$
σ_2	$\varnothing$	$\{\sigma_3\}$
σ_3	$\{\sigma_3\}$	$\{\sigma_3\}$

11. (For Exercise 5) $N = \{\sigma_0, \sigma_1, \sigma_2\}$, $T = \{a, b\}$,

$$\sigma_0 \to a\sigma_1, \ \sigma_0 \to b\sigma_0, \ \sigma_1 \to a\sigma_0, \ \sigma_1 \to b\sigma_2, \ \sigma_2 \to b\sigma_1, \ \sigma_2 \to a\sigma_0, \ \sigma_1 \to b$$

13. Yes

15. The string α is of the form $b^n a b^m$, where $n \geq 0$ and $m \geq 1$. A path representing this string terminating at F is $((\sigma*)^{n+1} C^m F)$. Any path starting at $\sigma*$ terminating at F is of the form $((\sigma*)^{n+1} C^m F)$ and thus represents $b^n a b^m$, where $n \geq 0$ and $m \geq 1$.

17. Yes

19. To reach σ_3 on a path from σ_0 we must have ended *bab*. Any string α that ends *bab* is accepted, since we can remain at σ_0 until we encounter the final *bab*, at which time we move successively to σ_1, σ_2, and σ_3.

21.

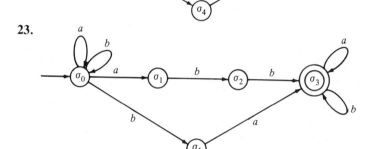

23.

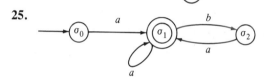

25.

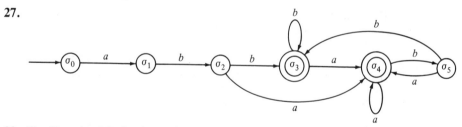

27.

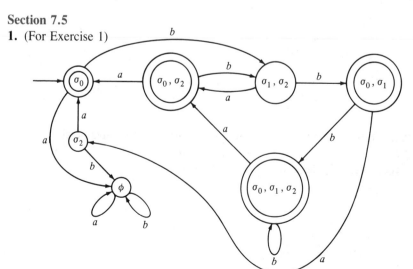

29. Use Exercise 36, Section 7.2.

Section 7.5

1. (For Exercise 1)

3.

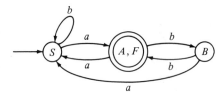

5.

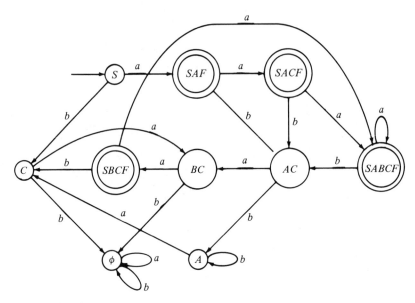

7. (For Exercise 21)

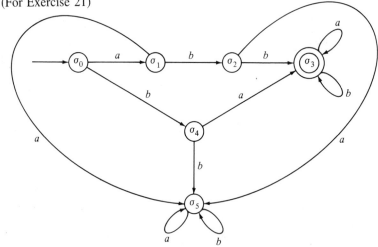

9. Exercise 19, Section 7.2, shows that Figure 7.5.4 accepts precisely the strings over $\{a, b\}$ that end bb. We may now use Example 7.5.7 to conclude that Figure 7.5.5 accepts precisely the strings over $\{a, b\}$ that start bb.

11.

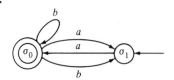

13.

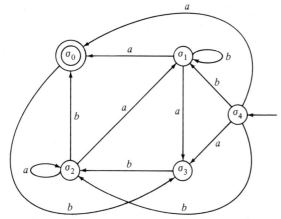

15. To find the nondeterministic finite-state automaton that accepts $Ac(A)^+$, allow an edge in A that terminates on an accepting state to optionally return to the starting state:

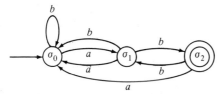

17. See the hint for Exercise 15.

19. Allow an edge in A_1 that terminates on an accepting state in A_1 to terminate optionally on the starting state in A_2. The accepting states of the nondeterministic finite-state automaton are the accepting states of A_2:

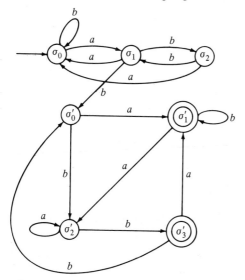

21. See the hint for Exercise 19.

23. Allow any terminating production, alternatively, to return to the start:

$$<S> ::= b<S> \mid a<A> \mid a<S> \mid a$$

$$<A> ::= a<S> \mid b<B>$$

$$<B> ::= b<A> \mid a<S> \mid b<S> \mid b$$

25. Suppose that L is regular. Then there exists a finite-state automaton A with $L =$ Ac(A). Suppose that A has k states. Consider the string $a^{k+1}bba^{k+1}$ and argue as in Example 7.5.6.

27. Consider

$$L_1 = \{a^n b^n c^k \mid n, k \in \{1, 2, \ldots\}\}$$

$$L_2 = \{a^k b^n c^n \mid n, k \in \{1, 2, \ldots\}\}.$$

Section A.1

1. T   **3.** T   **5.** F

7.

p	q	$(\bar{p} \vee \bar{q}) \vee p$
T	T	T
T	F	T
F	T	T
F	F	T

9.

p	q	$(p \wedge q) \wedge \bar{p}$
T	T	F
T	F	F
F	T	F
F	F	F

11.

p	q	r	$\overline{(p \wedge q)} \vee (r \wedge \bar{p})$
T	T	T	F
T	T	F	F
T	F	T	T
T	F	F	T
F	T	T	T
F	T	F	T
F	F	T	T
F	F	F	T

13. Exercise 1—Neither; Exercise 7—Tautology; Exercise 9—Contradiction
15. $p \wedge q$ **17.** $\overline{p \wedge q}$ **19.** $\overline{p \wedge (r \vee q)} \vee (\bar{p} \wedge r)$
21. Today is not Monday and either it is raining or it is hot.
23. (Today is Monday and either it is raining or it is hot) and (it is hot or either it is raining or today is Monday).
25. 1   **27.** 2   **29.** 2

Section A.2

1. If a person is a Cub, then the person is a great baseball player.
3. If better cars are built, then Buick will build them.
5. If the Cubs get a left-handed relief pitcher, then they will win the World Series.
7. If the chairperson gives the lecture, then the audience will go to sleep.
9. (For Exercise 1) If the person is a great baseball player, then the person is a Cub.
11. True   **13.** False   **15.** False   **17.** False   **19.** $p \to \bar{q}$
21. $(\bar{p} \wedge q) \to \bar{r}$ **23.** $p \leftrightarrow (q \wedge r)$
25. If today is Monday, then it is raining.
27. If today is not Monday, then either it is raining or it is hot.
29. If today is Monday and either it is raining or it is hot, then either it is hot or (it is raining or today is Monday).

31. Let p: $4 < 6$ and q: $9 > 12$.
Given statement: $p \rightarrow q$; false.
Converse: $q \rightarrow p$; if $9 > 12$, then $4 < 6$; true.
Contrapositive: $\bar{q} \rightarrow \bar{p}$; if $9 \leq 12$, then $4 \geq 6$; false.
33. Let p: $|1| < 3$ and q: $-3 < 1 < 3$.
Given statement: $q \rightarrow p$; true.
Converse: $p \rightarrow q$; if $|1| < 3$, then $-3 < 1 < 3$; true.
Contrapositive: $\bar{p} \rightarrow \bar{q}$; if $|1| \geq 3$, then either $-3 \geq 1$ or $1 \geq 3$; true.
35. b **37.** a, b, c **39.** b **41.** c **43.** b

45.

p	q	p *impl* q	q *impl* p
T	T	T	T
T	F	F	F
F	T	F	F
F	F	T	T

Since p *impl* q is true precisely when q *impl* p is true, p *impl* $q \equiv q$ *impl* p.

47.

p	q	$\overline{p \wedge q}$	$\bar{p} \vee \bar{q}$
T	T	F	F
T	F	T	T
F	T	T	T
F	F	T	T

Since $\overline{p \wedge q}$ is true precisely when $\bar{p} \vee \bar{q}$ is true, $\overline{p \wedge q} \equiv \bar{p} \vee \bar{q}$.

Section A.3

1. If three points are not collinear, then there is exactly one plane that contains them.
3. An *isosceles trapezoid* is a trapezoid with equal legs.
5. The diagonals of a rhombus are perpendicular to each other.
7. $x \cdot 0 = x \cdot (0 + 0) = x \cdot 0 + x \cdot 0$; therefore, $x \cdot 0 = 0$.
9. Suppose that $xy = 0$ and $x \neq 0$ and $y \neq 0$. Since $xy = 0 = x \cdot 0$ and $x \neq 0$; by the assumed theorem, with $a = x$, $b = y$, and $c = 0$, we deduce that $y = 0$. We now have the contradiction $y = 0$ and $y \neq 0$. Therefore, if $xy = 0$, then either $x = 0$ or $y = 0$.

11. Valid $p \rightarrow q$
$$\underline{p \qquad\quad}$$
$$\therefore q$$

13. Valid $p \leftrightarrow r$
$$\underline{r \qquad\quad}$$
$$\therefore p$$

15. Valid $p \rightarrow (q \vee r)$
$$\underline{\bar{q} \wedge \bar{r} \qquad\quad}$$
$$\therefore \bar{p}$$

17. Invalid. If 64K is better than no memory at all, then either we will buy a new computer or we will buy more memory. If we will buy a new computer, then we will not buy more memory. Therefore, if 64K is better than no memory at all, then we will buy a new computer.

19. Invalid. If we will not buy a new computer, then 64K is not better than no memory at all. We will buy a new computer. Therefore, 64K is better than no memory at all.

21. Invalid   **23.** Valid   **25.** Valid

27. Suppose that $p, p_2, \ldots, p_n$ are true. Since $p_1 \equiv p$, p_1 is also true. Since the argument

$$p_1, p_2, \ldots, p_n \; / \therefore c$$

is valid, c is true. We have shown that if $p, p_2, \ldots, p_n$ are true, then c is true. Therefore, the argument

$$p, p_2, \ldots, p_n \; / \therefore c$$

is valid.

Section A.4

3. Type (a) with
  C_1 = class of all isosceles triangles.
  C_2 = class of all equilateral triangles.
Negation: Some isosceles triangle is not equilateral.

5. Type (c) with
  C_1 = class of all expert bridge players.
  C_2 = class of all persons who can play a mean game of cribbage.
Negation: No expert bridge player can also play a mean game of cribbage.

7. Type (b) with
  C_1 = class of all persons over 30.
  C_2 = class of all trustworthy persons.
Negation: Some person over 30 is trustworthy.

9. Type (a) with
  C_1 = class of all violent movies.
  C_2 = class of all R-rated movies.
Negation: Some violent movie is not R-rated.

11. Type (c) with
  C_1 = class of all linear programming algorithms.
  C_2 = class of all polynomial-time algorithms.
Negation: No linear programming algorithm is a polynomial-time algorithm.

13. C_1 = class of all planar maps.
  C_2 = class of all maps that can be colored using at most four colors.
All C_1 is C_2. Negation: Some planar map cannot be colored using at most four colors.

15. C_1 = class of all continuous functions.
  C_2 = class of all functions differentiable at some point.
Some C_1 is not C_2. Negation: All continuous functions are differentiable at some point.

17. Valid   **19.** Invalid   **21.** Invalid   **23.** Valid   **25.** Valid

27. C_1 = class of all Cubs.

C_2 = class of all White Sox.

C_3 = class of all baseball players.

No C_1 is C_2.

All C_2 is C_3.

∴ Some C_3 is not C_1.

Invalid

29. C_1 = class of all trees.

C_2 = class of all graphs.

C_3 = class of all structures.

All C_1 is C_2.

Some C_3 is not C_2.

∴ Some C_3 is not C_1.

Valid

31. C_1 = class of all integers.

C_2 = class of all perfect numbers.

C_3 = class of all real numbers.

Some C_1 is not C_2.

All C_1 is C_3.

∴ Some C_3 is not C_2.

Valid

33. $(x - 1)^2(x + 2) = 0$ **35.** Apple's Lisa

37. $x = 3, y = 4, z = 5$

39. If $b \geq 1$, take $n = 1$. If $b < 1$, take $n = \lceil \log b/\log a \rceil + 1$ ($\log = \log_{10}$). Then, $n > \log b/\log a$, so $n \log a < \log b$. Thus $\log a^n < \log b$. Now $a^n < b$.

Section B.1

1. $\begin{pmatrix} 2 + a & 4 + b & 1 + c \\ 6 + d & 9 + e & 3 + f \\ 1 + g & -1 + h & 6 + i \end{pmatrix}$

3. $\begin{pmatrix} 5 & 7 & 7 \\ -7 & 10 & -1 \end{pmatrix}$

5. $\begin{pmatrix} 3 & 18 & 27 \\ 0 & 12 & -6 \end{pmatrix}$

7. $\begin{pmatrix} 9 & 8 & 5 \\ -14 & 16 & 0 \end{pmatrix}$

9. $\begin{pmatrix} 18 & 10 \\ 14 & -6 \\ 23 & 1 \end{pmatrix}$

11. $\begin{pmatrix} -11 & -6 \\ 18 & -8 \end{pmatrix}$

13. $\begin{pmatrix} 2a + 4c + e & 2b + 4d + f \\ 6a + 9c + 3e & 6b + 9d + 3f \\ a - c + 6e & b - d + 6f \end{pmatrix}$

15. $x = 1, y = -2, z = 8$

17.

	a	b	c	d
1	1	0	1	0
2	0	1	0	0
3	1	0	0	0

19.

	1	2	3	4	5
1	1	1	1	1	1
2	0	1	1	1	1
3	0	0	1	1	1
4	0	0	0	1	1
5	0	0	0	0	1

21. (a)

$$\begin{array}{c} \\ a \\ b \\ c \\ d \end{array} \begin{array}{cccc} w & x & y & z \\ \begin{pmatrix} 0 & 1 & 0 & 1 \\ 0 & 1 & 0 & 0 \\ 0 & 0 & 1 & 1 \\ 0 & 0 & 0 & 1 \end{pmatrix} \end{array} \qquad \begin{array}{c} \\ w \\ x \\ y \\ z \end{array} \begin{array}{ccc} 1 & 2 & 3 \\ \begin{pmatrix} 1 & 0 & 0 \\ 1 & 1 & 1 \\ 0 & 0 & 1 \\ 0 & 0 & 1 \end{pmatrix} \end{array}$$

(b)

$$\begin{array}{c} \\ a \\ b \\ c \\ d \end{array} \begin{array}{ccc} 1 & 2 & 3 \\ \begin{pmatrix} 1 & 1 & 1 \\ 1 & 1 & 1 \\ 0 & 0 & 1 \\ 0 & 0 & 1 \end{pmatrix} \end{array}$$

23. (a) Every column must contain at least one 1.
 (b) Every column must contain at most one 1.

25. Let $A = (b_{ij})$, $I_n = (a_{jk})$, $AI_n = (c_{ik})$. Then

$$c_{ik} = \sum_{j=1}^{n} b_{ij} a_{jk} = b_{ik} a_{kk} = b_{ik}.$$

Therefore, $AI_n = A$. Similarly, $I_n A = A$.

27. Let

$$A = \begin{pmatrix} a & b \\ c & d \end{pmatrix}.$$

Suppose that $AB = I_2$ for some 2×2 matrix

$$B = \begin{pmatrix} e & f \\ g & h \end{pmatrix}.$$

We have

$$\begin{pmatrix} ae + bg & af + bh \\ ce + dg & cf + dh \end{pmatrix} = AB = \begin{pmatrix} 1 & 0 \\ 0 & 1 \end{pmatrix}.$$

Now

$$(ad - bc)(eh - fg) = (ae + bg)(cf + dh) - (af + bh)(ce + dg)$$
$$= 1 \cdot 1 - 0 \cdot 0 = 1.$$

Thus $ad - bc \neq 0$.

If $\alpha = ad - bc \neq 0$, setting

$$B = \begin{pmatrix} d/\alpha & -b/\alpha \\ -c/\alpha & a/\alpha \end{pmatrix},$$

we obtain $AB = I_2 = BA$.

29. Let

$$A = (a_{ij}), \ B = (b_{pq}), \ AB = (c_{rs}), \ B^T A^T = (d_{uv}), \ A^T = (a'_{ji}),$$

$$B^T = (b'_{qp}), \ (AB)^T = (c'_{sr}).$$

Then

$$d_{uv} = \sum_{x=1}^{k} b'_{ux} a'_{xv} = \sum_{x=1}^{k} a_{vx} b_{xu} = c_{vu} = c'_{uv}.$$

Index